W0080629

8.5 × 5.5

525 gm

Fundamentals of
Heat and Mass
Transfer

Fundamentals of Heat and Mass Transfer

S.C. BHATIA
B.E. (Chemical), M.B.A.

CBS Publishers & Distributors Pvt. Ltd.

New Delhi • Bengaluru • Chennai • Kochi • Kolkata • Mumbai
Hyderabad • Nagpur • Patna • Pune • Vijayawada

ISBN: 81-239-0827-X

First Edition: 2002
Reprint: 2007, 2009, 2011, 2012, 2016

Copyright © Publisher

All rights reserved. No part of this book may be reproduced or transmitted in any form or by any means, electronic or mechanical, including photocopying, recording, or any information storage and retrieval system without permission, in writing, from the publisher.

Published by:
Satish Kumar Jain for CBS Publishers & Distributors Pvt. Ltd.,
4819/XI Prahlad Street, 24 Ansari Road, Daryaganj, New Delhi - 110002
delhi@cbspd.com, cbspubs@airtelmail.in • www.cbspd.com
Ph.: 23289259, 23266861, 23266867 • Fax: 011-23243014

Corporate Office: 204 FIE, Industrial Area, Patparganj, Delhi - 110 092
Ph: 49344934 • Fax: 011-49344935
E-mail: publishing@cbspd.com • publicity@cbspd.com

Branches:
• *Bengaluru:* 2975, 17th Cross, K.R. Road, Bansankari 2nd Stage,
 Bengaluru - 70 • Ph: +91-80-26771678/79 • Fax: +91-80-26771680
 E-mail: cbsbng@gmail.com, bangalore@cbspd.com
• *Chennai:* No. 7, Subbaraya Street, Shenoy Nagar, Chennai - 600030
 Ph: +91-44-26681266, 26680620 • Fax: +91-44-42032115
 E-mail: chennai@cbspd.com
• *Kochi:* Ashana House, 39/1904, A.M. Thomas Road, Valanjambalam,
 Ernakulum, Kochi • Ph: +91-484-4059061-65
 Fax: +91-484-4059065 • E-mail: cochin@cbspd.com
• *Kolkata:* 6-B, Ground Floor, Rameshwar Shaw Road, Kolkata - 700014
 Ph: +91-33-22891126/7/8 • E-mail: kolkata@cbspd.com
• *Mumbai:* 83-C, Dr. E. Moses Road, Worli, Mumbai - 400018
 Ph: +91-9833017933, 022-24902340/41 • E-mail: mumbai@cbspd.com

Representatives:

• Hyderabad: 0-9885175004   • Nagpur: 0-9021734563
• Patna: 0-9334159340   • Pune: 0-9623451994
• Vijayawada: 0-9000660880

Printed at:
J.S. Offset Printers, Delhi (India)

Preface

Heat and mass transfer is such a familiar and common occurrence in our daily life that we tend to take the process for granted and think only in terms of its manifestations and effects. Our physical intution enables us to interpret thermal phenomena, and our reactions to hot and cold objects on environments rapidly become voluntary. However, when the subject is considered from a quantitative point of view, it is evident that the actual mechanisms of heat and mass transfer and their operation in a given system are very complex. Analysis may be extremely difficult–even if possible–and an accurate description of some of the most ordinary cases may become quite involved. Because of this there is a tendency for introductory material on heat and mass transfer to be of a descriptive nature.

This reference text book, is therefore devoted to a basic, fundamental, and yet understandable exposition of the principles of heat and mass transfer. The treatment of the subject is based upon background in physics, mathematics, and thermodynamics which has been acquired by average engineering student. Although special effort has been made to provide a physical interpretation of the many quantities and processes, rather extensive analytical material is also included. This approach is necessary to give a representative picture and to encourage appreciation for both the experimental and analytical procedures employed.

Special attention has been given to the derivation and explanation of numerous dimensionless parameters and theoretical and semi-empirical equations in order to establish the student's confidence in them and develop his judgement of their range of applicability and limitations.

In an era when energy conservation and efficiency in conversion are of paramount importance to engineers, heat and mass transfer is becoming an increasingly important part of engineering design. It is hoped that by presenting the subject in an interesting and up to date manner, a service have been provided to those engineers who will face the challenge of efficient design of heat exchange equipment in their future careers.

The author hopes and strongly believes that this treatise on fundamentals of heat and mass transfer will contribute effectively to the existing literature in this field and will be a useful text for students of mechanical, electrical, chemical, metallurgical, nuclear, aeronautical engineering, AMIE, engineering services examination and the students associated with thermal power systems.

In addition, the text would also be of assistance to practicing engineers and other persons who need a concise reference to the basics of heat exchange equipment and energy related problems.

Chapter 1 is devoted to basic modes of heat transfer and presents an overview of the field of heat transfer and its relationship to thermodynamics. All heat transfer processes involve the transfer and conversion of energy, they must therefore obey the first as well as the second law of thermodynamics. In general, the temperature distribution throughout a body varies both with location and time. If this distribution as a function of the space coordinates and time is known or can be determined, the heat flow at any point in any direction can be specified by appropriate differentiation. Keeping this in mind, chapter 2 describes the derivation of the various differential fourier conduction equations governing the temperature distribution. Steady state conduction is the focus of chapter 3 while chapter 4 and 5 are devoted to conduction with heat generation and extended surfaces or fins. Before steady-state conditions can be reached in a process, some time must elapse after the heat transfer process is initiated to allow unsteady state conditions to disappear. Usteady state heat transfer is important because of the large number of heating and cooling problems occurring industrially and chapter 6 thus deals with unsteady state conduction. In radiant heat transfer the medium through which heat is transferred is usually not heated. Radiation heat transfer is the transfer of heat by electromagnetic radiation and thus chapter 7 is devoted to radiation.

Chapter 8 is devoted to dimensional analysis which deals with the process whereby all the important variables involved in a physical phenomenon are systematically organised into dimensionless groups which are less numerous than original variables. Thus this chapter presents the general methods of dimensional analysis and illustrates its applications to various problems of fluid mechanics and convection heat transfer.

When a fluid flows around an object or when the object moves through a body of fluid, their exists a thin layer of fluid close to the solid surface within which shear stresses significantly influence the velocity distribution. The fluid velocity varies from zero at the solid surface to the velocity of free stream flow at a certain distance away from the solid surface. This thin layer of changing velocity has been called the hydrodynamic boundry layer. Keeping this in mind, chapter 9 discusses the hydrodynamic and thermal boundry layers.

Chapter 10 is devoted to condensation and boiling which are the convective heat transfer processes that are associated with change in the phase of a fluid. Condensation refers to a change from the vapour to a liquid phase and boiling involves change from liquid to vapour phase of a fluid substance. Chapter 11 focuses on heat exchanger which is a device used to transfer heat between two or more fluids. Notable examples are boilers, automobile radiators, evaporators, water and air heaters or coolers. The heat transferred in the heat exchanger may be in the form of latent heat or sensible heat. In mass transfer, mass is being transferred from one phase to another distinct phase, the basic mechanism is the same whether the phases are gas, solid or liquid. Keeping this in view chapter 12 deals with mass transfer.

The appendices at the end of the book serve as quick references. Appendix I highlights the various notation/symbols used in the text while Appendix II gives details of various fundamental constants and conversion factors. Appendix III explains the physical properties of water such as latent heat of water, latent heat of vaporisation, vapour pressure of water, densities of liquid water, viscosity of liquid water, thermal conductivity and heat transfer properties of water etc. To reflect modern trends, SI units are used.

This reference textbook fills in the lacunae caused by the near total absence of books on heat and mass transfer in Indian context and is designed to cater to the needs of the Indian students who have, uptil now, been forced to rely on expensive and almost inaccessible American textbooks. The simple, lucid, and readable style will help the Indian students to comprehend the concepts and information presented, and ensures an easy grasp of the key points which are illustrated by flow diagrams, tables and equations.

The text is throughout supplemented with diagrams, figures and tables wherever needed. The treatment of all topics is in a cogent, lucid style aimed at enabling the reader to grasp the information quickly and easily. In the end of all chapters typical solved examples are given which will help to the students to understand the subject more thoroughly.

While painstaking care has gone into producing a useful and exhaustive reference textbook, the author would welcome any constructive criticism and a creative feedback from students, teachers and professionals in the industry and interaction that will certainly be mutually fruitful.

The author S C Bhatia, is a Chemical Engineer with management qualifications who has written several books on chemical and allied subjects, such as chemical process industries; perfumes, soaps, detergents and cosmetics; engineering chemistry, handbook of pollution and its control;

environmental chemistry etc. At present, he is a renowned consultant in the field of environment, waste heat recovery (energy conservation) and petrochemicals.

Acknowledgements are due to Mr. Santosh Kumar Shrivastava and Mr. Harinder Singh Negi, the computer operators who worked long hours to complete the task of bringing out the book on time. The author is also indebited to Dr. M M Verma, Ms. Namita Das (M.Sc. Chemistry) for editing and proof-reading of the book. Acknowledgements are also due to Maj. Gen. H C Dua and Mr. Madhuresh Kumar for proof-reading.

S C Bhatia

Contents at a Glance

Contents

CHAPTER 1

Basic Modes of Heat Transfer

INTRODUCTION

Heat and mass transfer is such a familiar and common occurrence in our daily life that we tend to take the process for granted and think only in terms of its manifestations and effects. Our physical intution enables us to interpret thermal phenomena, and our reactions to hot and cold objects on environments rapidly become voluntary. However, when the subject is considered from a quantitative point of view, it is evident that the actual mechanisms of heat and mass transfer and their operation in a given system are very complex. Analysis may be extremely difficult–even if possible–and an accurate description of some of the most ordinary cases may become quite involved. Because of this there is a tendency for introductory material on heat and mass transfer to be of a descriptive nature.

Heat transfer is the science of spontaneous irreversible process of heat propagation in space. By a process of heat propagation is meant the exchange in internal energy between individual elements, regions of the medium considered. Heat may be transmitted in three ways, by conduction, convection or thermal radiation.

Heat conduction is identified as the process of molecular transport of heat in bodies (or between them), due to temperature variation in the medium considered. Convection is possible only in a fluid medium. The term is applied to the transport of heat that occurs as volumes of liquid or gas fluid medium move from regions of one temperature to those of another temperature. The transport of heat is inseparably linked here with the movement of the medium itself.

Thermal radiation is the process of heat propagation by means of electromagnetic waves, depending only on the temperature and on the optical properties of an emitter, with its internal energy being converted into radiation energy. The process involving the conversion of the internal energy of a substance into radiation energy is referred to as (thermal) radiation heat transfer. The various basic processes of heat transfer—conduction, convection and radiation—are often combined both in nature and in engineering applications.

Pure conduction is found only in solids. Heat transfer by convection is always accompanied by conduction. The combined process of heat transfer by convection and conduction is referred to as convective heat transfer.

The subject of engineering analysis is usually the transfer of heat by convection between a stream of liquid or gas and a solid surface; this process of convection heat exchange is referred to as convection heat transfer or heat transfer.

Processes of conduction and convection can be accompanied by thermal radiation heat transfer. Heat transfer between media characterised by combined transport of heat by radiation and conduction is called radiation-conduction heat transfer. If heat is transported also by convection, a process of this kind is called radiation-convective heat transfer. Radiation-conduction and radiation-convection modes of heat transport are sometimes referred to as complex heat transfer.

In engineering and daily life processes are often encountered, involving heat transfer between various fluids separated by a solid wall. The process of heat transfer from a hot fluid to a cold one through a partition separating the two fluids is also simply referred to as heat transfer, and it may involve heat transfer by conduction, radiation and convection. The steam generating tubes of a boiler, for instance, receive heat from the products of fuel combustion by all three modes of heat transfer: conduction, convection, and radiation. Heat is transferred by conduction through the layer of ash covering the tube, the metal wall of the tube itself and the layer of scale covering its inner surface; from the inner surface heat is transferred to the liquid filling the tube both by convection and conduction.

Process of heat transfer may occur in various media, in pure substances and in mixtures, with and without changes of phase of the working medium, etc. and accordingly will differ in character and be described by different equations.

Many of the processes of heat transfer are accompanied by transfer of mass. As water evaporates, for instance, heat transfer is accompanied by transport of the vapour formed through an air-vapour mixture. The transport of steam generally occurs both through molecular interaction and convection. The combined molecular and convective transport of mass is called convection mass transfer. With mass transfer the process of heat transfer becomes more complicated. In addition, heat may be transported together with the mass of diffusing substances.

In the general case, in a mixture of various substances, transport of heat may be provoked by non-uniform distribution of other physical quantities, apart from the temperature. A difference in the concentration of mixture components, for instance, brings about an additional molecular transport of heat (the diffusion thermoeffect). The amount of heat transported due to such effects is usually comparatively small and it can be ignored.

Theoretical studies of heat transfer necessitate the introduction of some models, giving an idea of the medium in which the investigated processes take place. The gases, liquids and solids being investigated in the book are usually considered as a continuum, i.e. as a medium whose discrete structure can be ignored.

Homogeneous and heterogeneous continuous media are distinguished. In the first, the physical properties at different points are the same at equal temperatures and pressures, and in heterogeneous media the physical properties differ. Also distinguished are isotropic and anisotropic continuous media. The physical properties of an isotropic medium at any point of the medium are independent of the direction selected and, on the contrary, at a given point some properties of an anisotropic medium may happen to be a function of direction. Heat transfer in isotropic media is often encountered in practical applications and the process has been more exhaustively studied.

There may be single-phase and multi-phase continuous media. The properties of a single-phase medium, consisting of a pure substance or a mixture of substances, change in space continuously. The properties of a multi-phase medium, consisting of a number of single-phase parts, change at the interfaces in jumps. In single-phase and multi-phase media heat transfer occurs in different ways.

THERMODYNAMICS

Thermodynamics is that part of science which is concerned with the conditions that material systems may assume and the changes in conditions that may occur either spontaneously or as a result of interactions between systems. The word "thermodynamics" was derived from the Greek words thermé (heat) and dynamis (force). The formulation of the first and second laws of thermodynamics by the German scientist Rudolf Julius Clausius in 1850 lay the foundation for what is now called "classical" or "equilibrium thermodynamics". More recently, thermodynamics has been extended to include non-equilibrium states.

Basic Concepts

The description of physical phenomena is based on the concept of the state of a system and the changes that occur spontaneously or through interaction with other systems. The term system means any identifiable collection of matter that can be separated from everything else by a well-defined surface. Examples for thermodynamic systems are the water molecules in a container or, much more complex, a complete process plant. If the system boundaries permit the exchange of heat and work, but not of physical matter, the system is termed closed system, as compared to the open system, where mass transfer may occur. At any time, a system is in a condition called "state" which encompasses all that can be said about the results of any measurements or observations that can be performed on the system at that time. It is necessary to distinguish between quantities which depend on the path between states (such as heat and work) and those which depend solely on the state (such as temperature, internal energy or entropy).

A system is in equilibrium if there is no tendency for a change in state to occur, i.e. if all forces are in exact balance. The number of independent intensive (mass-independent) properties that must be arbitrarily fixed to establish the state of a system is named the degree of freedom. It can be calculated according to Gibbs

$$F = 2 - P + n$$

with P being the number of phases and n being the number of chemical species present.

A process is a change in the system from an initial state to a final state. Processes can be divided into two types, namely reversible processes which can be reversed at any time so that the system and the surrounding are returned to their original condition and irreversible processes in which the reversal can not be carried out without leaving some change in the system or the surroundings. Reversible processes are ideal processes in the absence of friction and finite temperature differences.

Four laws form the foundation of thermodynamics, even though the first and the second laws are considered the most important.

TEMPERATURE, HEAT AND THERMAL EQUILIBRIUM

The temperature of an object or fluid is that property which determines the direction of the flow of heat from that body or fluid to an adjacent body or fluid with which it is in contact. Thus, heat flows from a body or fluid of higher temperature to a body or fluid of lower temperature. Temperature is one of the main parameters of state which defines the thermal state of the system. The temperature of all parts of the system in thermodynamic equilibrium is the same. Based on the molecular-kinetic approach, the temperature of a system characterises the intensity of thermal motion of atoms, molecules and other particles forming the system.

For instance, for a system described by the laws of classical statistical physics the mean kinetic energy of thermal motion of particles is directly proportional to the absolute temperature of the system. In this regard we can say that the temperature characterises the thermal motions within a body.

In thermodynamics the reciprocal of the derivative of the entropy S of a body with respect to its energy E is called the absolute temperature T :

$$\frac{dS}{dE} = \frac{1}{T} \qquad \qquad ...(1.1)$$

Temperature, like entropy, is a purely statistical quantity and makes sense only for macroscopic bodies. According to the second law of thermodynamics, energy is transferred from bodies with higher temperature to bodies with lower temperature. The absolute temperature is always positive, $T > 0$. The least absolute temperature possible is the absolute zero. At absolute zero, the translatory and rotary motion of atoms and molecules comes to an end, and they are in a state of the so-called "zero vibrations" rather than in a state of rest. By Nernst theorem the entropy of any body becomes zero at absolute zero temperature. Absolute zero is unttainable. The entropy S is a dimensions quantity and from equation (1.1) it follows that temperature has the dimensions of energy and can be measured in Joules. The ratio Joules/Kelvins(K) called Boltzmann's constant k is equal to k = 1.38×10^{-23} Joules/K.

Actually, the temperature is usually measured in arbitrary units, degrees (Celsius degree, °C, Fahrenheit, °F, Réaumur, °R) or in "Kelvins" whose value is determined by the corresponding temperature scales.

Temperature scales are systems of sequentially numbered values corresponding to various temperatures. The temperature can be determined by measuring any quantity dependent on it and it is convenient to measure a physical property of a certain, so-called thermometric substance (for instance, the volume or pressure of a gas, the resistance of a conductor). To realise a temperature scale, we must select its origin and the dimension of the temperature unit (degree). For this purpose, we usually use two reference points—temperatures of transition of a substance from one aggregate state to another. Such temperature scales are called "practical". The first practical temperature scale was suggested by Fahrenheit in which one of the reference points was the temperature of a human body, accepted by Fahrenheit to be equal to 96 degrees (°F), the second, the temperature of ice melting, equal to 32 degrees (°F). A liquid mercury thermometer served as an interpolation device.

More accurate practical temperature scales were suggested by Celsius and Réaumur. In these scales the temperatures of melting of ice and boiling of water at atmosphere pressure were used as reference points. The temperature interval between these points in Celsius' scale (°C) was divided by 100, and in Réaumur's scale (°R) by 80 equal parts. In Fahrenheit's temperature scale, this temperature interval is equal to 180 (°F). In the absolute Kelvin (K) and Rankine (R) temperature scales, where the origin of scale is the absolute zero, the temperature interval is equal to 100 and 180 temperature units, respectively. The principle of constructing temperature scales suggested by Fahrenheit (the refence points and interpolation device) is used in the international temperature scales. For instance, the international temperature scale ITS-27, was realised using two points (0°C and 100°C), the unit of temperature is the degree Celsius (°C); in the ITS-90, one point, the temperature of the triple point of water, 273.16 K, was used; the unit of temperature is the Kelvin (K). The main interpolation device is a platinum resistance thermometer.

The so-called thermodynamic temperature scale, which is independent of the particular properties of a thermometric substance, can be realised on the basis of the second law of thermodynamics, by determining the ratio of temperatures from the ratio of temperatures in Carnot's cycle. In practice, to construct such a scale, relations are used which, whilst not contradicting the second law of thermodynamics, relate the thermodynamic temperature to some additive physical quantity which can be measured accurately enough. The most widespread are :

1. The gas thermometer based on the gas law

$$P \tilde{V} = RT, \qquad ...(1.2)$$

where R is the universal gas constant, and P, V, T are the pressure, molar volume and temperature of a working substance in an ideal gas state.

2. The acoustic thermometer based on measuring the velocity of sound, C, in a gas

$$C^2 = \frac{\gamma RT}{\tilde{M}}, \qquad ...(1.3)$$

where $\gamma = C_p/C_v$ is the specific heat ratio (for an ideal gas γ = constant), and $\tilde{M}$ is the molecular weight of the working substance.

3. The radiation thermometer based on measuring the total energy of heat radiation E(T) emitted by a blackbody at temperature T

$$E(T) = \sigma T^4, \qquad ...(1.4)$$

where σ is the Stefan-Boltzmann constant.

4. The thermal noise thermometer based on measuring the root-mean square of voltage noise $\overline{V}^2$ in current flow through a resistance Ω (ohm) at a temperature T (Nyquist' equation)

$$\overline{V}^2 = 4kT\Omega\,\Delta f, \qquad \qquad ...(1.5)$$

where Δf is the band width (Hz).

Presently the most exact values of thermodynamic temperature in a wide range of values can be obtained using the gas thermometer; however, near 4 K and above 200 K the noise and radiation thermometers approach the gas thermometer in accuracy. In 1848, J. Thomson (Lord Kelvin) proved that the temperatures determined from Carnot's cycle and by the gas thermometer are identical and represent the thermodynamic or absolute temperature. With the assumption for a dimension of a temperature unit made by Celsius, Thomson determined the value of temperature of ice melting on a new scale, − 273.15 K.

Since 1960, the unit of thermodynamic temperature (K) was determined as 1/273.16 of the temperature of the triple point of water, −273.16 K. When using a gas thermometer of constant volume ($\tilde{V}$ = const) equation (1.2) for determining the unknown temperature T_x takes the form

$$T_x = \frac{T_0}{P_0}\,P_x, \qquad \qquad ...(1.6)$$

where T_0 = 273.16 K, P_0 is the pressure of the working substance at a temperature T_x. In the ITS-90 the basic unit of temperature T_{90} is the Kelvin (K). The measurements allowed for measuring temperature in °C (T_{90}) are defined as

$$T_{90} = T_{90} - 273.15$$

Heat

Heat is less easily defined than work. In general terms, the temperature of a substance increases if heat is added to it. Therefore, heat was, for a long time, thought to be an invisible fluid named 'calorific' which flows along a temperature difference from one system to the other. We now know that this is false and that heat is not contained in a system but is manifested only as an interaction of the system with its surroundings as the system changes from one state to another. Like work, it can be considered as energy in transit. Heat is best defined through the first law of thermodynamics :

$$\Delta U = MC_v\Delta T = Q - W = Q - \int pdV \qquad ...(1.7)$$

$$\Delta H = MC_p \Delta T = \Delta U + \int V dp + \int p dV = Q + \int V dp \qquad ...(1.8)$$

where U is internal energy, M mass, T temperature, C_v and C_p specific heat capacity at constant volume and pressure respectively, and Q is heat, W work, P pressure, V volume, and H enthalpy.

Therefore,

$$Q = \Delta U \text{ or } V = \text{constant and } Q = \Delta H \text{ for } P = \text{constant} \qquad ...(1.9)$$

If phase change occurs as a result of heat transfer, the latent heat of evaporation and/or the latent heat of solidification have to be included in equation (1.9).

The sign convention is that heat transfer is positive into the systems and negative out of the system.

Thermal Equilibrium

If the only process which a body undergoes is the addition or removal of heat the effect of this addition or removal is to cause a definite change of state (since in this case $dq = dU$) and thereby a definite change of entropy. The equation

$$dS = \frac{dq}{T}$$

is therefore applicable to processes of pure heat transfer, even when they are not carried out under reversible conditions, provided that the temperature remains sensibly uniform throughout the body.

Consider the transfer, under the above conditions, of a quantity of heat dq from a body at a temperature T_2 to another body at the temperature T_1. The overall entropy change is

$$dS = dS_1 + dS_2$$

$$= \frac{dq}{T_1} - \frac{dq}{T_2}$$

$$= dq(T_2 - T_1)/T_1 T_2.$$

Since dS must be positive or zero, $T_2 > T_1$, and thus the heat flows from the hotter to the cooler body, in agreement with experience. The creation of entropy in the system continues for as long as T_2 exceeds T_1 and the state of equilibrium requires equality of these temperatures. This is in accordance with the meaning of temperature. A reversible transfer of heat thus requires that there shall be only an infinitesimal temperature difference.

Thus if T_2-T_1 is an infinitesimal, the increase of entropy in the above equation becomes equal to $dqdT/T^2$ and is of the second order of smallness.

The above theorem, like all those which have appeared so far, is based on the supposition that there is no exchange of matter between the bodies. If such an exchange takes place the notion of heat flow becomes ambiguous and we can speak only of the total energy flow.

RELATION OF HEAT TRANSFER TO THERMODYNAMICS

Whenever a temperature gradient exists within a system, or when two systems at different temperatures are brought into contact, energy is transferred. The process by which the energy transport takes place is known as heat transfer. The thing in transit, called heat, cannot be measured or observed directly, but the effects it produces are amenable to observation and measurement. The flow of heat, as the performance of work, is a process by which the internal energy of a system is changed.

The branch of science which deals with the relation between heat and other forms of energy is called thermodynamics. Its principles, like all laws of nature, are based on observations and have been generalised into laws which are believed to hold for all processes occurring in nature, because no exceptions have ever been found. The first of these principles, the first law of thermodynamics, states that energy can be neither created nor destroyed but only changed from one form to another. It governs all energy transformations quantitatively but places no restrictions on the direction of the transformation. It is known, however, from experience that no process is possible whose sole result is the net transfer of heat from a region of lower temperature to a region of higher temperature. This statement of experimental truth is known as the second law of thermodynamics.

All heat transfer processes involve the transfer and conversion of energy. They must therefore obey the first as well as the second law of thermodynamics. This, however, would be an erroneous conclusion because classical thermodynamics is restricted primarily to the study of equilibrium states, including mechanical and chemical as well as thermal equilibriums, and is therefore, by itself, of little help in determining quantitatively the transformations that occur from a lack of equilibrium in engineering processes. Since heat flow is the result of temperature non-equilibrium, its quantitative treatment must be based on other branches of science. The same reasoning applies to other types of transport processes such as mass transfer and diffusion.

Limitations of Classical Thermodynamics

Classical thermodynamics deals with the states of systems from a macroscopic view and makes no hypotheses about the structure of matter.

To perform a thermodynamic analysis it is necessary to describe the state of a system in terms of gross characteristics, such as pressure, volume, and temperature, which can be measured directly and involve no special assumptions regarding the structure of matter. These variables or thermodynamic properties are of significance for the system is in equilibrium. Thus, classical thermodynamics is not concerned with the details of a process but rather with equilibrium states and the relations among them. The processes employed in a thermodynamic analysis are idealised processes, devised to give information concerning equilibrium states.

From a thermodynamic viewpoint, the amount of heat transferred during a process simply equals the difference between the energy change of the system and the work done. It is evident that this type of analysis considers neither the mechanism of heat flow nor the time required to transfer the heat. It simply prescribes how much heat supply to, or reject from, a system during a process between specified end states without taking care of whether, or how, this could be accomplished. The reason for this lack of information obtainable from a thermodynamic analysis is the absence of time as a variable. The question of how long it would take to transfer a specified amount of heat, although it is of great practical importance, does not usually enter into the thermodynamic analysis.

Engineering Heat Transfer

From an engineering viewpoint, the determination of the rate of heat transfer at a specified temperature difference is the key problem. To estimate the cost, the feasibility, and the size of equipment necessary to transfer a specified amount of heat in a given time, a detailed heat transfer analysis must be made. The dimensions of boilers, heaters, refrigerators, and heat exchangers depend not only on the amount of heat to be transmitted, but also on the rate at which the heat is to be transferred under given conditions. The successful operation of equipment components such as turbine blades or the walls of combustion chambers depends on the possibility of cooling certain metal parts by removing heat continuously at a rapid rate from a surface. Also, in the design of electric machines, transformers, and bearings, a heat transfer analysis must be made to avoid conditions that will cause overheating and damage the equipment. These examples show that in almost every branch of engineering, heat transfer problems are encountered which are not capable of solution by thermodynamic reasoning alone, but require an analysis based on the science of heat transfer.

In heat transfer, as in other branches of engineering, the successful solution of a problem requires assumptions and idealisations. It is almost impossible to describe physical phenomena exactly, and in order to express a problem in the form of an equation that can be solved it is necessary to make some approximations. In electric circuit calculations, for example, it is usually assumed that the values of the resistances, capacitances, and inductances are independent of the current flowing through them. This assumption simplifies the analysis but may in certain cases severely limit the accuracy of the results.

It is important to keep the assumptions, idealisations, and approximations made in the course of an analysis in mind when the final results are interpreted. Sometimes insufficient information on physical properties makes it necessary to use engineering approximations to solve a problem. For example, in the design of machine parts for operation at elevated temperatures it may be necessary to estimate the proportional limit of the fatigue strength of the material from low-temperature data. To assure satisfactory operation of the part, the designer should apply a factor of safety to the results obtained from the analysis. Similar approximations are also necessary in heat transfer problems. Physical properties, such as the thermal conductivity or the viscosity, change with temperature, but if suitable average values are selected, the calculations can be considerably simplified without introducing an appreciable error in the final result. When heat is transferred from a fluid to a wall, as in a boiler, a scale forms under continued operation and reduces the rate of heat flow. To assure satisfactory operation over a long period of time, a factor of safety must be applied to provide for this contingency.

When it becomes necessary to make an assumption or approximation in the solution of a problem, the engineer must rely on ingenuity and past experience. There are no simple guides to new and unexplored problems, and an assumption valid for one problem may be misleading in another. Experience has shown, however, that the first requirement for making sound engineering assumptions or approximations is a complete and thorough physical understanding of the problem at hand. In the field of heat transfer, this means familiarity not only with the laws and physical mechanisms of heat flow, but also with those of fluid mechanics, physics, and mathematics.

Heat transfer can be defined as the transmission of energy from one region to another as a result of a temperature difference between them. Since differences in temperatures exist all over the universe, the phenomena of heat flow are as universal as those associated with gravitational attractions. Unlike gravity, however, heat flow is governed not by a unique relationship but rather heat transfer generally recognises three distinct modes of heat transmission: conduction, radiation, and convection. Strictly speaking, only conduction and radiation should be classified as heat transfer processes, because only these two mechanisms depend for their operation on the mere existence of a temperature difference. The last of the three, convection, does not strictly comply with the definition of heat transfer because it depends for its operation on mechanical mass transport also. But since convection also accomplishes transmission of energy from regions of higher temperature to regions of lower temperature, the term "heat transfer by convection" has became generally accepted.

However, it should be emphasised that in most natural situations heat is transferred not by one, but by several mechanisms operating simultaneously.

CONDUCTION

Whenever, a temperature gradient exists in a solid medium, heat will flow from the higher-temperature to the lower-temperature region. The rate at which heat is transferred by conduction, q_k, is proportional to the temperature gradient dT/dx times the area A through which heat is transferred, or

$$q_k \propto A \frac{dT}{dx}$$

In this relation $T(x)$ is the local temperature and x is the distance in the direction of the heat flow. The actual rate of heat flow depends on the thermal conductivity k, which is a physical property of the medium. For conduction through a homogeneous medium, the rate of heat transfer is then

$$q_k = -kA \frac{dT}{dx} \qquad \qquad ..(1.10)$$

The minus sign is a consequence of the second law of thermodynamics, which requires that heat must flow in the direction from higher to lower temperature. The temperature gradient, as shown in Fig. 1.1, will be negative if the temperature decreases with increasing values of x. Therefore, if heat transferred in the positive x directions is to be a positive quantity, a negative sign must be inserted on the right side of Eq. (1.10).

Equation (1.10) defines the thermal conductivity. It is called Fourier's law of conduction in honour of the French scientist J.B.J. Fourier, who proposed it in 1822. The thermal conductivity in eq. (1.10) is a material property that indicates the amount of heat that will flow per unit time across a unit area when the temperature gradient is unity. In the SI system the area is in square metres (m^2), the temperature in kelvin (K), x in metres (m), and the rate of heat flow in watts (W). The thermal conductivity, therefore, has the units of watts per metre per kelvin (W/m K).

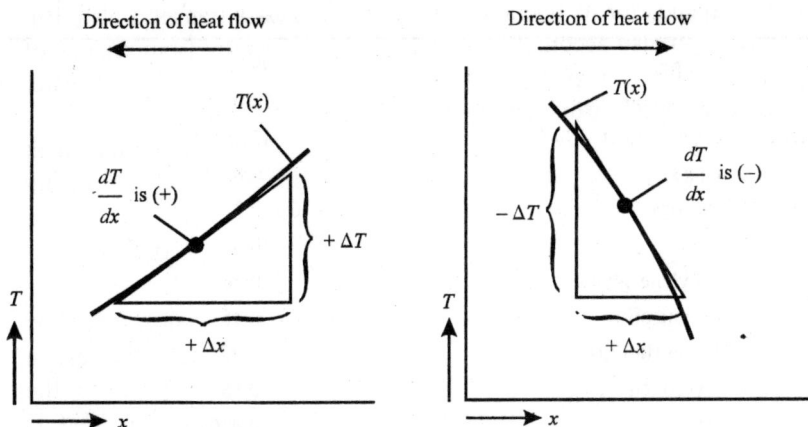

Fig. 1.1. Sketch illustrating sign convention for conduction heat flow.

In the English system, which is still widely used by engineers in the United States, the area is expressed in square feet (ft^2), x in feet (ft), the temperature in degrees Fahrenheit (°F), and the rate of heat flow in Btu/h. Thus, k has the units Btu/h ft °F. The conversion constant for k between the SI and English systems is

$$1 \text{ W/m K} = 0.578 \text{ Btu/h ft °F}$$

Orders of magnitude of the thermal conductivity for various types of materials are presented in Table 1.1. Although in general the thermal conductivity varies with temperature, in many engineering problems the variation is sufficiently small to be neglected.

Plane Walls

For the simple case of steady-state heat flow through a plane wall, the temperature gradient and the heat flow do not vary with time and the cross-sectional area along the heat flow path is uniform. The variables in Eq. (1.10) can then be separated, and the resulting equation is

$$\frac{q_k}{A} \int_0^L dx = -\int_{T_{\text{hot}}}^{T_{\text{cold}}} k\,dT = -\int_{T_1}^{T_2} k\,dT$$

Table 1.1. Thermal conductivities of some metals, non-metallic solids, liquids, and gases.

Material	Thermal conductivity at 300 K (W/m K)
Copper	399.0
Aluminium	273.0
Carbon steel, 1% C	43.0
Glass	0.81
Plastics	0.2–0.3
Water	0.6
Ethylene glycol	0.26
Engine oil	0.15
Freon (liquid)	0.07
Hydrogen	0.18
Air	0.026

The limits of integration can be checked by inspection of Fig. 1.2, where the temperature at the left face ($x = 0$) is uniform at T_{hot} and the temperature at the right face ($x = L$) is uniform at T_{cold}.

If k is independent of T, we obtain, after integration, the following expression for the rate of heat conduction through the wall :

$$q_k = \frac{Ak}{L}(T_{hot} - T_{cold}) = \frac{\Delta T}{L / Ak}$$

In this equation ΔT, the difference between the higher temperature T_{hot} and the lower temperature T_{cold}, is the driving potential that causes the flow of heat. The quantity L/Ak is equivalent to a thermal resistance R_k that the wall offers to the flow of heat by conduction :

$$R_k = \frac{L}{Ak} \qquad \text{...(1.12)}$$

The reciprocal of the thermal resistance is referred to as the thermal conductance K_k, defined by

$$K_k = \frac{Ak}{L} \qquad \text{...(1.13)}$$

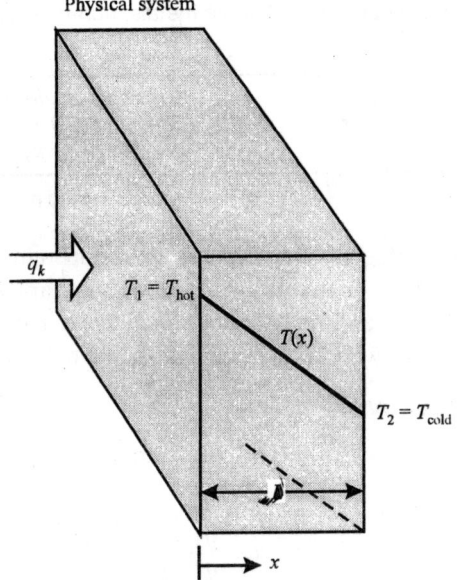

Physical system

q_k

$T_1 = T_{hot}$

$T(x)$

$T_2 = T_{cold}$

x

Thermal circuit

T_1 T_2

$R_k = \dfrac{L}{Ak}$

q_k

Fig. 1.2. Temperature distribution for steady-state conduction through a plane wall.

The ratio k/L in Eq. (1.13), the thermal conductance per unit area, is called the unit thermal conductance for conduction heat flow. The subscript k indicates that the transfer mechanism is conduction. The thermal conductance has the units of watts per kelvin temperature difference (Btu/h °F in the engineering system) and the thermal resistance has the units kelvin per watt (h °F/Btu in the engineering system) The concepts of resistance and conductance are helpful in the analysis of thermal systems, where several modes of heat transfer occur simultaneously.

For many materials, the thermal conductivity can be approximated as a linear function of temperature over limited ranges of temperature :

$$k(T) = k_0(1 + \beta_k T) \qquad\qquad ...(1.14)$$

where β_k is an empirical constant and k_0 is the value of conductivity at a reference temperature. In such cases, integration of Eq. (1.1) gives

$$q_k = \frac{k_0 A}{L}\left[(T_1 - T_2) + \frac{\beta_k}{2}(T_1^2 - T_2^2)\right] \qquad ...(1.15)$$

or

$$q_k = \frac{k_{av} A}{L}(T_1 - T_2) \qquad ...(1.16)$$

where k_{av} is the value of k at the average temperature $(T_1 + T_2)/2$.

The temperature distributions for a constant value of thermal conductivity $(\beta_k = 0)$ and for thermal conductivity increasing $(\beta_k > 0)$ and decreasing $(\beta_k < 0)$ with temperature are shown in Fig. 1.3.

Thermal Conductivity

According to Fourier's law, Eq. (1.10), the thermal conductivity is defined as

$$k \equiv \frac{q_k / A}{dT / dx}$$

For engineering calculations we generally use experimentally measured values of thermal conductivity, although for gases at moderate temperatures the kinetic theory of gases may be used to predict the experimental values accurately. Theories have also been proposed for other materials to calculate thermal conductivities, but in the case of liquids and solids, theories are not adequate to predict the thermal conductivity with satisfactory accuracy. Table 1.1 lists values of thermal conductivity for several materials. Note that the best conductors are pure metals and the poorest ones are gases. In between lie alloys, non-metallic solids, and liquids.

The mechanism of thermal conduction in a gas can be explained on a molecular level from basic concepts of the kinetic theory of gases. The kinetic energy of a molecule is related to its temperature. Molecules in a high-temperature region have higher velocities than those in a lower-temperature region. But molecules are in continuous random motion, and as they collide with one another they exchange energy as well as momentum. When a molecule moves from a higher-temperature region to a lower-temperature region, it transports kinetic energy from the higher- to the lower-temperature part of the system. Upon collision with slower molecules, it gives up some of this energy and increases the energy of molecules with a lower energy content. In this manner thermal energy is transferred from higher- to lower-temperature regions in a gas by molecular action.

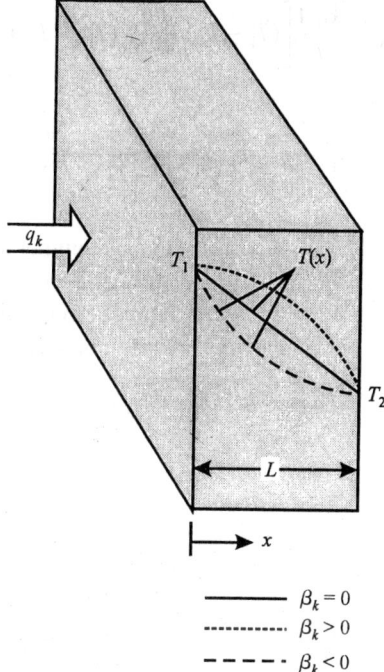

Fig. 1.3. Temperature distribution in conduction through a plane wall with constant and variable thermal conductivity.

In accordance with the above simplified concept, the faster molecules move, the faster they will transport energy. Consequently, the transport property that we have called thermal conductivity should be dependent on the temperature of the gas. A somewhat simplified analytical treatment indicates that the thermal conductivity of a gas is proportional to the square root of the absolute temperature. At moderate pressures the space between molecule is large compared to the size of a molecule; thermal conductivity of gases is therefore essentially independent of pressure. Fig. 1.4. shows how the thermal conductivities of some typical gases vary with temperature.

The basic mechanism of energy conduction in liquids is qualitatively similar to that in gases. However, molecular conditions in liquids are more difficult to describe and the details of the conduction mechanisms in liquids are not as well understood. For most liquids, the thermal conductivity decreases with increasing temperature, but water is a notable exception. The thermal conductivity of liquids is insensitive to pressure, except near the critical point. As a general rule, the thermal conductivity of liquids decreases with increasing molecular weight. For engineering purposes, .

values of the thermal conductivity of liquids are taken from tables as a function of temperature in the saturated state.

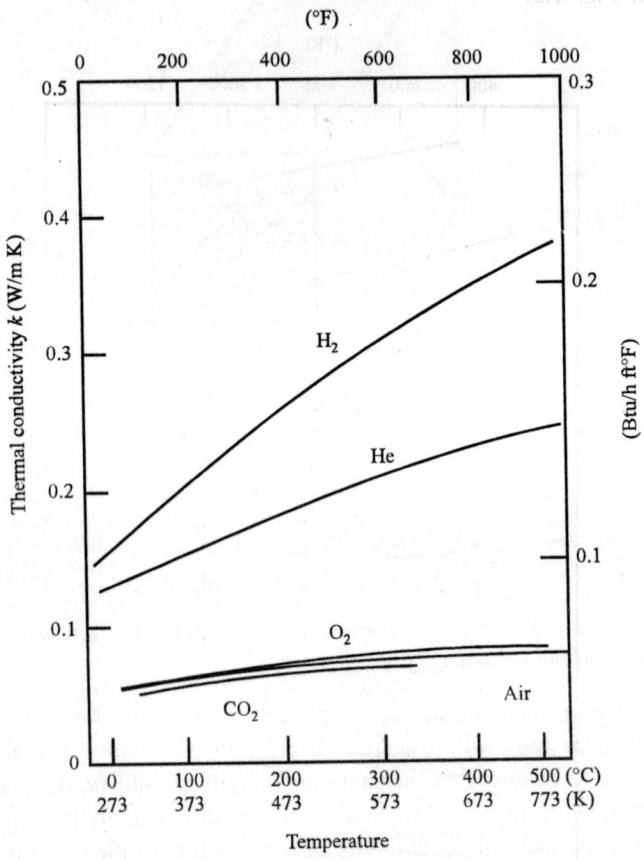

Fig. 1.4. Variation of thermal conductivity with temperature for gases.

According to current theories, solid materials consist of free electrons and of atoms in a periodic lattice arrangement. Thermal energy may thus be conducted by two mechanisms: migration of free electrons and lattice vibration.

These two effects are additive, but, in general, the transport due to electrons is more effective than the transport due to vibrational energy in the lattice structure. Since electrons transport electric charge in a manner similar to the way in which they carry thermal energy from a higher- to a lower-temperature region, good electrical conductors are usually also good heat conductors, whereas good electrical insulators are poor heat conductors. In non-metallic solids there is little or no electronic transport and the conductivity is therefore determined primarily by lattice vibration.

This explains why these materials have a lower thermal conductivity than metals. Thermal conductivities of some typical metals and alloys are shown in Fig. 1.5.

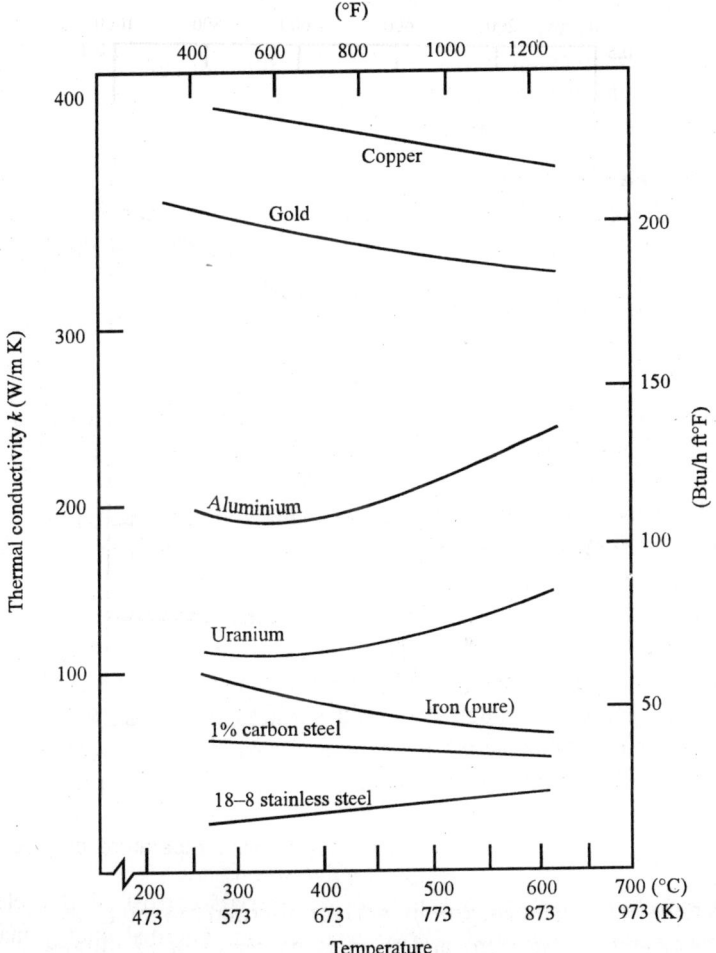

Fig. 1.5. Variation of thermal conductivity with temperature for typical metals and alloys.

An important group of solid materials for heat transfer design are thermal insulators. These materials are solids, but their structure contains air spaces that are sufficiently small to suppress gaseous motion and thus take . advantage of the low thermal conductivity of gases in reducing heat transfer.

Although we usually speak of a thermal conductivity of thermal insulators, in reality the transport through an insulator is comprised of conduction as well as radiation across the interstices filled with gas. In good insulators the spaces containing the air are sealed from each other, as in cellular foams made from plastic or glass.

The thermal conductivity values of insulation systems is always an effective value that accounts for conduction, radiation, and sometimes also convection within the material.

Contact Resistance

When different conducting surfaces are placed in contact, as shown in Fig. 1.6, a thermal resistance is present at the interface of the solids. The interface resistance, frequently called the contact resistance, is developed when two materials will not fit tightly together and a thin layer of fluid is trapped between them.

Examination of an enlarged view of the contact between the two surfaces shows that the solids touch only at peaks in the surface and that the valleys in the mating surfaces are occupied by a fluid (possibly air), a liquid, or a vacuum.

The interface resistance is primarily a function of surface roughness, the pressure holding the two surfaces in contact, the interface fluid, and the interface temperature.

At the interface, the mechanism of heat transfer is complex. Conduction takes place through the contact points of the solid, while heat is transferred by convection and radiation across the trapped interfacial fluid.

If the heat flux through two solid surfaces in contact is q/A and the temperature difference across the fluid gap separating the two solids is ΔT_i, the interface resistance R_i is defined by

$$R_i = \frac{\Delta T_i}{q/A} \qquad ...(1.17)$$

When two surfaces are in perfect thermal contact, the interface resistance approaches zero and there is no temperature difference across the interface. For imperfect thermal contact, a temperature difference occurs at the interface.

Table 1.2 shows the influence of contact pressure on the thermal contact resistance between metal surfaces under vacuum conditions. It is apparent that an increase in the pressure can reduce the contact resistance appreciably. As shown in Table 1.3, the interfacial fluid also affects the thermal resistance. Putting a viscous liquid such as glycerine on the interface reduces the contact resistance between two aluminium surfaces by a factor of 10 at a given pressure.

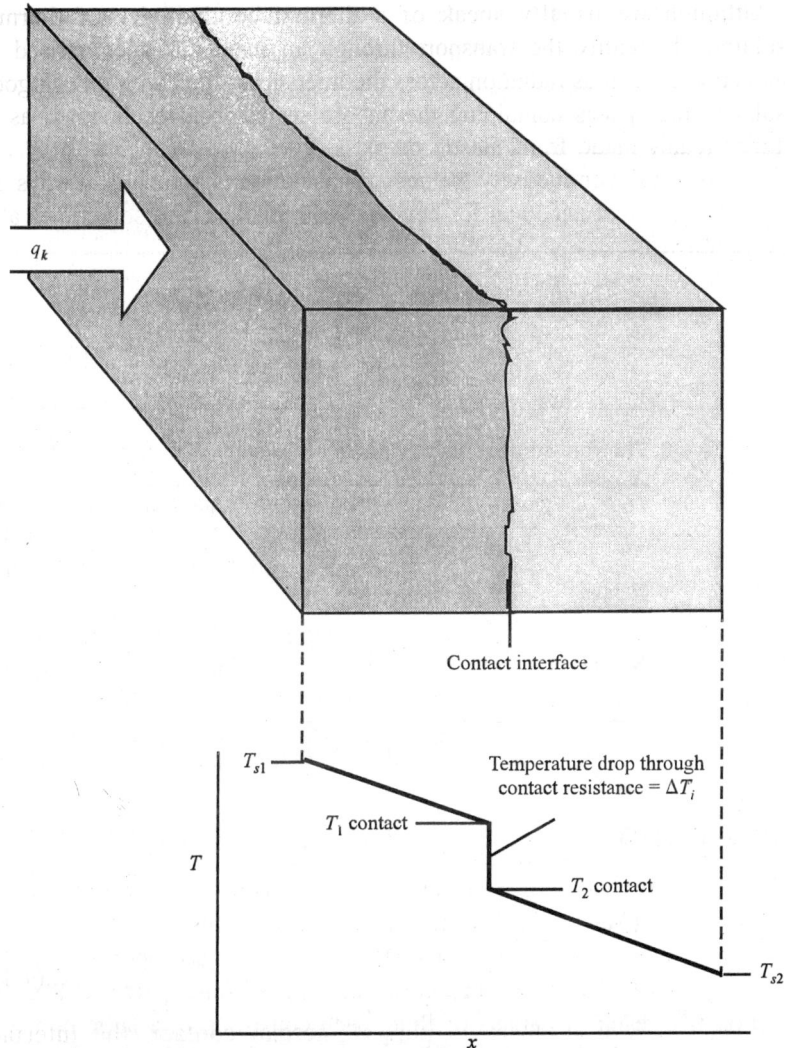

Fig. 1.6. Schematic diagram illustrating physical contact and temperature distribution through a contact interface. (a) Physical model of contact interface; and (b) temperature profile through solids A and B and contact interface.

We should, therefore, always be aware of the existence of the interface resistance and the resulting temperature difference across the interface. Particularly with rough surfaces and low bonding pressures, the temperature drop across the interface can be significant and cannot be ignored. The subject of interface resistance is complex, and no single theory or set of empirical data accurately describes the interface resistance for surfaces of engineering importance.

Table 1.2. Approximate range of thermal contact resistance for metallic interfaces under vacuum conditions.

| | Thermal resistance, $R_i(m^2\ K/W \times 10^4)$ | |
| | Contact pressure | Contact pressure |
Interface material	100 kN/m²	10,000 kN/m²
Stainless steel	6–25	0.7–4.0
Copper	1–10	0.1–0.5
Magnesium	1.5–3.5	0.2–0.4
Aluminium	1.5–5.0	0.2–0.4

Table 1.3. Thermal contact resistance for aluminium-aluminium interface[a].

Fluid	Thermal resistance, $R_i\ (m^2\ K/W)$
Air	2.75×10^{-4}
Helium	1.05×10^{-4}
Hydrogen	0.720×10^{-4}
Silicone oil	0.525×10^{-4}
Glycerine	0.265×10^{-4}

[a] 10 μm surface roughness under 10^5 N/m² contact pressure with different interfacial fluids.

CONVECTION

The convective mode of heat transfer actually consists of two mechanisms operating simultaneously. The first is the energy transfer due to molecular motion, that is, the conductive mode. But superimposed upon this mode is energy transfer by the macroscopic motion of fluid parcels. The fluid motion is a result of parcels of fluid, each consisting of a large number of molecules, moving by virtue of an extraneous force. This extraneous force may be due to a density gradient, as in natural convection, or due to a pressure difference generated by a pump or a fan, or possibly to a combination of the two.

Fig. 1.7 shows a plate at surface temperature T_s and a fluid at temperature T_∞ flowing parallel to the plate. As a result of viscous forces the velocity of the fluid will be zero at the wall and will increase to U_∞ as shown. Since the fluid is not moving at the interface, heat is transferred at that location only by conduction. If we knew the temperature gradient and the thermal conductivity at this interface, we could calculate the rate of heat transfer from equation (1.10), or

$$q_c = -k_{\text{fluid}} A \frac{\partial T}{\partial y}\bigg|_{\text{at } y = 0} \qquad \qquad \ldots(1.18)$$

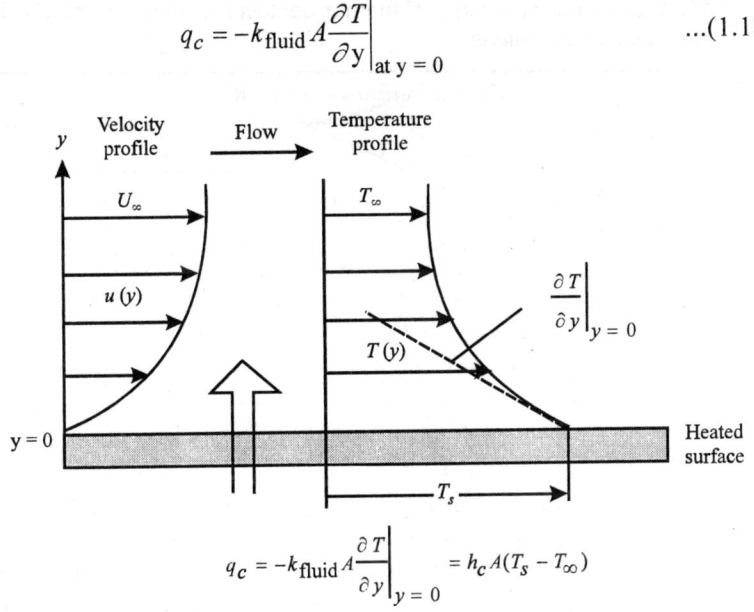

$$q_c = -k_{\text{fluid}} A \frac{\partial T}{\partial y}\bigg|_{y = 0} = h_c A(T_s - T_\infty)$$

Fig. 1.7. Velocity and temperature profile for convection heat transfer from a heated plate with flow over its surface.

But the temperature gradient at the interface depends on the rate at which the macroscopic as well as the microscopic motion of the fluid carries the heat away from the interface. Consequently, the temperature gradient at the fluid plate interface depends on the nature of the flow field, particularly the free-steam velocity U_∞.

The situation is quite similar in natural convection. The principal difference is that in forced convection the velocity far from the surface approaches the free-stream value imposed by an external force, whereas in natural convection the velocity at first increases with increasing distance from the heat transfer surface and then decreases, as shown in Fig. 1.8. The reason for this behaviour is that the action of viscosity diminishes rather rapidly with distance from the surface while the density difference decreases more slowly. Eventually, however, the buoyant force also decreases as the fluid density approaches the value of the unheated surrounding fluid. This interaction of forces will cause the velocity to reach a maximum and then approach zero far from the heated surface. The temperature fields in free and forced convection have similar shapes, and in both cases the heat transfer mechanism at the fluid-solid interface is conduction.

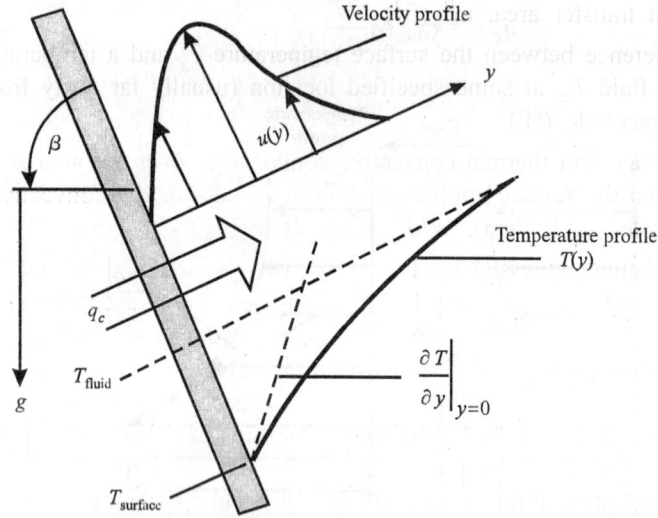

Fig. 1.8. Velocity and temperature distribution for free convection over a heated flat plate inclined at angle β from the horizontal.

The preceding discussion indicates that the convection heat transfer coefficient will depend on the density, viscosity of the fluid as well as on its thermal properties (thermal conductivity and specific heat). Whereas in forced convection the velocity is usually imposed on the system by a pump or a fan and can be directly specified, in free convection the velocity will depend on the temperature difference between the surface and the fluid, the coefficient of thermal expansion of the fluid (which determines the density change per unit temperature difference), and the body force field, which in systems located on the earth is simply located on the earth is simply the gravitational force.

In later chapters we will develop methods for relating the temperature gradient at the interface to the external flow conditions. But for the time being we shall use a simpler approach to calculate the rate of convection heat transfer, is shown below.

Irrespective of the details of the mechanism, the rate of heat transfer by convection between a surface and a fluid may be calculated from the relation

$$q_c = \bar{h}_c A \Delta T \qquad \qquad ...(1.19)$$

where

q_c = rate of heat transfer by convection, W (Btu/h)

A = heat transfer area, m^2 (ft^2)

ΔT = difference between the surface temperature T_s and a temperature of the fluid T_∞ at some specified location (usually far away from the surface), K (°F)

$\bar{h}_c$ = average unit thermal convective conductance over the area A (often called the surface coefficient of heat transfer or the convective heat transfer coefficient), W/m^2 K (Btu/h ft^2 °F)

The relation expressed by Eq. (1.19) was originally proposed by the British scientist Isaac Newton in 1701. Engineers have used this equation for many years, even though it is a definition of $\bar{h}_c$ rather than a phenomenological law of convection. Evaluation of the convective heat transfer coefficient is difficult because convection is a very complex phenomenon. The methods and techniques available for a quantitative evaluation of $\bar{h}_c$ will be presented in later chapters. At this point it is sufficient to note that the numerical value of $\bar{h}_c$ in a system depends on the geometry of the surface and the velocity, as well as on the physical properties of the fluid and often even on the temperature difference ΔT. In view of the fact that these quantities are not necessarily constant over a surface, the convective heat transfer coefficient may also vary from point to point. For this reason we must distinguish between a local and an average convective heat transfer coefficient. The local coefficient h_c is defined by

$$dq_c = h_c \, dA(T_s - T_\infty) \qquad \text{...(1.20)}$$

while the average coefficient $\bar{h}_c$ can be defined in terms of the local value by

$$\bar{h}_c = \frac{1}{A} \iint_A h_c dA \qquad \text{...(1.21)}$$

For most engineering applications, we are interested in average values. For general orientation, typical values of the order of magnitude of average convective heat transfer coefficients encountered in engineering practice are presented in Table 1.4.

Using equation (1.19), we can define the thermal conductance for convective heat transfer K_c as

$$K_c = \bar{h}_c A \quad \text{(W/K)} \qquad \text{...(1.22)}$$

Table 1.4. Order of magnitude of convective heat transfer coefficients $\bar{h}_c$.

Fluid	$W/m^2 K$	$Btu/h\ ft^2\ °F$
Air, free convection	6–30	1–5
Superheated steam or air, forced convection	30–300	5–50
Oil, forced convection	60–1,800	10–300
Water, forced convection	300–6,000	50–2,000
Water, boiling	3,000–60,000	500–10,000
Steam, condensing	6,000–120,000	1,000–20,000

and the thermal resistance to convective heat transfer R_c, which is equal to the reciprocal of the conductance, as

$$R_c = \frac{1}{\bar{h}_c A} \quad (K/W) \qquad \qquad ...(1.23)$$

RADIATION

The quantity of energy leaving a surface as radiant heat depends on the absolute temperature and the nature of the surface. A perfect radiator or blackbody emits radiant energy from its surface at a rate q_r given by

$$q_r = \sigma A_1 T_1^{\,4} \qquad \qquad ...(1.24)$$

The heat flow rate q_r will be in watts if the surface area A_1 is in square metres and the surface temperature T_1 is in kelvins; σ is a dimensional constant with a value of 5.67×10^{-8} W/m^2 K^4. (In the engineering system the heat flow rate will be in Btu's per hour if the surface area is in square feet, the surface temperature in degrees Rankine (R), and σ is 0.1714×10^{-8} Btu/h ft^2 R^4.) The constant σ is the Stefan-Boltzmann constant; it was named after two Austrian scientists, J. Stefan, who in 1879 discovered Eq. (1.24) experimentally, and L. Boltzmann, who in 1884 derived it theoretically.

Inspection in Eq. (1.24) shows that any blackbody surface above a temperature of absolute zero radiates heat at a rate proportional to the fourth power of the absolute temperature. While the rate of radiant heat emission is independent of the conditions of the surroundings, a net transfer of radiant heat requires a difference in the surface temperature of any two bodies between which the exchange is taking place. If the blackbody radiates to an enclosure (Fig. 1.9) that is also black, that is, absorbs all the radiant energy incident upon it, the net rate of radiant heat transfer is given by

$$q_r = A_1\sigma\ (T_1^4 - T_2^4) \qquad\qquad ...(1.25)$$

where T_2 is the surface temperature of the enclosure in kelvins.

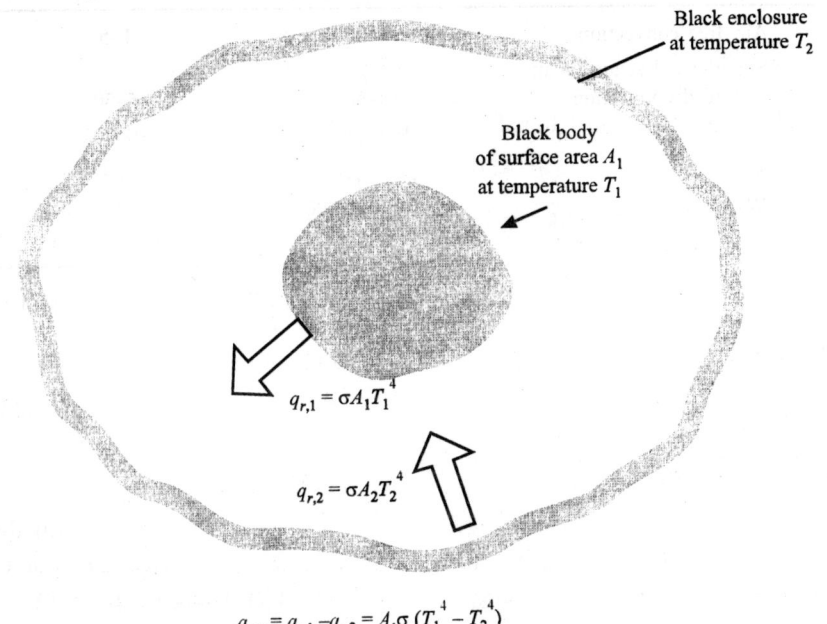

Black enclosure
at temperature T_2

Black body
of surface area A_1
at temperature T_1

$q_{r,1} = \sigma A_1 T_1^4$

$q_{r,2} = \sigma A_2 T_2^4$

$$q_{net} = q_{r,1} - q_{r,2} = A_1\sigma\ (T_1^4 - T_2^4)$$

Fig. 1.9. Schematic diagram of radiation between body 1 and enclosure 2.

Real bodies do not meet the specifications of an ideal radiator but emit radiation at a lower rate than blackbodies. If they emit, at a temperature equal to that of a blackbody, a constant fraction of blackbody emission at each wavelength, they are called gray bodies. A gray body A_1 at T_1 emits radiation at the rate $\varepsilon_1\sigma A_1 T_1^4$, and the net rate of heat transfer between a gray body at a temperature T_1 and a surrounding black enclosure at T_2 is

$$q_r = A_1\varepsilon_1\sigma(T_1^4 - T_2^4) \qquad\qquad ...(1.26)$$

where ε_1 is the emittance of the gray surface and is equal to the ratio of the emission from the gray surface to the emission from a perfect radiator at the same temperature.

If neither of two bodies is a perfect radiator and if the two bodies have a given geometric relationship to each other, the net heat transfer by radiation between them is given by

$$q_r = A_1\mathscr{F}_{1-2}\sigma(T_1^4 - T_2^4) \qquad\qquad ...(1.27)$$

where $\mathscr{F}_{1-2}$ is a dimensionless modulus that modifies the equation for perfect radiators to account for the emittances and relative geometries of the actual bodies.

In many engineering problems, radiation is combined with other modes of heat transfer. The solution of such problems can often be simplified by using a thermal conductance K_r, or a thermal resistance R_r, for radiation. The definition of K_r is similar to that of K_k, the thermal conductance for conduction. If the heat transfer by radiation is written

$$q_r = K_r(T_1 - T'_2) \qquad\qquad ...(1.28)$$

the radiation conductance, by comparison with eq. (1.21), is given by

$$K_r = \frac{A_1 \mathscr{F}_{1-2}\sigma(T_1^4 - T_2^4)}{T_1 - T'_2} \quad \text{W}/\text{K (Btu}/\text{h ft}^2{}^\circ\text{F)} \qquad ...(1.29)$$

The unit thermal radiation conductance, or radiation heat transfer coeficient, $\bar{h}_r$, is then

$$\bar{h}_r = \frac{K_r}{A_1} = \frac{\mathscr{F}_{1-2}\sigma(T_1^4 - T_2^4)}{T_1 - T'_2} \quad \text{W}/\text{m}^2\,\text{K (Btu}/\text{h ft}^2{}^\circ\text{F)} \quad ...(1.30)$$

where T'_2 is any convenient reference temperature whose choice is often dictated by the convection equation. Similarly, the thermal resistance for radiation is

$$R_r = \frac{T_1 - T'_2}{A_1 \mathscr{F}_{1-2}\sigma(T_1^4 - T_2^4)} \qquad\qquad ...(1.31)$$

COMBINED HEAT TRANSFER SYSTEMS

In the preceding sections the three basic mechanisms of heat transfer have been treated separately. In practice, however, heat is usually transferred by several of the basic mechanisms occurring simultaneously. For example, in the winter, heat is transferred from the roof of a house to the colder ambient environment not only by convection but also by radiation, while the heat transfer through the roof from the interior to the exterior surface is by conduction. Heat transfer between the panes of a double-glazed window occurs by convection and radiation acting in parallel, while the transfer through the panes of glass is by conduction with some radiation passing directly through the entire window system. In this section we will examine combined heat transfer problems. We will set up and solve these problems by dividing the heat transfer path into sections that can be connected in series, just like an electrical circuit, with heat being transferred in each section by one or more mechanisms acting in parallel. Table 1.5 summarises the basic relations for the rate equation of each of the three basic heat transfer mechanisms to aid in setting up the thermal circuits or solving combined heat transfer problems.

Table 1.5. Three modes of heat transfer.

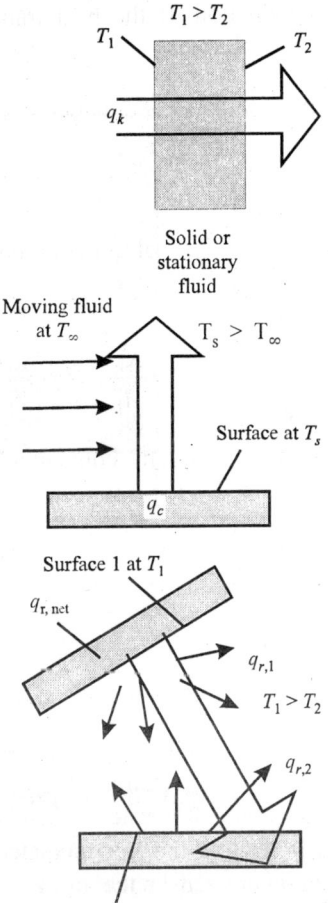

One-dimensional conduction heat transfer through a stationary medium

$$q_k = \frac{kA}{L}(T_1 - T_2)$$

$$R_k = \frac{L}{kA}$$

Convection heat transfer from a surface to a moving fluid

$$q_c = \bar{h}_c A(T_s - T_\infty)$$

$$R_c = \frac{1}{\bar{h}_c A}$$

Net radiation heat transfer from surface 1 to surface 2

$$q_r = A_1 \mathscr{F}_{1-2}\sigma(T_1^4 - T_2^4)$$

$$R_r = \frac{T_1 - T_2}{A_1 \mathscr{F}_{1-2}\sigma(T_1^4 - T_2^4)}$$

Plane Walls in Series and Parallel

If heat is conducted through several plane walls in good thermal contact, as through a multilayer wall of a building, the rate of heat conduction is the same through all sections. However, as shown in Fig. 1.10 for a three-layer system, the temperature gradients in the layers are different. The rate of heat conduction through each layer is q_k, and from Eq. (1.10) we get

$$q_k = \left(\frac{kA}{L}\right)_A (T_1 - T_2) = \left(\frac{kA}{L}\right)_B (T_2 - T_3) = \left(\frac{kA}{L}\right)_C (T_3 - T_4) \quad ...(1.32)$$

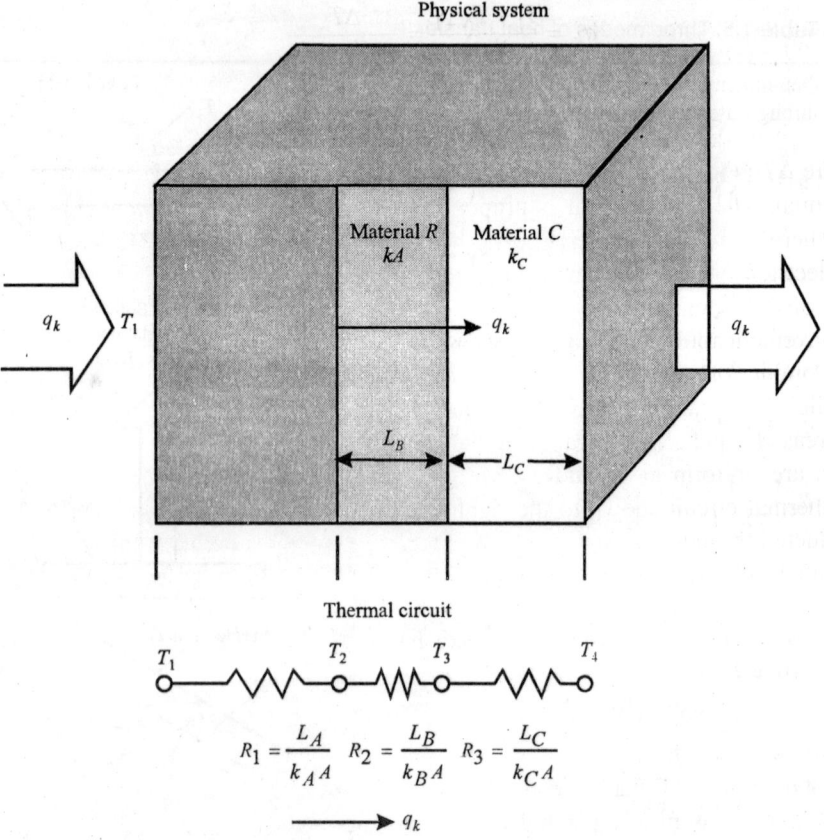

Fig. 1.10. Conduction through a three-layer system in series.

Eliminating the intermediate temperatures T_2 and T_3 in Eq. (1.32), q_k can be expressed in the form

$$q_k = \frac{T_1 - T_4}{(L/kA)_A + (L/kA)_B + (L/kA)_C}$$

Similarly, for N layers in series we have

$$q_k = \frac{\Delta T}{(L/kA)_n} = \frac{T_1 - T_{N+1}}{\sum\limits_{n=1}^{n=N}(L/kA)_n} \quad ...(1.33)$$

where T_1 is the outer-surface temperature of layer 1 and T_{N+1} is the outer-surface temperature of layer N. Using the definition of thermal resistance from Eq. (1.13), Eq. (1.33) becomes

$$q_k = \frac{T_1 - T_{N+1}}{\sum\limits_{n=1}^{n=N} R,k,n} = \frac{\Delta T}{\sum\limits_{n=1}^{n=N} R,k,n} \quad ...(1.34)$$

where ΔT is the overall temperature difference, often called the temperature potential. The flow of heat is proportional to the temperature potential.

There is an analogy between the flow of heat and electricity. The flow of electricity is directly proportional to the voltage potential divided by the sum of the electrical resistances in the circuit. This analogy will be found a convenient tool, especially for visualising more complex situations.

Conduction can occur in a section with two different materials in parallel. For example, Fig. 1.11. shows a slab with two different materials of areas A_A and A_B in parallel. If the temperatures over the left and right faces are uniform at T_1 and T_2, we can analyse the problem in terms of the thermal circuit shown to the right of the physical system. Since heat is conducted through the two materials along separate paths between the same potential, the total rate of heat flow is the sum of the flows through A_1 and A_2:

$$q_k = q_1 + q_2 = \frac{T_1 - T_2}{(L/kA)_A} + \frac{T_1 - T_2}{(L/kA)_B} = \frac{T_1 - T_2}{R_1 R_2 / (R_1 + R_2)} \quad ...(1.35)$$

Note that the total heat transfer area is the sum of A_A and A_B and that the total resistance equals the product of the individual resistances divided by their sum, as in any parallel circuit.

A more complex application of the thermal network approach is illustrated in Fig. 1.12, where heat is transferred through a composite structure involving thermal resistances in series and in parallel. For this system the resistance of the middle layer, R_2 in Fig. 1.12, becomes

$$R_2 = \frac{R_B R_C}{R_B + R_C}$$

and the rate of heat flow is

$$q_k = \frac{\Delta T_{overall}}{\sum\limits_{n=1}^{n=3} R_n} \quad ...(1.36)$$

where N = number of layers in series (three)

R_n = thermal resistance of nth layer

$\Delta T_{overall}$ = temperature difference across two outer surfaces

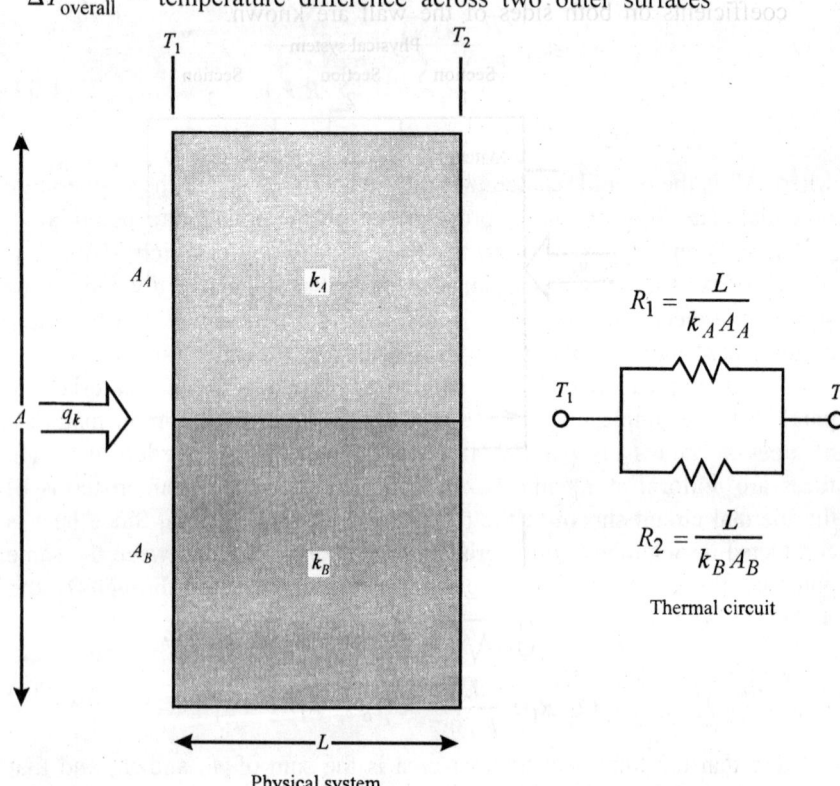

Fig. 1.11. Heat conduction through a wall section with two paths in parallel.

By analogy to Eqs. (1.13) and (1.14), Eq. (1.36) can also be used to obtain an overall conductance between the two outer surfaces :

$$K_k = \left(\sum_{n=1}^{n=N} R_n \right)^{-1}$$ ...(1.37)

Convection and Conduction in Series

In the preceding section we have treated conduction through composite walls when the surface temperatures on both sides are specified. The more common problem encountered in engineering practice, however, is heat being transferred between two fluids separated by a wall with the fluid temperatures specified. In such a situation the surface temperatures are not known, but they can be calculated if the convection heat transfer coefficients on both sides of the wall are known.

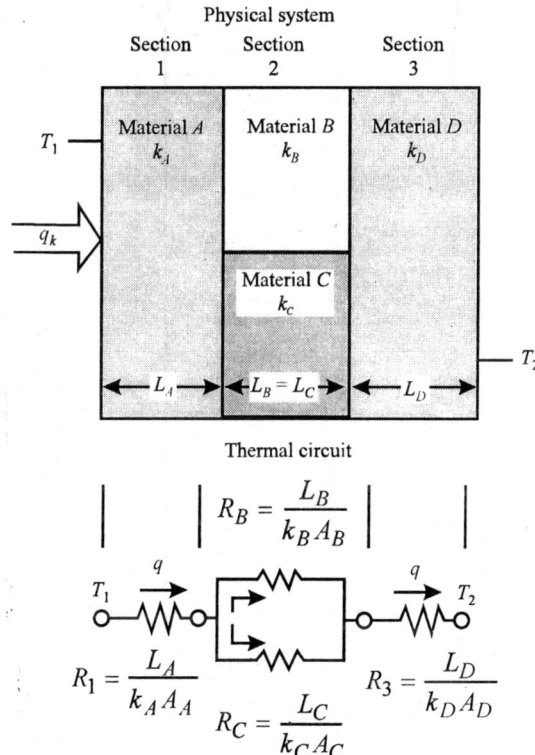

Fig. 1.12. Conduction through a wall consisting of series and parallel thermal paths.

Convection heat transfer can easily be integrated into a thermal network. From Eq. (1.24), the thermal resistance for convection heat transfer is

$$R_c = \frac{1}{\overline{h}_c A}$$

Fig. 1.13 shows a situation where heat is transferred between two fluids separated by a wall. According to the thermal network shown below the physical system, the rate of heat transfer from the hot fluid h at temperature T_h to the cold fluid c at temperature T_c is

$$q = \frac{T_h - T_c}{\sum\limits_{n=1}^{n=3} R_i} = \frac{\Delta T}{R_1 + R_2 + R_3} \qquad ...(1.38)$$

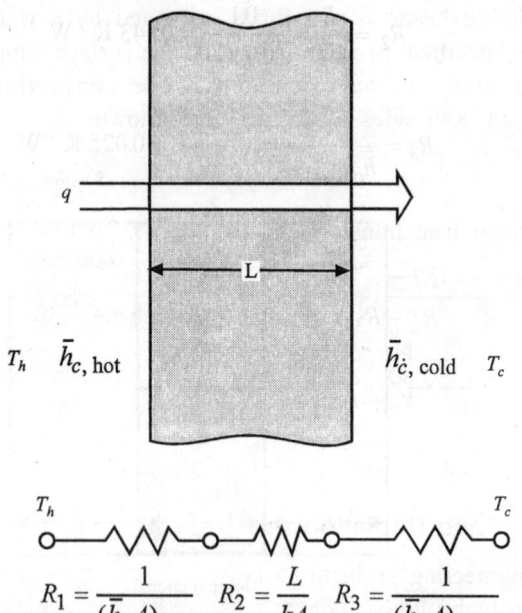

T_h $\bar{h}_{c,\text{hot}}$ $\bar{h}_{\dot{c},\text{cold}}$ T_c

T_h T_c

$$R_1 = \frac{1}{(\bar{h}_c A)_{\text{hot}}} \qquad R_2 = \frac{L}{kA} \qquad R_3 = \frac{1}{(\bar{h}_c A)_{\text{cold}}}$$

Fig. 1.13. Thermal circuit with conduction and convection in series.

where

$$R_1 = \frac{1}{(\bar{h}_c A)_{\text{hot}}}$$

$$R_2 = \frac{L}{kA}$$

$$R_3 = \frac{1}{(\bar{h}_c A)_{\text{cold}}}$$

Solved Examples

Example 1.1. A 0.1-m-thick brick wall (k = 0.7 W/m K) is exposed to a cold wind at 270 K through a convection heat transfer coefficient of 40 W/m² K. On the outer side is calm air at 330 K, with a free-convection heat transfer coefficient of 10 W/m² K. Calculate the rate of heat transfer per unit area (i.e., the heat flux).

Solution. The three resistances are

$$R_1 = \frac{1}{\bar{h}_{c,\text{hot}} A} = \frac{1}{(10)(1)} = 0.10 \text{ K / W}$$

$$R_2 = \frac{L}{kA} = \frac{0.1}{(0.7)(1)} = 0.143 \text{ K / W}$$

$$R_3 = \frac{1}{\bar{h}_{c,\,cold}\, A} = \frac{1}{(40)(1)} = 0.025 \text{ K / W}$$

and the rate of heat transfer per unit area is from eq. (1.37).

$$\frac{q}{A} = \frac{\Delta T}{R_1 + R_2 + R_3} = \frac{(330-270)\text{K}}{(0.10+0.143+0.025)\text{ K / W}} = 223.9 \text{ W}$$

The same approach as used in Example 1.1 can also be used for composite walls, and Fig. 1.14 shows the structure, temperature distribution, and equivalent network for a wall with three layers and convection on both surfaces.

Convection and Radiation in Parallel

In many engineering problems a surface loses or receives thermal energy simultaneously by convection and radiation. For example, the roof of a house heated from the interior is at a higher temperature than the ambient air and thus loses heat by convection as well as radiation. Since both heat flows emanate from the same potential, that is, the roof, they act in parallel. Similarly, the gases in a combustion chamber contain species that emit and absorb radiation. Consequently, the wall of the combustion chamber receives heat by convection as well as radiation. Fig. 1.15 illustrates the concurrent heat transfer from a surface to its surrounding by convection and radiation. The total rate of heat transfer is the sum of the rates of heat flow by convection and radiation, or

$$\begin{aligned} q &= q_c + q_r \\ &= \bar{h}_c A(T_1 - T_2) + h_r A(T_1 - T_2) \\ &= (\bar{h}_c + h_r)A(T_1 - T_2) \end{aligned} \qquad \text{...(1.39)}$$

where $\bar{h}_c$ is the average convection heat transfer coefficient between area A_1 and the ambient air at T_2, and, as shown previously, the radiation heat transfer coefficient between A_1 and the surroundings at T_2 is

$$\bar{h}_r = \frac{\varepsilon_1 \sigma(T_1^4 - T_2^4)}{T_1 - T_2} \qquad \text{...(1.40)}$$

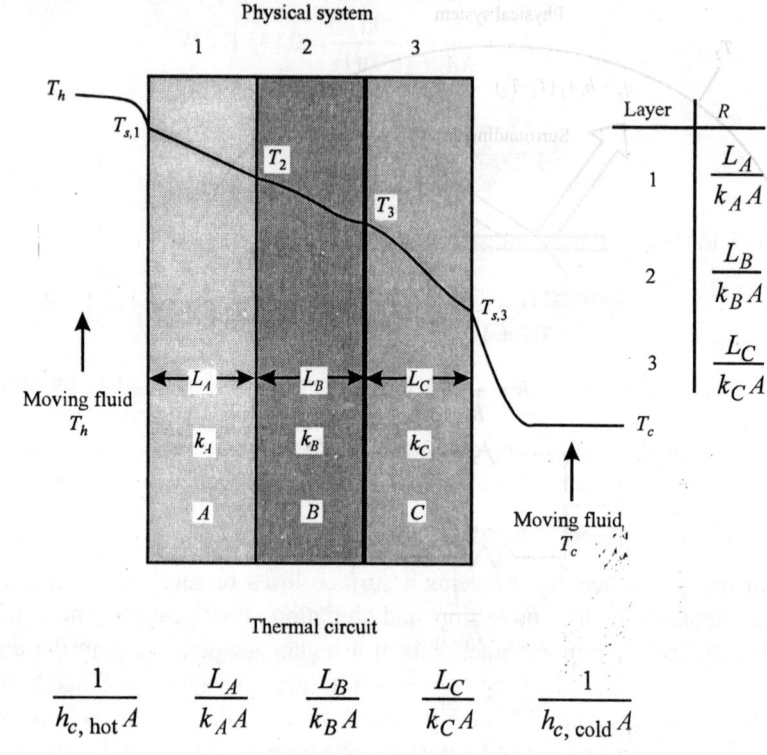

Fig. 1.14. Schematic diagram and thermal circuit for composite three-layer wall with convection over both exterior surfaces.

The analysis of combined heat transfer, especially at boundaries of a complicated geometry or in unsteady-state conduction, can often be simplified by using an effective unit surface conductance that combines convection and radiation. The combined unit surface conductance, or unit surface conductance for short, is defined by

$$\bar{h} = \bar{h}_c + h_r \qquad \qquad ...(1.41)$$

The unit surface conductance specifies the average total rate of heat flow between a surface and an adjacent fluid and the surroundings per unit surface area and unit temperature difference between the surface and the fluid. Its units are $W/m^2 \ K$.

Physical system

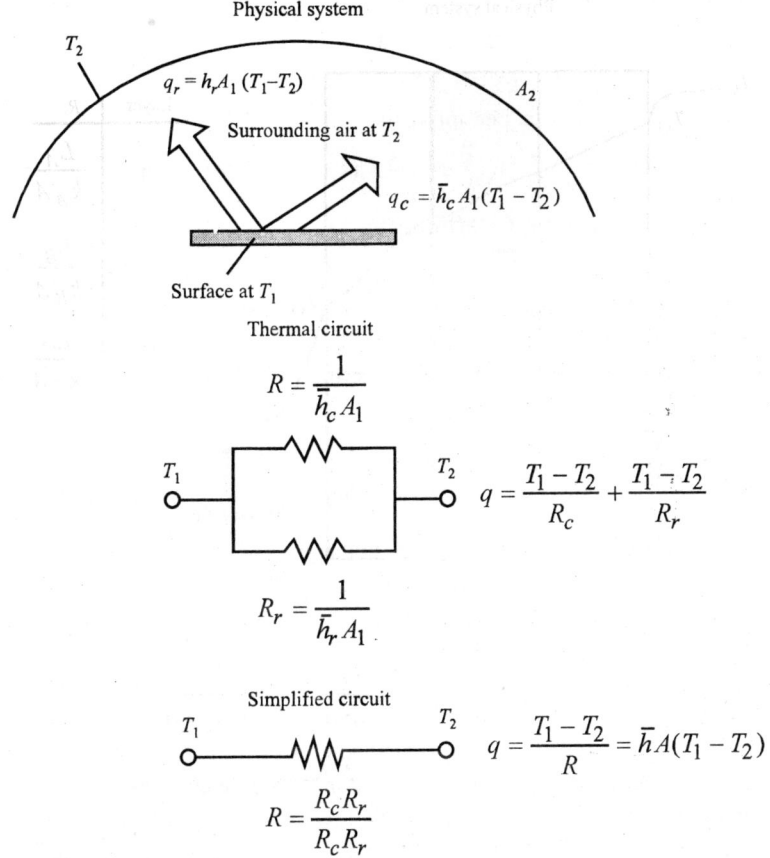

Fig. 1.15. Thermal circuit with convection and radiation acting in parallel.

Example 1.2. A 0.5 m diameter pipe ($\varepsilon = 0.9$) carrying steam has a surface temperature of 500 K. The pipe is located in a room at 300 K and the convection heat transfer coefficient between the pipe surface and the air in the room is 20 W/m² K. Calculate the combined unit surface conductance and the rate of heat loss per metre of pipe length.

Solution. This problem may be idealised as a small object (the pipe) inside a large black enclosure (the room). Noting that

$$\frac{T_1^4 - T_2^4}{T_1 - T_2} = (T_1^2 + T_2^2)(T_1 + T_2)$$

the radiation heat transfer coefficient is, from Eq. (1.40),

$$h_r = \sigma\varepsilon(T_1^2 + T_2^2)(T_1 + T_2) = 13.9 \ \text{W/m}^2 \ \text{K}$$

The combined unit surface conductance is, from Eq. (1.41),

$$h = \bar{h}_c + h_r = 20 + 13.9 = 33.9 \text{ W/m}^2 \text{ K}$$

and the rate of heat loss per metre is

$$q = \pi D L h (T_{pipe} - T_{air}) = \pi(0.5)(1)(33.9)(200) = 10,650 \text{ W}$$

Overall Heat Transfer Coefficient

We noted previously that a common heat transfer problem is to determine the rate of heat flow between two fluids, gaseous or liquid, separated by a wall. If the wall is plane and heat is transferred only by convection on both sides, the rate of heat transfer in terms of the two fluid temperatures is given by Eq. (1.38), or

$$q = \frac{T_h - T_c}{(1/h_c A)_h + (L/kA) + (1/h_c A)_c} = \frac{\Delta T}{R_1 + R_2 + R_3}$$

where the subscripts h and c denote hot and cold and T_h and T_c are the temperatures of the hot and cold fluids, respectively.

In Eq. (1.38) the rate of heat flow is expressed only in terms of an overall temperature potential and the heat transfer characteristics of individual sections in the heat flow path. From these relations it is possible to evaluate quantitatively the importance of each individual thermal resistance in the path. Inspection of the order of magnitudes of the individual terms in the denominator often indicates means of simplifying a problem. When one term dominates quantitatively, it is sometimes permissible to neglect the rest. As we gain facility in the techniques of determining individual thermal resistances and conductances, there will be numerous occasions where such approximations will be illustrated. There are, however, certain types of problems, notably in the design of heat exchangers, where it is convenient to simplify the writing of Eq. (1.38) by combining the individual resistances or conductances of the thermal system into one quantity, called the overall unit conductance, the overall transmittance, or the overall coefficient of heat transfer U. The use of an overall coefficient is a convenience in notation, and it is important not to lose sight of the significance of the individual factors that determine the numerical value of U.

Writing Eq. (1.38) in terms of an overall coefficient gives

$$q = UA\Delta T_{total} \qquad \qquad ...(1.42)$$

where

$$UA = \frac{1}{R_1 + R_2 + R_3} = \frac{1}{R_{\text{total}}} \qquad \text{...(1.43)}$$

The overall coefficient U may be based on any chosen area. This becomes particularly important in heat transfer through the walls of tubes in a heat exchanger, and to avoid misunderstandings the area basis of an overall coefficient should always be stated.

An overall heat transfer coefficient can also be obtained in terms of individual resistances in the thermal circuit when convection and radiation transfer heat to and/or from one or both surfaces of the wall. In general, radiation will not be of any significance when the fluid is a liquid, but can play an important role in convection to or from a gas when the temperatures are high or the convection heat transfer coefficient is small, for instance, in free convection. The integration of radiation into an overall heat transfer coefficient will be illustrated below.

The schematic diagram in Fig. 1.16 shows the heat transfer from hot products of combustion in the chamber of a rocket motor through a wall that is liquid-cooled on the outside by convection. In the first section of this system heat is transferred by convection and radiation in parallel. Hence, the rate of heat flow to the interior surface of the wall is the sum of the two heat flows.

$$q = q_c + q_r$$

$$= \bar{h}_c A(T_g - T_{sg}) + h_r A(T_g - T_{sg})$$

$$= (\bar{h}_c + h_{r1}) A(T_g - T_{sg}) = \frac{T_g - T_{sg}}{R_1} \qquad \text{...(1.44)}$$

where $T_g = T_1$ = temperature of the hot gas in the interior

$T_{sg} = T_2$ = temperature of the hot wall surface

$h_{r1} = \dfrac{\sigma A(T_g^4 - T_{sg}^4)}{T_g - T_{sg}}$ = the radiation heat transfer coefficient in the

first section (ε is assumed unity)

$\bar{h}_c$ = convection heat transfer coefficient from gas to wall

$R_1 = \dfrac{1}{(h_r + \bar{h}_{c1})A}$ = combined thermal resistance of first section

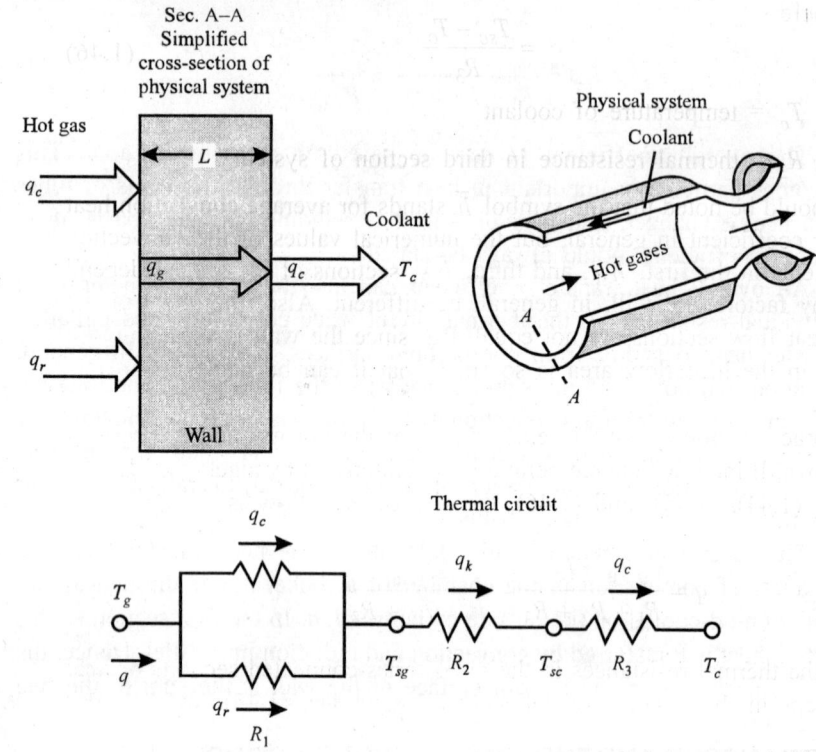

Fig. 1.16. Heat transfer from combustion gases to a liquid coolant in a rocket motor.

In the steady state, heat is conducted through the shell, the second section of the system, at the same rate as to the surface and

$$q = q_k = \frac{kA}{L}(T_{sg} - T_{sc})$$

$$= \frac{T_{sg} - T_{sc}}{R_2} \qquad \text{...(1.45)}$$

where

T_{sc} = surface temperature at wall on coolant side

R_2 = thermal resistance of second section

After passing through the wall, the heat flows through the third section of the system by convection to the coolant. The rate of heat flow in the last step is

$$q = q_c = \bar{h}_{c3}A(T_{sc} - T_c)$$

$$= \frac{T_{sc} - T_c}{R_3} \qquad \qquad ...(1.46)$$

where T_c = temperature of coolant

R_3 = thermal resistance in third section of system

It should be noted that the symbol $\bar{h}_c$ stands for average convection heat transfer coefficient in general, but the numerical values of the convection coefficients in the first, $\bar{h}_{c1}$, and third, $\bar{h}_{c3}$, sections of the system depend on many factors and will, in general, be different. Also, the areas of the three heat flow sections are not equal. But since the wall is very thin, the change in the heat flow area is so small that it can be neglected in this system.

In practice, often only the temperatures of the hot gas and the coolant are known. If intermediate temperatures are eliminated by algebraic addition of Eqs. (1.44), (1.45), and (1.46), the rate of heat flow is

$$q = \frac{T_g - T_c}{R_1 + R_2 + R_3} = \frac{\Delta T_{total}}{R_1 + R_2 + R_3} \qquad ...(1.47)$$

where the thermal resistances in the three series-connected sections or heat flow steps in the system are defined in Eqs. (1.44), (1.45), and (1.46).

HEAT TRANSFER AND THE LAW OF ENERGY CONSERVATION

In addition to the heat transfer rate equations we shall also often use the first law of thermodynamics, the law of conservation of energy, in analysing a system. Although, as mentioned previously, a thermodynamic analysis alone cannot predict the rate at which the transfer will occur in terms of the degree of thermal non-equilibrium, the basic laws of thermodynamics must be obeyed and any physical law that must be satisfied by a process or a system provides an equation that can be used for analysis. We have already used the second law of thermodynamics to indicate the direction of heat flow. We will now demonstrate how the first law of thermodynamics can be applied in the analysis of heat transfer problems.

First Law of Thermodynamics

The first law of thermodynamics states that energy cannot be created or destroyed, but may be transformed from one form to another or transferred as heat or work. To apply the law of conservation of energy, we first need to identify a control volume. A control volume is a fixed region in space bounded by a control surface through which heat, work, and mass can pass. The conservation of energy requirement for an open system in a form useful for heat transfer analysis is: *The rate at which thermal and mechanical energies enter a control volume plus the rate at which energy is generated within that volume minus the rate at which thermal and mechanical energies leave the control volume must equal the rate at which energy is stored inside this volume.*

If the sum of the energy inflow and the generation exceeds the outflow, there will be an increase in the amount of energy stored in the control volume, whereas when the outflow exceeds the inflow and generation there will be a decrease in energy storage. But when there is no generation and the rate of energy inflow is equal to the rate of outflow, steady state exists and there is no change in the energy stored in the control volume.

Referring to Fig. 1.17, the energy conservation requirements may be expressed in the form

$$(e\dot{m})_{\text{in}} + q + \dot{q}_G\,(e\dot{m})_{\text{out}} - W_{\text{out}} = \frac{\partial E}{\partial t} \qquad \text{...(1.48)}$$

where $(e\dot{m})_{\text{m}}$ is the rate of energy inflow, $(e\dot{m})_{\text{out}}$ is the rate of energy outflow, q is the net rate of heat transfer into the control volume ($q_{\text{in}} - q_{\text{out}}$), W_{out} is the net rate of work output, q_G is the rate of energy generation within the control volume, and $\partial E/\partial t$ is the rate of energy storage inside the control volume.

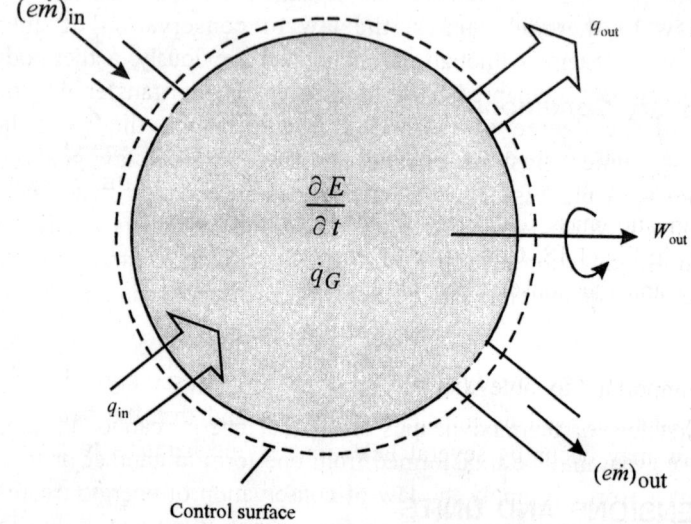

Fig. 1.17. Control volume for first law of thermodynamics or conservation of energy.

The specific energy carried by the mass flow, e, across the surface may contain potential and kinetic as well as thermal (internal) forms. But for most heat transfer problems the potential and kinetic energy terms are negligible. The inflow and outflow energy terms may also include work interactions, but these phenomena are of significance only in extremely high-speed flow processes.

Observe that the inflow and outflow rate terms are surface phenomena and are, therefore, proportional to the surface area. The internal energy generation term $\dot{q}_G$ is encountered when a form of energy (such as chemical, electrical, or nuclear energy) is converted to thermal energy within the control volume. The generation term is therefore a volumetric phenomenon and its rate is proportional to the volume within the control surface. Energy storage is also a volumetric phenomenon associated with the internal energy of the mass in the control volume. But the process of energy generation is quite different from that of energy storage, although both will contribute to the rate of energy storage.

Equation (1.48) can be simplified when there is no transport of mass across the boundary. Such a system is called a closed system, and eq. (1.48) for such conditions becomes

$$q + \dot{q}_G - W_{out} = \frac{\partial E}{\partial t} \qquad \text{...(1.48a)}$$

where the right side represents the rate of energy storage or the rate of increase in internal energy. Note that E is the total internal energy stored in the system and equals the product of the specific internal energy and the mass of the system.

Boundary Conditions

There are many situations in which the conservation of energy requirement is applied at the surface of a system. In these cases the control surface contains no mass and the volume it encompasses approaches zero, as shown in Fig. 1.18. Consequently, there can be no storage or generation of energy and the conservation requirement reduces to

$$q_{net} = q_{in} - q_{out} = 0 \qquad \text{...(1.49)}$$

It is important to note that in this form the conservation law holds for steady-state as well as transient conditions and that the heat inflow and outflow may occur by several heat transfer mechanisms in parallel.

DIMENSIONS AND UNITS

This section introduces systems of units and defines the system used in this book. It is important not to confuse the meaning of the terms units and dimensions. Dimensions are our basic concepts of measurements such as length, time, and temperature. For example, the distance between two points is a dimension called length. Units are the means of expressing dimensions numerically, for instance, metre or foot for length, second or hour for time. Before numerical calculations can be made, dimensions must be quantified by units.

Several different systems of units are in use throughout the world. The SI systems (Système International d'Unités) has been adopted by the International Organisation for Standardisation and is recommended by most US national standard organisations. We will therefore use the SI systems of units in this book. But in the United States the English system of units is still widely used. It is, therefore, important to be able to change from one set of units to another. To be able to communicate with engineers who are still in the habit of using the English systems, several examples in the book will be worked in the English system.

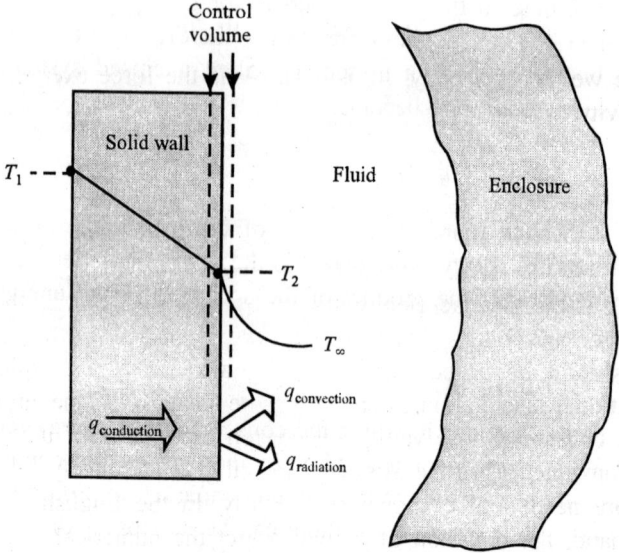

Fig. 1.18. Application of conservation of energy law at the surface of a system.

The basic SI units are those for length, mass, time, and temperature. The unit of force, the newton, is obtained from Newton's second law of motion, which states that force is proportional to the time rate of change of momentum. For a given mass, Newton's law can be written in the form

$$F = \frac{1}{g_c} ma \qquad \qquad ...(1.50)$$

where F is the force, m is the mass, a is the acceleration, and g_c is a constant whose numerical value and units depend on those selected for F, m, and a.

In the SI system the unit of force, the newton, is defined as

$$1 \text{ newton} = \frac{1}{g_c} \times 1 \text{ kg} \times 1 \text{ m/s}^2$$

Thus we see that

$$g_c \equiv 1 \text{ kg m/newtons s}^2$$

In the English system we have the relation

$$1 \text{ lb}_f = \frac{1}{g_c} \times 1 \text{ lb}_m \times g \text{ ft/s}^2$$

The numerical value of the conversion constant g_c is determined by the acceleration imparted to a 1-lb mass by a 1-lb force, or

$$g_c = 32.174 \text{ ft lb}_m/\text{lb}_f \text{ s}^2$$

The weight of a body, W, is defined as the force exerted on the body by gravity.

Thus,

$$W = \frac{g}{g_c} m$$

where g is the local acceleration due to gravity. Weight has the dimensions of a force and 1 kg_{mass} will weigh 1 kg_{force} at sea level.

It should be noted that g and g_c are not similar quantities. The gravitational acceleration g depends on the location and the altitude, whereas g_c is a constant whose value depends on the system of units. One of the great conveniences of the SI system is that g_c is identically equal to one and therefore need not be shown specifically. In the English system, on the other hand, the omission of g_c will affect the numerical answer and it is therefore imperative that it be included and clearly displayed in analysis and especially in numerical calculations.

In the SI systems with the fundamental units of metre, kilogram, second, and kelvin, the units for both force and energy or heat are derived units. The joule (newton metre) is the only energy unit in the SI system, and the watt (joule per second) is the corresponding unit of power. In the engineering system of units, on the other hand, the Btu is the unit for heat or energy. It is based on thermal phenomena and is defined as the energy required to raise 1 lb_m of water 1°F at 68°F.

The conversions between the Btu and other common energy units are given below :

$$1 \text{ Btu} = 1054.35 \text{ J (or newton metre)}$$

$$1 \text{ Btu} = 778.16 \text{ ft lb}_f$$

The SI unit of temperature is the kelvin, but use of the Celsius temperature scale is widespread and generally considered permissible. The kelvin is based on the thermodynamic scale, while zero on the Celsius scale (0°C) corresponds to the freezing temperature of water and is equivalent to 273.15 K on the thermodynamic scale. Thus, the relation between a temperature in kelvin and degrees Celsius or Centigrade is

$$K = 273.15 + {}^\circ C$$

Note, however, that temperature differences are numerically equivalent in K and °C, since 1 K is equal to 1°C.

In the engineering system of units the temperature is usually expressed in degrees Fahrenheit (°F), or, on the thermodynamic temperature scale, in degrees Rankine (°R). Here 1 K is equal to 1.8°R and conversions for other temperature scales are given below :

$$^\circ F = 1.8 {}^\circ C + 32$$

$$^\circ R = {}^\circ F + 459.69$$

$$^\circ R = 1.8 \text{ K}$$

$$^\circ C = \frac{{}^\circ F - 32}{1.8}$$

SOLVED EXAMPLE

Example 1.1. It is necessary to determine by experiment the distribution of temperature in a long shaft $d = 400$ mm in a time interval $\tau = 2.5$ h after the shaft was charged into a furnace.

The thermal conductivity and thermal diffusivity of steel are respectively: $\lambda = 42$ W/m.°C, $a = 1.18 \times 10^{-5}$ m²/s. In the furnace the local coefficient of heat transfer to the shaft is equal to $\alpha = 116$ W/m².°C.

It is decided to conduct the investigation in a small furnace using for a test specimen a geometrically similar model of the shaft fabricated from alloyed steel. For the model $\lambda_m = 16$ W/m.°C; $a_m = 0.53 \times 10^{-5}$ m²/s, $\alpha_m = 150$ W/m².°C.

Determine the diameter d_m of the model and how soon after the model is charged into the furnace it is necessary to measure the distribution of temperature in the model.

Solution. The temperature fields of the shaft and its model will be similar when the dimensionless terms for the shaft and model will be the same:

$$Bi_m = Bi \text{ and } Fo_m = Fo.$$

For the shaft the dimensionless terms

$$Bi = \frac{\alpha r}{\lambda} = \frac{116 \times 0.2}{42} = 0.552;$$

$$Fo = \frac{a\tau}{r^2} = \frac{1.18 \times 10^{-5} \times 9 \times 10^3}{0.2^2} = 2.66$$

From the condition $Bi_m = Bi$ we find the diameter for the shaft model

$$d_m = 2r_m = 2\frac{\lambda_m}{\alpha_m}Bi = 2\frac{16}{150}0.552 = 0.1175 \text{ m}.$$

From the condition $Fo_m = Fo$ we find the unknown time interval:

$$\tau_m = \frac{r_m^2}{a_m}Fo = \frac{(0.05875)^2}{0.53 \times 10^{-5}}2.66 = 1735 \text{ s.}$$

PROBLEMS

Problem 1.1. Determine the diameter of the model of the shaft d_m and the required value of the local heat-transfer coefficient α_m at which, under the conditions assumed in problem, the temperature fields shall become similar in time $t_m = 15$ min after the model of the shaft is charged into the furnace.

Also determine the relations of the linear dimensions, time and temperatures for the shaft and model, if it is known that at the moment of charging their temperatures and the temperature inside the furnace were respectively equal to:

$$t_0 = 10°C, \ t_{0m} = 20°C, \ t_f = 1000°C, \ t_{fm} = 200°C.$$

Answer. $d_m = 85$ mm; $\alpha_m = 208$ W/m².°C;

$$r/r_m = 4.7 \ ; \ \tau/\tau_m = 10; \ t = 5.5 \ t_m - 100.$$

Problem 1.2. The hot water flowing through a tube having a diameter $d = 16$ mm and length $l = 2.1$ m transfers heat through the wall of the tube to the medium to which the outside surface of the tube is exposed. The rate of water flow through the tube $G = 0.0091$ kg/s; the water inlet temperature

t_{f1} = 87.2°C; the water outside temperature t_{f2} = 29°C; the mean temperature of the wall of the tube t_w = 15.3°C.

Calculate the magnitudes of the dimensionless terms Nu, Re and Pe, assuming as the reference temperature the arithmetic mean temperature of the water. Relate the heat-transfer coefficient to the arithmetic mean difference between the temperatures of the water and wall.

Answer. Nu = 11.9; Re = 1485; Pe = 4600.

CHAPTER 2

Fourier Conduction Equation

INTRODUCTION

Conduction is primarily a molecular phenomenon requiring temperature gradient as the driving force. The functions obtained by successive differentiations, which are employed in the development of infinite series and in the solution of numerical equations, correspond also to physical properties. The first of these functions, or the fluxion properly so-called, expresses in geometry the inclination of the tangent of a curved line, and in dynamics the velocity of a moving body when the motion varies; in the theory of heat it measures the quantity of heat which flows at each point of a body across a given surface. Mathematical analysis has therefore necessary relations with sensible phenomena; its object is not created by human intelligence; it is a pre-existent element of the universal order, and is not in any way contingent or fortuitous; it is imprinted throughout all nature.

In general, the temperature distribution throughout a body varies both with location and time. If this distribution as a function of the space coordinates and time is known or can be determined, the heat flow at any point in any direction can be specified by appropriate differentiation. In this chapter we shall derive the differential equation governing the temperature distribution in a homogeneous solid. The determination of solutions satisfying this equation and the boundary conditions of particular problems will be treated in this chapter.

DERIVATION OF THE FOURIER EQUATION

The fundamental law of heat conduction for one direction is given by

$$\Delta Q = -kA \frac{dt}{dx} \Delta \tau \qquad \qquad ...(2.1)$$

We shall see now how its application in three mutually perpendicular directions leads to the differential equation of heat conduction.

Assume that the temperature is varying with time and location in the isotropic medium shown in Fig. 2.1. Select the cartesian system of coordinates, ox, oy, and oz for reference and consider a small rectangular parallelepiped of edges Δx, Δy, and Δz parallel, respectively, to these three axes. Let t denote the temperature at the centre of this element of volume; then, we may express its value on any face of the parallelepiped—this being so small that the temperature is effectively uniform over one face—as being greater or less than this mean temperature t by a small amount. Taking Δx, Δy, and Δz sufficiently small, this magnitude for the $\Delta y \Delta z$ faces can be expressed as

$$\frac{1}{2}\frac{\partial t}{\partial x}\Delta x \qquad \text{...(2.2)}$$

where $\partial t/\partial x$ is the temperature gradient in the x direction and the distance of $\Delta y \Delta z$ from the centre is evidently $\frac{1}{2}\Delta x$. The temperatures of the left and right hand faces may then be written

$$t_L = t - \frac{1}{2}\frac{\partial t}{\partial x}\Delta x, \quad t_R = t + \frac{1}{2}\frac{\partial t}{\partial x}\Delta x \qquad \text{...(2.3)}$$

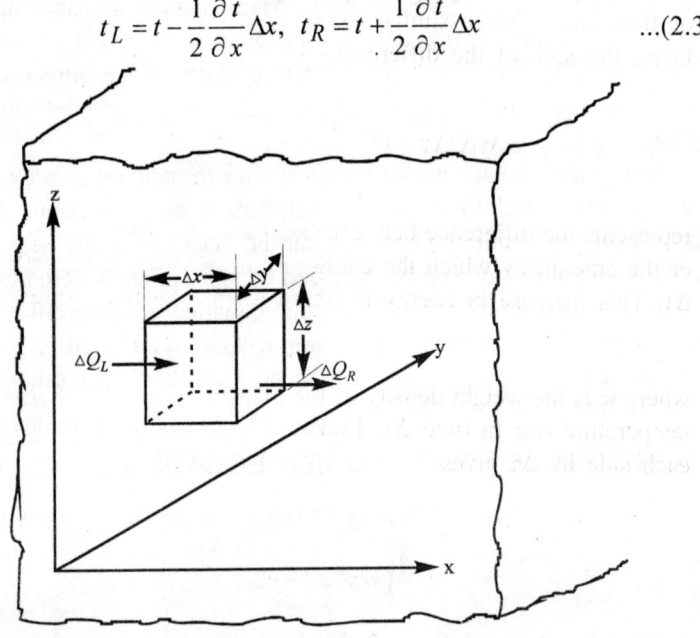

Fig. 2.1. Elementary parallelepiped in medium through which heat is flowing.

We now set up a "heat balance" equation for the element $\Delta x \Delta y \Delta z$. By applying equation (2.1) the flow of heat during the time interval $\Delta \tau$ in the positive x direction through the left-hand face $\Delta y \Delta z$ is

$$\Delta Q_L = -k \Delta y \Delta z \frac{\partial}{\partial x} \left(t - \frac{1}{2} \frac{\partial t}{\partial x} \Delta x \right) \Delta \tau \qquad \ldots (2.4)$$

and through the right-hand face in the same direction

$$\Delta Q_R = -k \Delta y \Delta z \frac{\partial}{\partial x} \left(t + \frac{1}{2} \frac{\partial t}{\partial x} \Delta x \right) \Delta \tau \qquad \ldots (2.5)$$

Subtracting the heat flow out the right-hand face from that in the left gives

$$\Delta Q_L - \Delta Q_R = -k \Delta y \Delta z \left\{ \frac{\partial}{\partial x} \left(t - \frac{1}{2} \frac{\partial t}{\partial x} \Delta x \right) - \frac{\partial}{\partial x} \left(t + \frac{1}{2} \frac{\partial t}{\partial x} \Delta x \right) \right\} \Delta \tau$$

$$= k \Delta y \Delta z \frac{\partial^2 t}{\partial x^2} \Delta x \Delta \tau \qquad \ldots (2.6)$$

This represents a gain in energy of the element due to the x component of flow alone. Since similar expressions hold for the other two pairs of faces, the sum of the differences for the three coordinate directions,

$$k \frac{\partial^2 t}{\partial x^2} \Delta x \Delta y \Delta z \Delta \tau + k \frac{\partial^2 t}{\partial y^2} \Delta x \Delta y \Delta z \Delta \tau + k \frac{\partial^2 t}{\partial z^2} \Delta x \Delta y \Delta z \Delta \tau \qquad \ldots (2.7)$$

represents the difference between the total inflow and total outflow of heat, or the amount by which the energy in the element is increased in the time $\Delta \tau$. This increase in energy is stored within the element and is equal to

$$wc \Delta x \Delta y \Delta z \Delta t \qquad \ldots (2.8)$$

where w is the weight density of the material, c its specific heat, and Δt the temperature rise in time $\Delta \tau$. Equating these two expressions and dividing each side by $\Delta \tau$ gives

$$k \left(\frac{\partial^2 t}{\partial x^2} + \frac{\partial^2 t}{\partial y^2} + \frac{\partial^2 t}{\partial z^2} \right) = wc \frac{\Delta t}{\Delta \tau} \qquad \ldots (2.9)$$

As the time interval $\Delta t \to 0$ this becomes

$$\frac{k}{wc}\left(\frac{\partial^2 t}{\partial x^2} + \frac{\partial^2 t}{\partial y^2} + \frac{\partial^2 t}{\partial z^2}\right) = \frac{\partial t}{\partial \tau} \qquad \text{...(2.10)}$$

Equation (2.10) is known as the Fourier equation. It expresses the conditions which govern the flow of heat in a body and consequently any solution to a particular problem must satisfy it.

THERMAL DIFFUSIVITY

During steady-state heat conduction the only property of a substance which determines the temperature distribution is its thermal conductivity.

When the temperature changes with time, the thermal storage capacity (the product of the density and specific heat) of a substance, in addition to its thermal conductivity, influences the temperature variation. In this case equation (2.10) shows that the behaviour of different substances will vary as the ratio of the thermal conductivity to the thermal capacity. It is therefore convenient to define this combination of the properties

$$\frac{k}{wc} = \alpha \qquad \text{...(2.11)}$$

as a single property; it is called the thermal diffusivity and is denoted by α.

POTENTIAL FIELD EQUATION

Introducing the commonly used symbol ∇^2 (del squared) to indicate the second partial differentiation, equation (2.10) becomes

$$\alpha \nabla^2 t = \frac{\partial t}{\partial \tau} \qquad \text{...(2.12)}$$

Equation (2.12) should be recognised as the equation governing a potential field written in terms of temperature. Similar equations are satisfied by other potential field phenomena such as electricity, magnesium, diffusion, and ideal fluid flow. Because of this, solutions to problems in one field are applicable to analogous systems in the others. Also an experimental solution of a problem in one field may be obtained from an analogous system in another.

PRESENCE OF A HEAT SOURCE

When there is heat release or generation (e.g., due to chemical or atomic reactions or electrical current flow) within a substance, equation (2.12) must be modified to account for it. This can be done by including the heat generated in the heat balance equation for the small element of Fig. 2.1. If the rate of heat generation throughout the body is uniform and equal to G

Btu/ht ft^3, we would have to add a term, G/wc, to the left side of equation (2.12) giving

$$\alpha \nabla^2 t + \frac{G}{wc} = \frac{\partial t}{\partial \tau} \qquad \ldots (2.13)$$

In the general case where G is not uniform but varies with time and location, the equation would have the same form, but G would represent a function of t and x, y, and z.

THERMAL CONDUCTIVITY

Thermal conduction is the transfer of heat from hotter to cooler parts of a body resulting in equalising of temperature. In contrast to heat transfer by convection, thermal conduction has nothing to do with microscopic displacements in the body, but is a result of a direct energy transfer between particles, such as molecules, atoms, and electrons, with higher energy and ones with lower energy. Contrary to heat transfer by radiation, there is no thermal conduction in vacuum.

The basic law of thermal conduction is the Fourier law which states that the heat flux density $\bar{q}$ is proportional to the temperature gradient T in a isotropic body: $\bar{q} = - \lambda \operatorname{grad} T$. The constant of proportionally λ is the thermal conductivity. The minus sign indicates that the temperature decreases in the direction of heat transport and, hence, the temperature gradient is a negative quantity.

Deviations from the Fourier law can be observed at extremely high values of gradT, e.g. in powerful shock waves, at low temperatures for liquid helium HeII. and at high temperatures on the order of tens of thousands of degrees when energy transfer in gases is mainly due to radiation. In highly rarefield media, in which molecules collide with the walls of the vessel rather than with one another, the concept of local temperature is meaningless and the Fourier law is inapplicable. In this case, we deal not with thermal conduction in a gas, but with heat exchange between the bodies in it.

Among solid, anisotropic substances (e.g. crystals, sedimentary rocks, lamellar and pyrolytic materials) occur for which the heat flux density vector $\bar{q}$ does not coincide with the normal to an isothermal surface. The simplest assumption generalising the Fourier hypothesis is that each component of the vector $\bar{q}$ at the point (x, y, z) is a linear combination of all the components of the temperature gradient in it

$$q_x = \lambda_{11} \frac{\partial T}{\partial x} + l_{12} \frac{\partial T}{\partial y} + \lambda_{13} \frac{\partial T}{\partial z} \text{ etc.,}$$

the coefficients of thermal conductivity $\lambda_{LK}\varepsilon$ of anisotropic body form a tensor in the 2nd dimension. For crystals, it is found, within the measurement error, that thermal conductivity in mutually opposite directions is the same.

In multicomponent gas mixtures, one has to take into account the so-called "cross effects" such as mass flows brought about by the temperature gradient (thermal diffusion or the Soret effect), energy flows due to density gradient (diffusion thermal effect or the Dufour effect). The heat flux density in a v-component gas mixture is written down as a sum

$$\vec{q} = -\lambda \operatorname{grad} T + \sum_{i=1}^{v} n_i h_i \vec{V}_i + \frac{KT}{n} \sum_{i=1}^{v} \sum_{i \neq j}^{v} \frac{n_i D_j^T}{m_i D_{ij}} (\vec{V}_i - \vec{V}_j),$$

where n and n_i are the total number of molecules and the number of molecules of a given species per unit volume, h_i and m_i the enthalpy and the mass of a single particle of the i-th species, $\vec{V}_i$ the diffusion rate, D_{ij} the coefficient of binary diffusion, D_j^T the coefficient of thermal diffusion (Dufour effect), and K is Boltzmann's constant. Thus, in multicomponent gas mixtures energy transfer is also accomplished, in addition to convection and heat conduction, by diffusion flow of molecules relative to the bulk velocity and by the Dufour effect. In this case the apparent thermal conductivity may be far different from the molecular thermal conductivity λ.

If we assume that the coefficient of thermal conductivity λ as well as the coefficient of heat capacity c and the density ρ of a substance do not depend on other parameters, then in the absence of internal heat sources the temperature inside the body is described by the differential equation of thermal conductivity

$$\frac{\partial T}{\partial \tau} = \frac{\lambda}{\rho c} \nabla^2 T$$

where $\nabla^2 T = \dfrac{\partial^2 T}{\partial x^2} + \dfrac{\partial^2 T}{\partial y^2} + \dfrac{\partial^2 T}{\partial z^2}$ is the Laplacian, τ the time, and x, y, z the Cartesian coordinates.

The group $\kappa = (\lambda/\rho c)$ is known as thermal diffusivity. It characterises the velocity of propagation of isothermal surfaces in a body.

Solving equations of thermal conduction enables the establishment of the temperature distribution $T(\tau, x, y, z)$ to a certain degree of accuracy for any body. The degree of accuracy depends on the initial and boundary conditions. Solutions of the equation of thermal conductance in a general form, including the variable coefficients λ, c, and ρ, using high-speed computers, in principle presents no difficulties. But this specifies higher requirements for reliability in determining the variable coefficient of thermal conductivity λ for the given substances. Numerous theoretical and experimental investigations have led to the discovery of some specific features and regularities.

The highest thermal conductivity is inherent in metals (Table 2.1). Amongst the metals silver shows the highest thermal conductivity and bismuth the lowest. In the temperature range above ambient, λ for nearly all pure metals falls with increasing temperature. λ is greatly affected by the presence of additives and impurities. Thus, the coefficient of thermal conductivity for steel containing 1% carbon is 40% lower than for pure iron. For metals it depends to a great extent on their treatment. Quenching and cold treatment of metals decrease λ, while pre-heating up to a high temperature increases it.

Table 2.1. Values of the coefficient of thermal conductivity for various substances at atmospheric pressure and moderate temperatures.

Substance	t, °C	λ, W/mK	Substance	t, °C	λ, W/mK
Metals			*Liquid*		
Silver	0	429	Mercury	0	7.82
Copper	0	403	Water	20	0.599
Iron	0	86.5	Acetone	16	0.190
Tin	0	68.2	Ethyl alcohol	20	0.167
Lead	0	35.6			
Non-metallic materials			*Gases*		
Sodium chloride	0	6.9	Hydrogen	0	0.1655
Tourmaline	0	4.6	Helium	0	0.1411
Glass	18	0.4–1.0	Oxygen	0	0.0239
Wood	18	0.16–0.25	Nitrogen	-3	0.0237
Asbestos	18	0.12	Air	4	0.0226

Thermal conductivity of non-metallic liquids under normal conditions is much lower than that of metals and ranges from 0.1 to 0.6 W/mK. In the interval between the melting point and the boiling point, thermal conductivity of liquids may change by a factor of 1.1 to 1.6.

Finally, the lowest thermal conductivity is observed in gases (under normal conditions it is from 0.006 to 0.1 W/mK). Hydrogen and helium are distinguished among gases for the highest thermal conductivity.

The coefficients of thermal conductivity presented in the table evidence that this parameter varies widely. It is determined using various techniques based on the molecular kinetic theory, the phenomenological approaches of the generalised conductivity theory, and generalisation of experimental data.

THERMAL CONTACT RESISTANCE

When a junction is formed by pressing two similar or dissimilar metallic materials together, only a small fraction of the nominal surface area is actually in contact because of the non-flatness and roughness of the contacting surfaces. If a heat flux is imposed across the junction, the uniform flow of heat is generally restricted to conduction through the contact spots, as shown in Fig. 2.2. The limited number and size of the contact spots results in an actual contact area which is significantly smaller than the apparent contact area. This limited contact area causes a thermal resistance, the contact resistance or thermal contact resistance.

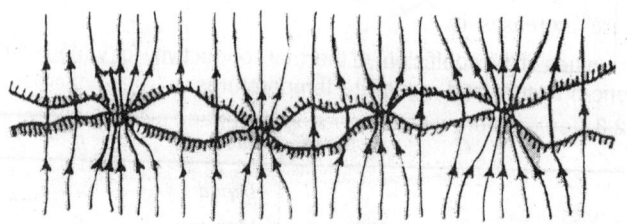

Fig. 2.2. Magnified view of two materials in contact.

The presence of a fluid or solid interstitial medium between the contacting surfaces may contribute to or restrict the heat transfer at the junction, depending upon the thermal conductivity, thickness, and hardness (in the case of a solid) of the interstitial medium. If there is a significant temperature difference between the surfaces composing the junction, heat exchange by radiation also may occur across the gaps between the contacting surfaces.

When a metallic junction is placed in a vacuum, conduction through the contact spots is the primary mode of heat transfer, and the contact resistance is generally greater than when the junction is in the presence of air or other fluid. In a vacuum, the temperature distribution in the contacting materials, with the resulting temperature difference at the junction, is shown in Fig. 2.3 for both flat and cylindrical junctions.

This temperature difference is used to define the contact resistance at the junction, such that :

$$1/S\alpha_c = (T_1 - T_2)/\dot{Q} = \Delta T/\dot{Q}$$

or $\qquad 1/\alpha_c = (T_1 - T_2)/\dot{q} = \Delta T/\dot{q}$ $\qquad$ ...(2.14)

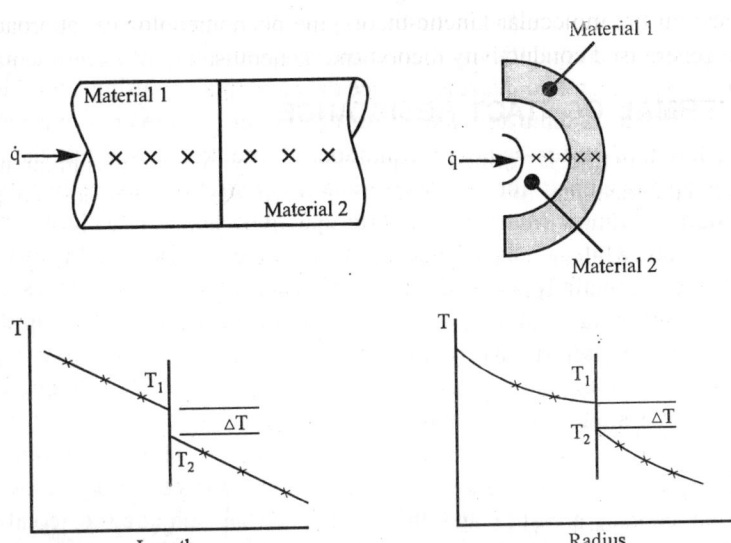

Fig. 2.3. Temperature distribution across flat and cylindrical contacting solids.

where T_1 and T_2 are the temperatures of the bounding contact surfaces, S is the area across which the heat is transferred, and α_c is the heat transfer coefficient for the junction, or the thermal contact conductance. This contact conductance or joint conductance is often reported in the literature and is defined as :

$$\alpha_c = \dot{q}/\Delta T \qquad ...(2.15)$$

The magnitude of the contact conductance is a function of a number of parameters including the thermophysical and mechanical properties of the material in contact, the characteristics of the contacting surfaces, the presence of gaseous or non-gaseous interstitial media, the apparent contact pressure, the mean junction temperature, and the conditions surrounding the junction, as noted by Fletcher (1988).

In view of the significant number of parameters affecting the contact conductance or contact resistance, it has not been possible to develop a single analytical expression for the prediction of the contact resistance at a junction between two materials, except for cases of highly idealised single and multiple contacts. An overview of the idealised models has been reported by Sridhar and Yovanovich (1994). An analytical expression for predicting the contact conductance of non-flat or machined metallic surfaces in contact has been developed by Lambert and Fletcher (1995) for a wide range of metallic materials and test conditions. Despite the availability of these models, a majority of the contact resistance information is determined experimentally in order to provide a measure of the thermal performance of a specific configuration or system.

Most experimental contact resistance data are obtained using a traditional cut-bar, vertical column test facility in a vacuum or ambient environment over a range of steady-state test conditions. More specialised test facilities have been developed for use with such configurations as bolted joints, periodic or sliding contacts, concentric cylinders, and full scale or partial scale models, while some configurations are studied by electrolytic analogue techniques. Essentially all of these experimental facilities may be used for evaluation of metallic and non-metallic materials in contact, or metallic and non-metallic materials with gaseous or non-gaseous interstitial media between the contacting surfaces, over a wide range of test parameters.

The force applied to the nominal contact area of the junction provides the apparent contact pressure on the junction. The mean junction temperature, T_m, is the average of the contacting surface temperatures. The apparent contact pressure and the mean junction temperature, combined with the thermophysical and mechanical properties of the contacting materials and the surface characteristics, are the primary factors in determining the magnitude of the contact resistance. High junction loads and high temperatures result in low contact resistances, whereas light junction loads and low temperatures lead to high contact resistances.

The surface finish, or roughness and flatness of the contacting surfaces, can significantly affect the magnitude of the contact resistance. If the axial force on the contacting surfaces is increased, the surface roughness peaks or asperities may deform plastically or elastically, depending upon the material properties, leading to increased contact area and decreased contact resistance. An elevated temperature at the junction may also cause plastic and/or elastic deformation of the roughness asperities, especially for softer materials, with an associated increase in the actual contact area and a decrease in the contact resistance. Typical contact resistance values for aluminium 2024-T4 samples in contact at moderate test conditions in a vacuum environment are shown in Fig. 2.4, to demonstrate the effect of surface finish and mean junction temperature on contact resistance (Fletcher, 1991).

Some additional factors which may affect the contact resistance are the direction of the heat flux, surface scratches or cracks, non-uniform loading which causes uneven contact pressure, relative motion or slipping between the surfaces, and the presence of oxides or contaminants on the contacting surfaces.

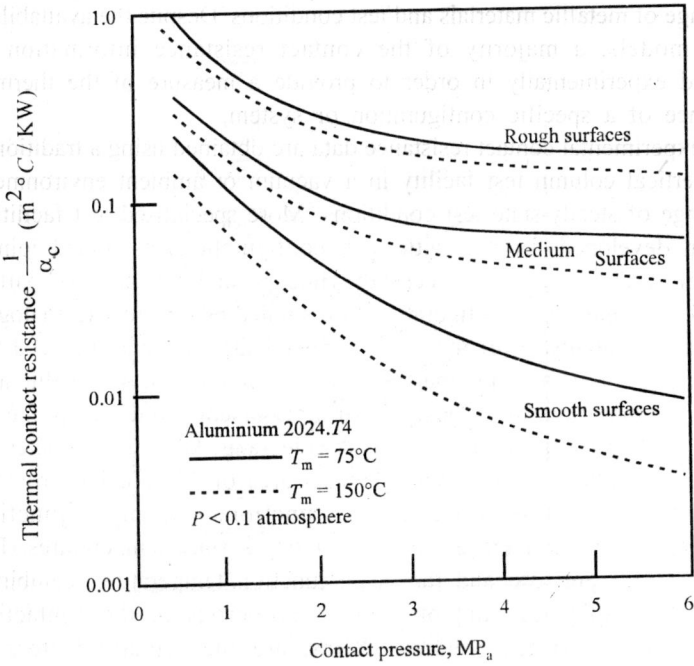

Fig. 2.4. Thermal contact resistance as a function of apparent contact pressure, mean junction temperature and surface finish.

The use of interstitial or thermal control materials for thermal enhancement or thermal isolation of metallic junctions further effects the contact resistance. Although there are variations in material thickness and composition, the contact resistance for representative interstitial materials is shown in Fig. 2.5. These interstitial materials have been categorised as greases and oils; metallic foils and screens; ceramic composites and cements; and synthetic and natural sheets (Fletcher, 1972). While metallic foils and greases are often used for thermal enhancement, most of the interstitial materials are generally used for thermal isolation.

Surface treatments, or coatings and films, may also be used for thermal enhancement or thermal isolation. Metallic coatings provide modest to significant thermal enhancement, depending upon the metal used and the method of application. Ceramic coatings provide modest to excellent thermal isolation depending upon the choice of material. Ceramic coatings may also provide hard, corrosion resistant coatings that are not electrically conducting. Care must be taken to assure that galvanic corrosion will not occur with the choice of materials for some applications.

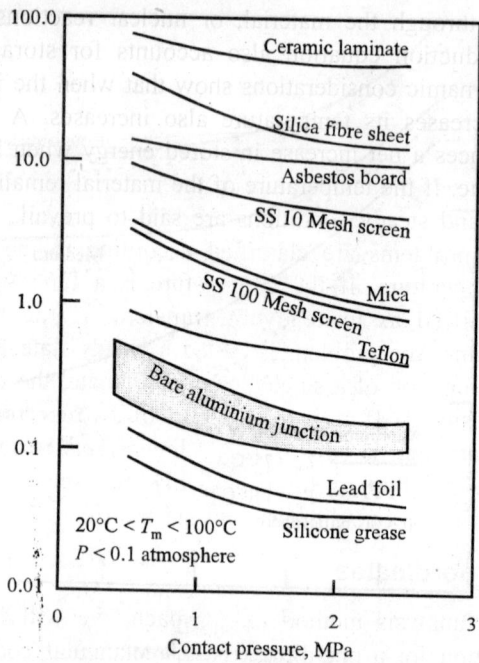

Fig. 2.5. Contact resistance for selected interstitial materials for thermal enhancement or thermal isolation.

Contact resistance can be an advantage or disadvantage. Proper consideration of contact resistance and its characteristics can lead to improved thermal design and overall thermal control of components and systems.

CONDUCTION EQUATION

In this section the general conduction equation is derived. A solution of this equation, subject to given initial and boundary conditions, yields the temperature distribution in a solid system. Once the temperature distribution is known, the heat transfer rate in the conduction mode can be evaluated by applying Fourier's law, Eq. (1.1) of Chapter 1.

The conduction equation is a mathematical expression of the conservation of energy in a solid substance. To derive this equation we perform an energy balance on an elemental volume of material in which heat is being transferred only by conduction. Heat transfer by radiation occurs in a solid only if the material is transparent or translucent.

The energy balance includes the possibility of heat generation in the material. Heat generation in a solid can occur by chemical reactions, electric currents passing through the material, or nuclear reactions. The general form of the conduction equation also accounts for storage of internal energy. Thermodynamic considerations show that when the internal energy of a material increases its temperature also increases. A solid material therefore experiences a net increase in stored energy when its temperature increases with time. If the temperature of the material remains constant, no energy is stored and steady conditions are said to prevail.

Heat transfer problems are classified according to the variables that influence the temperature. If the temperature is a function of time, the problem is classified as unsteady or transient. If the temperature is independent of time, the problem is called a steady-state problem. If the temperature is a function of a single space coordinate, the problem is said to be one-dimensional. If it is a function of two or three coordinate dimensions, the problem is two- or three-dimensional, respectively. If the temperature is a function of time and only one space coordinate, the problem is classified as one-dimensional and transient.

Rectangular Coordinates

To illustrate the analysis method of approach, we will first derive the conduction equation for a one-dimensional, rectangular coordinate system as shown in Fig. 2.6. We will assume that the temperature in the material is only a function of the x coordinate and time, or $T = T(x, t)$, and that the conductivity k, density ρ, and specific heat c of the solid are all constant.

The principle of conservation of energy for the control volume of Fig. 2.6 can be stated as follows :

$$\begin{array}{l}\text{Rate of heat conduction} \\ \text{into control volume}\end{array} + \begin{array}{l}\text{rate of heat generation} \\ \text{inside control volume}\end{array} =$$

$$\begin{array}{l}\text{rate of heat conduction} \\ \text{out of control volume}\end{array} + \begin{array}{l}\text{rate of energy storage} \\ \text{inside control volume}\end{array} \quad ...(2.16)$$

We will use Fourier's law to express the two conduction terms and define the symbol $\dot{q}_G$ as the rate of energy generation per unit volume inside the control volume. Then the word (Eq. (2.16) can be expressed in mathematical form :

$$kA\frac{\partial T}{\partial x}\bigg|_x + \dot{q}_G A\Delta x = -kA\frac{\partial T}{\partial x}\bigg|_{x+\Delta x} + \rho A\Delta x c\frac{\partial T(x+\Delta x/2,t)}{\partial t} \quad ..(2.17)$$

Dividing Eq. (2.17) by the control volume $A\Delta x$ and rearranging, we obtain

$$k \frac{(\partial T / \partial x)_{x+\Delta x} - (\partial T / \partial x)_x}{\Delta x} + \dot{q}_G = \rho c \frac{\partial T(x + \Delta x / 2, t)}{\partial t} \qquad ...(2.18)$$

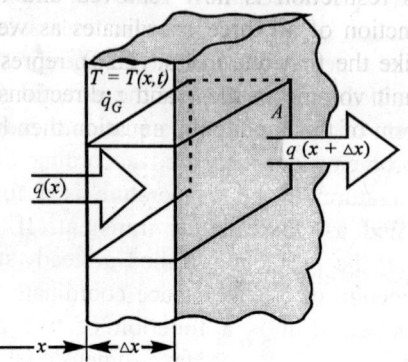

Fig. 2.6. Control volume for one-dimensional conduction in rectangular coordinates.

In the limit as $\Delta x \to 0$, the first term on the left side of Eq. (2.18) can be expressed in the form

$$\frac{\partial T}{\partial x}\bigg|_{x+dx} = \frac{\partial T}{\partial x}\bigg|_x + \frac{\partial}{\partial x}\left(\frac{\partial T}{\partial x}\right)_x dx = \frac{\partial T}{\partial x}\bigg|_x + \frac{\partial^2 T}{\partial x^2}\bigg|_x dx \qquad ...(2.19)$$

The right side of Eq. (2.18) can be expanded in a Taylor series as

$$\frac{\partial T}{\partial t}\left[\left(x + \frac{\Delta x}{2}\right), t\right] = \frac{\partial T}{\partial t}\bigg|_x + \frac{\partial^2 T}{\partial x \partial t}\bigg|_x \frac{\Delta x}{2} + ...$$

Equation (2.17) then becomes, to the order of Δx,

$$k \frac{\partial^2 T}{\partial x^2} + \dot{q}_G = \rho c \frac{\partial T}{\partial t} \qquad ...(2.20)$$

Physically, the first term on the left side represents the net rate of heat conduction into the control volume per unit volume. The second term on the left side is the rate of energy generation per unit volume inside the control volume. The right side represents the rate of increase in internal energy inside the control volume per unit volume. Each term has dimensions of energy per unit time and volume with the units (W/m^3) in the SI system and $(Btu/h\ ft^3)$ in the engineering system.

Equation (2.20) applies only to unidimensional heat flow because it was derived on the assumption that the temperature distribution is one-dimensional. If this restriction is now removed and the temperature is assumed to be a function of all three coordinates as well as time, or $T = T(x, y, z, t)$, terms like the first one in Eq. (2.20), representing the net rate of conduction per unit volume in the y and z directions, will appear. The three-dimensional form of the conduction equation then becomes (Fig. 2.7).

$$\frac{\partial^2 T}{\partial x^2} + \frac{\partial^2 T}{\partial y^2} + \frac{\partial^2 T}{\partial z^2} + \frac{\dot{q}_G}{k} = \frac{1}{\alpha} \cdot \frac{\partial T}{\partial t} \qquad ...(2.21)$$

where α is the thermal diffusivity, a group of material properties defined as

$$\alpha = \frac{k}{\rho c} \qquad ...(2.22)$$

The thermal diffusivity has units of (m^2/s) in the SI system and (ft^2/s) in the engineering system. Numerical values of the thermal conductivity, density, specific heat, and thermal diffusivity for several engineering materials are presented in Appendix 2.

Solutions to the general conduction equation in the form of Eq. (2.21) can only be obtained for simple geometric shapes and easily specific boundary conditions. However, as shown in the next chapter, solutions by numerical methods can be obtained quite easily for complex shapes and realistic boundary conditions without a great deal of effort, a procedure used in engineering practice today for the majority of conduction problems. But a basic understanding of analytic solutions is important in writing computer programs, and in the rest of this chapter we will examine problems for which simplifying assumptions can eliminate some terms from Eq. (2.21) and reduce the complexity of the solution.

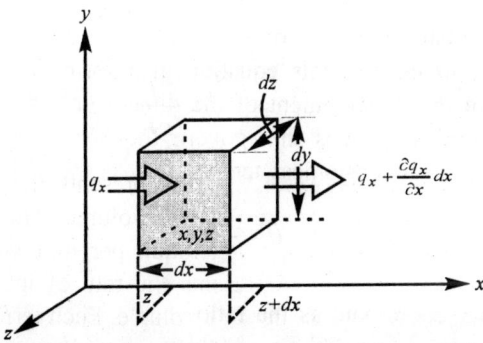

Fig. 2.7. Differential control volume for three-dimensional conduction in rectangular coordinates.

If the temperature of a material is not a function of time, the system is in the steady state and does not store any energy. The steady form of a three dimensional conduction equation in rectangular coordinates is

$$\frac{\partial^2 T}{\partial x^2} + \frac{\partial^2 T}{\partial y^2} + \frac{\partial^2 T}{\partial z^2} + \frac{\dot{q}_G}{k} = 0 \qquad ...(2.23)$$

If the system is in the steady state and no heat is generated internally, the conduction equation further simplifies to

$$\frac{\partial^2 T}{\partial x^2} + \frac{\partial^2 T}{\partial y^2} + \frac{\partial^2 T}{\partial z^2} = 0 \qquad ...(2.24$$

Equation (2.24) is known as the Laplace equation, in honour of the French mathematician Pierre Laplace. It occurs in a number of areas in addition to heat transfer, for instance, in diffusion of mass or in electromagnetic fields. The operation of taking the second derivatives of the potential in a field has therefore been given a shorthand symbol, ∇^2, called the Laplacian operator. For the rectangular coordinate system Eq. (2.24) becomes

$$\frac{\partial^2 T}{\partial x^2} + \frac{\partial^2 T}{\partial y^2} + \frac{\partial^2 T}{\partial z^2} = \nabla^2 T = 0 \qquad ...(2.25)$$

Since the operator ∇^2 is independent of coordinate system, the above form will be particularly useful when we want to study conduction in cylindrical and spherical coordinates.

Dimensionless Form

The conduction equation in the form of Eq. (2.21) is dimensional. It is often more convenient to express this equation in a form where each term is dimensionless. In the development of the dimensionless equation we will identify dimensionless groups that govern the heat conduction process. Begin by defining a dimensionless temperature as the ratio

$$\theta = \frac{T}{T_r} \qquad ...(2.26)$$

a dimensionless x coordinate as the ratio

$$\xi = \frac{x}{L_r} \qquad ...(2.27)$$

and a dimensionless time as the ratio

$$\tau = \frac{t}{t_r} \qquad\qquad ...(2.28)$$

where the symbols T_r, L_r, and t_r represent a reference temperature, a reference length, and a reference time, respectively. Although the choice of reference quantities is somewhat arbitrary, the values selected should be physically significant. The choice of dimensionless groups varies from problem to problem, but the form of the dimensionless groups should be structured so that they limit the dimensionless variables between convenient extremes, such as zero and one. The value for L_r should therefore be selected as the maximum x dimension of the system for which the temperature distribution is sought. Similarly, a dimensionless ratio of temperature differences that varies between zero and unity is often preferable to a ratio of absolute temperatures.

If the definitions of the dimensionless temperature, x coordinate, and time are substituted into Eq. (2.20), we obtain the conduction equation in the non-dimensional form

$$\frac{\partial^2 \theta}{\partial \xi^2} + \frac{\dot{q}_G L_r^2}{kT_r} = \frac{L_r^2}{\alpha t_r} \cdot \frac{\partial \theta}{\partial \tau} \qquad\qquad ...(2.29)$$

The reciprocal of the dimensionless group $(L_r^2/\alpha t_r)$ is called the Fourier number, designated by the symbol Fo:

$$\text{Fo} = \frac{\alpha t_r}{L_r^2} \qquad\qquad ...(2.30)$$

In a physical sense, the Fourier number is the ratio of the rate of heat transfer by conduction to the rate of energy storage in the system. It is an important dimensionless group in transient conduction problems and will be encountered frequently. The choice of reference time and length in the Fourier number depends on the specific problem, but the basic form is always a thermal diffusivity multiplied by time divided by the square of a characteristic length.

The other dimensionless group appearing in Eq. (2.29) is a ratio of internal heat generation per unit time to heat conduction through the volume per unit time. We will use the symbol $\dot{Q}_G$ to represent this dimensionless heat generation number :

$$\dot{Q}_G = \frac{\dot{q}_G L_r^2}{kT_r} \qquad\qquad ...(2.31)$$

The one-dimensional form of the conduction equation expressed in dimensionless form now becomes

$$\frac{\partial^2 \theta}{\partial \xi^2} + \dot{Q}_G = \frac{1}{\text{Fo}} \frac{\partial \theta}{\partial t} \qquad ...(2.32)$$

If steady-state prevails, the right side of Eq. (2.32) becomes zero.

Cylindrical and Spherical Coordinates

Equation (2.21) was derived for a rectangular coordinate system. Although the generation and energy storage terms are independent of the coordinate system, the heat conduction terms depend on geometry and, therefore, on the coordinate system. The dependence on the coordinate system used to formulate the problem can be removed by replacing the heat conduction terms with the Laplacian operator.

$$\nabla^2 T + \frac{\dot{q}_G}{k} = \frac{1}{\alpha} \frac{\partial T}{\partial t} \qquad ...(2.33)$$

The differential form of the Laplacian is different for each coordinate system, and the Laplacian operators in rectangular, cylindrical, and spherical coordinate systems are derived in the appendix.

For a general transient three-dimensional problem in the cylindrical coordinates shown in Fig. 2.8, $T = T(r, \phi, z, t)$ and $\dot{q}_G = \dot{q}_G(r, \phi, z, t)$. If the Laplacian is substituted into Eq. (2.33), the general form of the conduction equation in cylindrical coordinates becomes

$$\frac{1}{r} \frac{\partial}{\partial r}\left(r \frac{\partial T}{\partial r}\right) + \frac{1}{r^2} \frac{\partial^2 T}{\partial \phi^2} + \frac{\partial^2 T}{\partial z^2} + \frac{\dot{q}_G}{k} = \frac{1}{\alpha} \frac{\partial T}{\partial t} \qquad ...(2.34)$$

If the heat flow in a cylindrical shape is only in the radial direction, $T = T(r, t)$, the conduction equation reduces to

$$\frac{1}{r} \frac{\partial}{\partial r}\left(r \frac{\partial T}{\partial r}\right) + \frac{\dot{q}_G}{k} = \frac{1}{\alpha} \frac{\partial T}{\partial t} \qquad ...(2.35)$$

Furthermore, if the temperature distribution does not vary with time, the conduction equation becomes

$$\frac{1}{r} \frac{d}{dr}\left(r \frac{\partial T}{\partial r}\right) + \frac{\dot{q}_G}{k} = 0 \qquad ...(2.36)$$

In this case the equation for the temperature contains only a single variable r and is therefore an ordinary differential equation.

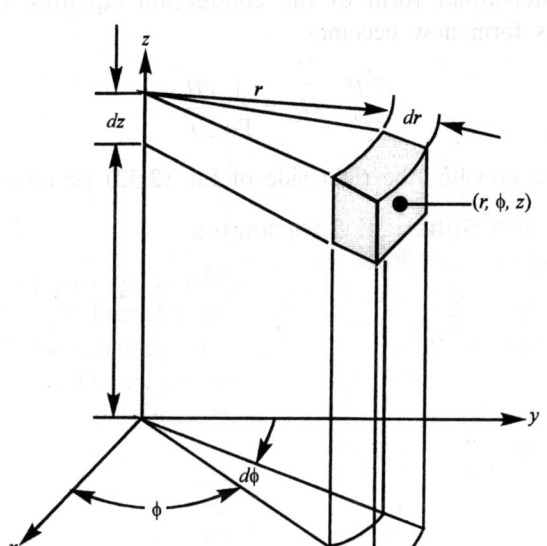

Fig. 2.8. Cylindrical coordinate system for the general conduction equation.

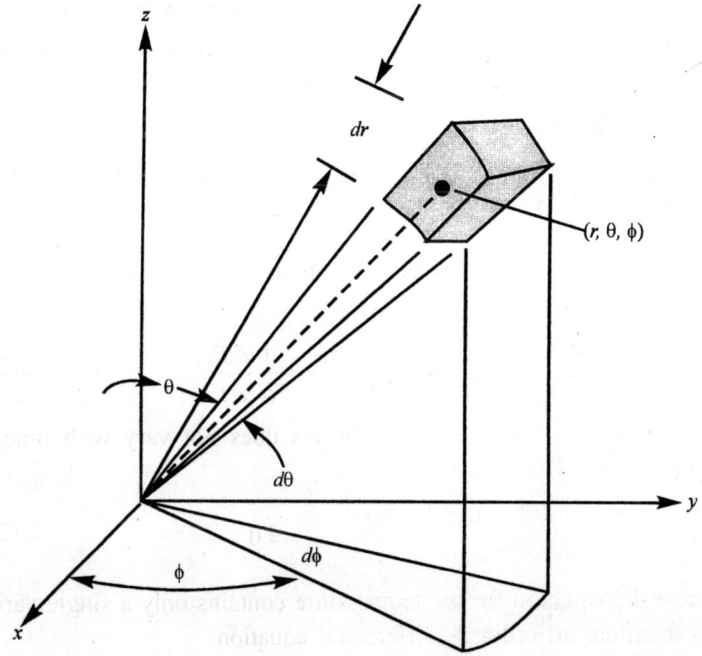

Fig. 2.9. Spherical coordinate system for the general conduction equation.

When no internal energy generation is present and the temperature is a function of the radius only, the steady-state conduction equation for cylindrical coordinates is

$$\frac{d}{dr}\left(r\frac{dT}{dr} \right) = 0 \qquad\qquad ...(2.37)$$

For spherical coordinates, as shown in Fig. 2.9, the temperature is a function of the three space coordinates r, θ, ϕ and time t, or $T = T(r, \theta, \phi, r)$. The general form of the conduction equation in spherical coordinates is then

$$\frac{1}{r^2}\frac{\partial}{\partial r}\left(r^2\frac{\partial T}{\partial r} \right) + \frac{1}{r^2 \sin \theta}\frac{\partial}{\partial \theta}\left(\sin\theta \frac{\partial T}{\partial \theta} \right)$$

$$+ \frac{1}{r^2 \sin \theta}\frac{\partial^2 T}{\partial \phi^2} + \frac{\dot{q}_G}{k} = \frac{1}{\alpha}\cdot\frac{\partial T}{\partial t} \qquad ...(2.38)$$

CHAPTER 3

Steady State Conduction

INTRODUCTION

One-dimensional steady-state heat flow includes the common cases of flow through a wall or along an insulated rod, when the two faces of the wall or ends of the rod are held at temperatures t_1 and t_2. In addition, heat flow through cylindrical and spherical walls and along thin rods, with heat losses to the surroundings, can be treated from this point of view. The heat flow may be from an external source, may be generated within the wall, or may be the result of a combination of these two sources.

Heat transfer problems are classified according to the variables that influence the temperature. If the temperature is a function of time, the problem is classified as unsteady or transient. If temperature is independent of time the problem is called a steadystate problem.

HEAT TRANSFER THROUGH A PLAIN WALL ($q_v = 0$)

With steady-state, or stationary conduction, the temperature of a body remains constant in time, i.e. $\partial t / \partial t = 0$.

The differential equation of steady-state conduction has the following form :

$$a\nabla^2 t + \frac{q_v}{c\rho} = 0 \qquad \text{...(3.1)}$$

or

$$\nabla^2 t + \frac{q_v}{\lambda} = 0 \qquad \text{...(3.1')}$$

In the absence of inner heat sources, i.e. ($q_v = 0$), Eq. (3.1) acquires a simpler form :

$$\nabla^2 t = 0 \qquad \text{...(3.2)}$$

or

$$\frac{\partial^2 t}{\partial x^2} + \frac{\partial^2 t}{\partial y^2} + \frac{\partial^2 t}{\partial z^2} = 0 \qquad \text{...(3.2')}$$

This chapter is concerned with heat transfer by conduction through bodies of the simplest geometrical form. The two variants of bodies, without inner heat sources ($q_v = 0$) and those with such sources ($q_v \neq 0$), are considered separately. Let us first consider the transfer of heat through a plain wall with $q_v = 0$.

Boundary Conditions of the First Kind

Let us consider a homogeneous, isotropic wall of thickness δ with constant thermal conductivity λ. The external surfaces of the wall are maintained at constant temperatures t_{w1} and t_{w2}.

With the given conditions temperature will change only in the direction normal to the plain wall. Should the Ox-axis be directed as shown in Fig. 3.1, temperature will remain constant in the direction of the Oy- and Oz-axes, i.e.,

$$\frac{\partial t}{\partial y} = \frac{\partial t}{\partial z} = 0$$

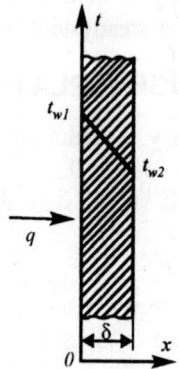

Fig. 3.1. Plain homogeneous wall.

Accordingly the temperature shall be a function of only one coordinate x and for the case being considered the differential equation of conduction acquires the following appearance :

$$\frac{d^2 t}{dx^2} = 0 \qquad \text{...(3.3)}$$

The boundary conditions for this problem may be formulated as follows :

$$\left.\begin{array}{ll} \text{with} & x = 0, \quad t = t_{w1} \\ \text{with} & x = \delta, \quad t = t_{w2} \end{array}\right\} \qquad ...(3.4)$$

Equation (3.3) and the conditions (3.4) give a full mathematical description of the problem. Its solution consists in finding the temperature distribution in the wall, i.e. the function $t = f(x)$, and in deriving the formula for determining the heat flow through the wall per unit time.

The law of temperature distribution in wall thickness may be found by double integration of Eq. (3.3).

The first integration gives :

$$\frac{dt}{dx} = C_1 \qquad ...(3.5)$$

The second integration yields :

$$t = C_1 x + C_2 \qquad ...(3.6)$$

From eq. (3.6) it follows that with constant thermal conductivity variation in wall temperature is linear.

The constants C_1 and C_2 of eq. (3.6) are determined from the boundary conditions :

With $x = 0$, $\quad t = t_{w1}$ and $C_2 = t_{w1}$

With $x = \delta$, $\quad t = t_{w2}$ and $C_1 = -\dfrac{t_{w1} - t_{w2}}{\delta}$

Substituting the constants C_1 and C_2 in eq. (3.6), we get the temperature distribution or temperature profile in a plain wall :

$$t = t_{w1} - \frac{t_{w1} - t_{w2}}{\delta} x \qquad ...(3.7)$$

If the reference temperature is taken as the smallest given temperature of the wall t_{w2}, we can present Eq. (3.7) in dimensionless form.

Let us denote :

$\Delta t = t - t_{w2}$ as the current temperature difference or excess temperature;

$\Delta t_0 = t_{w1} - t_{w2}$ as the full temperature difference or maximum excess temperature.

Upon introduction of these terms in Eq. (3.7), we get

$$\Delta t = \Delta t_0 - \frac{\Delta t_0}{\delta} x \qquad \qquad ...(3.8)$$

or

$$\frac{\Delta t}{\Delta t_0} = 1 - \frac{x}{\delta}$$

If we then denote $\dfrac{\Delta t}{\Delta t_0} = \Theta$ as the dimensionless temperature difference or dimensionless excess temperature, and $\dfrac{w}{\delta} = X$ as the dimensionless coordinate, we obtain :

$$\Theta = 1 - X \qquad \qquad ...(3.8')$$

The equation of temperature field (3.8') is universally applicable, i.e. it makes it possible to present the temperature distribution across the wall as a single straight line for any given value of t_{w1}, t_{w2} and δ (Fig. 3.2). The dimensionless equations are often very convenient in problems of conduction.

The quantity of heat flowing through unit area of a wall per unit time in the direction of the Ox-axis is determined by Fourier law according to which

$$q = -\lambda \frac{\partial t}{\partial x}$$

Since $\dfrac{\partial t}{\partial x} = C_1 = \dfrac{-(t_{w1} - t_{w2})}{\delta}$, then after substituting the value of $\dfrac{\partial t}{\partial x}$ into the expression of the Fourier law we get :

$$q = \frac{\lambda}{\delta}(t_{w1} - t_{w2}) \qquad \qquad ...(3.9)$$

From Eq. (3.9) it follows that the quantity of heat passing through unit area of the wall per unit time is directly proportional to the thermal conductivity λ and the difference between the temperatures of the external surfaces of the wall $(t_{w1} - t_{w2})$, and inversely proportional to the wall thickness δ. It should be noted here that the rate of heat flow depends not on the absolute values of the temperatures, but on their difference, $t_{w1} - t_{w2} = \Delta t$ which is generally referred to as the temperature difference.

The ratio λ/δ W/m^2.K is called the thermal conductance of the wall and its reciprocal δ/λ (m^2.K/W) the thermal resistance of the wall. The latter represents the temperature drop across the wall related to the unit rate of heat flow. This rate of heat flow being known, we can easily calculate the total quantity of heat Q_τ transferred through the surface of the wall F during a time interval τ :

$$Q_\tau = qF\tau = \frac{\lambda}{\delta}(t_{w1} - t_{w2})F\tau \qquad \text{...(3.10)}$$

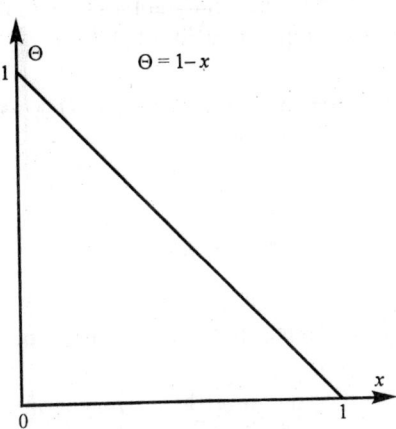

Fig. 3.2. Dimensionless temperature field in a plain wall $\Theta = 1 - x$.

From Eq. (3.9) we find :

$$\frac{t_{w1} - t_{w2}}{\delta} = \frac{q}{\lambda}$$

Introducing this expression into the temperature field equation (3.7), we obtain

$$t = t_{w1} - \frac{q}{\lambda} x \qquad \text{...(3.11)}$$

It follows from Eq. (3.11) that, other conditions being equal, the temperature drop within the wall is the larger, the greater the rate of heat flow.

Expressions (3.7) and (3.9) are based the on the assumption of constant thermal conductivity λ. In fact λ is variable.

Let us now consider the case where thermal conductivity is a function of temperature only :

$$\lambda = \lambda (t)$$

For many materials thermal conductivity has a nearly linear dependence on temperature

$$\lambda = \lambda_0 (1 + bt)$$

where λ_0 is the thermal conductivity at 0°C.

From Fourier law

$$q = -\lambda(t)\frac{dt}{dx} = -\lambda_0(1+bt)\frac{dt}{dx} \qquad \text{...(a)}$$

Separating the variables and integrating (a) within the limits from $x = 0$ and $x = \delta$ and in the temperature interval from t_{w1} to t_{w2}, we obtain:

$$q\delta = \lambda_0\left[1 + b\frac{t_{w1}+t_{w2}}{2}\right](t_{w1}-t_{w2}) \qquad \text{....(b)}$$

The factor

$$\lambda_0\left(1 + b\frac{t_{w1}+t_{w2}}{2}\right)$$

in expression (b) is the integral mean thermal conductivity, i.e.

$$\lambda_m = \frac{1}{t_{w1}-t_{w2}}\int_{t_{w2}}^{t_{w1}}\lambda(t)dt \qquad \text{...(3.12)}$$

Thus the rate of heat flow q, W/m^2, at the wall surface is

$$q = \frac{\lambda_m}{\delta}(t_{w1}-t_{w2}) \qquad \text{...(3.13)}$$

It follows from Eq. (3.13) that when λ is temperature-dependent, the rate of heat flow q may be found approximately as an integral mean value between temperatures t_{w1} and t_{w2} on the assumption λ = constant.

Integrating (a) between $x = 0$ and any given coordinate x and in the temperature interval from t_{w1} to t we obtain the expression for the temperature field :

$$t = \sqrt{\left(\frac{1}{b}+t_{w1}\right)^2 - \frac{2qx}{\lambda_0 b}} - \frac{1}{b} \qquad \text{...(3.14)}$$

It follows from this equation that the variation in wall temperature is not linear but follows a curve whose shape depends on the sign and numerical value of factor 'b'.

Let us now examine heat conduction through a composite plain wall built up of n homogeneous layers, assuming perfect contact between layers and equal temperatures at the surfaces in contact.

In steady-state conduction the quantity of heat passing through any isothermal surface of a heterogeneous wall is identical, i.e.

$$\frac{\partial q}{\partial x} = 0$$

Given the temperatures of the external wall surfaces, and thickness and heat conductivity of each layer, we can obtain the following system of equations:

$$\left. \begin{array}{l} q = \dfrac{\lambda_1}{\delta_1}(t_{w1} - t_{w2}) \\[2ex] q = \dfrac{\lambda_2}{\delta_2}(t_{w2} - t_{w3}) \\[1ex] \dotfill \\[1ex] q = \dfrac{\lambda_n}{\delta_n}(t_{wn} - t_{w(n+1)}) \end{array} \right\} \qquad \text{...(c)}$$

From equations (c) we determine the temperature difference for each layer and summation of the left and right sides of these equations gives :

$$t_{w1} - t_{w(n+1)} = q\left(\frac{\delta_1}{\lambda_1} + \frac{\delta_2}{\lambda_2} + ... + \frac{\delta_n}{\lambda_n}\right)$$

Hence, the rate of heat flow is

$$q = \frac{t_{w1} - t_{w(n+1)}}{\dfrac{\delta_1}{\lambda_1} + \dfrac{\delta_2}{\lambda_2} + ... + \dfrac{\delta_n}{\lambda_n}} = \frac{t_{w1} - t_{w(n+1)}}{\displaystyle\sum_{i=1}^{i=n} \frac{\delta_i}{\lambda_i}} \qquad \text{...(3.15)}$$

The value $\displaystyle\sum_{i=1}^{i=n} \frac{\delta_i}{\lambda_i}$ equalling the sum of the thermal resistances of each of n layers, is called the full thermal resistance of conduction of the composite wall.

Comparison of the process of heat transfer through composite and plain homogeneous walls is made more convenient by introducing the concept of equivalent thermal conductivity λ_{eq}. It is equal to the thermal conductivity coefficient of a homogeneous wall the thickness of which Δ equals the thickness of a composite wall $\sum\limits_{i=n}^{i=n} \delta_i$ and the thermal resistance of the composite wall concerned, i.e.

$$\frac{\sum\limits_{i=1}^{i=n} \delta_i}{\lambda_{eq}} = \sum\limits_{i=1}^{i=n} \frac{\delta_i}{\lambda_i}$$

Hence

$$\lambda_{eq} = \frac{\sum\limits_{i=1}^{i=n} \delta_i}{\sum\limits_{i=1}^{i=n} \frac{\delta_i}{\lambda_i}} \qquad ...(3.16)$$

It follows from Eq. (3.16) that equivalent thermal conductivity λ_{eq} depends not only on the thermal and physical properties of the layers, but also on their thickness. The interface temperatures between any two neighbouring layers are :

$$t_{w2} = t_{w1} - q\frac{\delta_1}{\lambda_1}$$

and

$$t_{w3} = t_{w2} - q\left(\frac{\delta_1}{\lambda_1} + \frac{\delta^2}{\lambda_2}\right) \qquad \left.\begin{array}{c} \\ \\ \\ \\ \\ \end{array}\right\} \qquad ...(3.17)$$

$$t_{w(i+1)} = t_{w1} - q\sum\limits_{i=1}^{i} \frac{\delta_i}{\lambda_i}$$

Within each layer temperature changes according to either eq. (3.7) or eq. (3.14), and the temperature profile of a composite wall as a whole is a broken line.

Boundary Conditions of the Third Kind (Heat Transfer)

The transport of heat from one moving medium (liquid or gas) to another through a plain or composite solid wall of any form separating the two media is referred to as heat transfer. The process involves the transfer of heat from the hotter fluid to the wall, transfer of heat by conduction through the wall and transfer of heat from the separating wall to the colder moving medium.

Let us consider heat transfer through homogeneous and composite plain walls.

Let a plain homogeneous wall be of thickness δ (Fig. 3.3), and the given data include: thermal conductivity λ of the wall; ambient temperatures t_{f1} and t_{f2}; and local heat-transfer coefficients α_1 and α_2. Let us further assume that t_{f1}, t_{f2}, α_1 and α_2 are constant and do not change along the surface; this allows us to consider temperature variations of the fluids and wall solely in the direction normal to the wall.

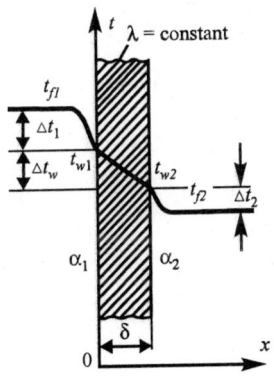

Fig. 3.3. Heat transfer through a plain homogeneous wall.

It is necessary to find the amount of heat flow in the given conditions from the hot fluid to the cold, and the temperatures at the wall surfaces.

The rate of heat flow from the hot fluid to the wall is determined by the equation

$$q = \alpha_1(t_{f1} - t_{w1}) \qquad \qquad ...(3.18)$$

In steady-state conditions the same amount of heat will be transferred through the wall by conduction :

$$q = \frac{\lambda}{\delta}(t_{w1} - t_{w2}) \qquad \qquad ...(3.19)$$

The same amount of heat is transferred from the surface of the wall to the cold fluid

$$q = \alpha_2 (t_{w2} - t_{f2}) \qquad \qquad ...(3.20)$$

Equations (3.18), (3.19) and (3.20) may be presented in the form

$$\left. \begin{aligned} q\,\frac{1}{\alpha_1} &= t_{f1} - t_{w1} \\[4pt] q\,\frac{\delta}{\lambda} &= t_{w1} - t_{w2} \\[4pt] q\,\frac{1}{\alpha_2} &= t_{w2} - t_{f2} \end{aligned} \right\} \qquad ...(3.21)$$

Putting together equalities (3.21) termwise, we obtain

$$q\left(\frac{1}{\alpha_1} + \frac{\delta}{\lambda} + \frac{1}{\alpha_2}\right) = t_{f1} - t_{f2}$$

Hence, the rate of heat flow, W/m²

$$q = \frac{t_{f1} - t_{f2}}{\dfrac{1}{\alpha_1} + \dfrac{\delta}{\lambda} + \dfrac{1}{\alpha_2}} \qquad ...(3.22)$$

Denote :

$$\frac{1}{\dfrac{1}{\alpha_1} + \dfrac{\delta}{\lambda} + \dfrac{1}{\alpha_2}} = k \qquad ...(3.23)$$

This quantity is in W/m².K.

Taking into account Eq. (3.23), we can write Eq. (3.22) in the following form :

$$q = k\,(t_{f1} - t_{f2}), \ \text{W/m}^2 \qquad ...(3.24)$$

The factor k has the same dimension as α, and is called the overall heat-transfer coefficient. It defines the intensity of heat transfer from one fluid to another through a wall separating them, and is numerically equal to the quantity of heat passing through unit area of wall surface in unit time at a temperature difference of 1°C.

The reciprocal of this coefficient is called the overall thermal resistance to heat transfer.

The overall thermal resistance of a single-layer homogeneous wall is

$$R = \frac{1}{k} = \frac{1}{\alpha_1} + \frac{\delta}{\lambda} + \frac{1}{\alpha_2} \qquad ...(3.25)$$

It will be seen from eq. (3.25) that the overall thermal resistance is the sum of separate thermal resistances $\dfrac{1}{\alpha_1}, \dfrac{\delta}{\lambda}$ and $\dfrac{1}{\alpha_2}$ where

$\dfrac{1}{\alpha_1} = R_1$ is the thermal resistance to heat transfer from the hot fluid to the wall;

$\dfrac{\delta}{\lambda} = R_w$ is the thermal resistance to heat transfer by conduction across the wall; and

$\dfrac{1}{\alpha_2} = R_2$ is the thermal resistance to heat transfer from the wall to the cold fluid.

Since the overall thermal resistance of a composite wall is the sum of the separate thermal resistances of the component layers, it is quite clear that in this case the thermal resistance of each component layer should be taken into account. If the wall consists of layers the overall thermal resistance to heat transfer through such a wall will be :

$$R = \frac{1}{k} = \frac{1}{\alpha_1} + \frac{\delta_1}{\lambda_1} + \frac{\delta_2}{\lambda_2} + \ldots + \frac{\delta_n}{\lambda_n} + \frac{1}{\alpha_2}$$

or

$$R = \frac{1}{\alpha_1} + \sum_{i=1}^{i=n} \frac{\delta_i}{\lambda_i} + \frac{1}{\alpha_2}$$

Hence

$$k = \frac{1}{\dfrac{1}{\alpha_1} + \displaystyle\sum_{i=1}^{i=n} \dfrac{\delta_i}{\lambda_i} + \dfrac{1}{\alpha_2}} \qquad \ldots (3.26)$$

The rate of heat flow through a composite wall consisting of n layers is :

$$q = \frac{1}{\dfrac{1}{\alpha_1} + \displaystyle\sum_{i=1}^{i=n} \dfrac{\delta_i}{\lambda_i} + \dfrac{1}{\alpha_2}} = k(t_{f1} - t_{f2}) \qquad \ldots (3.27)$$

Equation (3.27) for a composite wall is similar to eq. (3.24) for a plain homogeneous wall, differing only in the expression of the overall heat-transfer coefficient k. Comparison of equations (3.26) and (3.23) shows that the latter is a particular case of the former when $n = 1$.

The heat flow Q, W, through surface F of a wall is

$$Q = qF = k\Delta t F \qquad \qquad ...(3.28)$$

The surface temperatures of a homogeneous wall can be found from Eq. (3.21). It follows that

$$t_{w1} = t_{f1} - q\frac{1}{\alpha_1}$$

$$t_{w2} = t_{f1} - q\left(\frac{1}{\alpha_1} + \frac{\delta}{\lambda}\right)$$

or

$$t_{w2} = t_{f2} + q\frac{1}{\alpha_2}$$

Comparison of Eqs. (3.15) and (3.27) shows that heat transfer through a composite wall with boundary conditions of the first kind is a special case of heat transfer with boundary conditions of the third kind.

From the foregoing, the interface temperature between any two layers of a composite wall i and $i + 1$ with boundary conditions of the third kind, may be determined by the equation

$$t_{w(i+1)} = t_{f1} - q\left(\frac{1}{\alpha_1} + \sum_{i=1}^{t}\frac{\delta_i}{\lambda_i}\right) \qquad ...(3.29)$$

Along with eq. (3.29) interface temperatures can also be determined graphically. Let us consider the graphical method for determining interface temperatures of a non-homogeneous composite wall based on the linear relationship between the temperature difference in a wall and thermal resistance :

$$t_{f1} - t_{f2} = q\frac{1}{k}$$

or, for any layer of the system

$$t_{wi} - t_{w(i+1)} = q\frac{\delta_i}{\lambda_i}$$

This relationship allows us to construct an imaginary wall in which the thickness of the separate layers is proportional to the respective thermal resistance, and the external thermal resistances to heat transfer, $1/\alpha_1$ and $1/\alpha_2$ are accounted for by introducing two conditional boundary layers of corresponding thickness.

The essence of the method will be explained on the model of a three-layer wall.

The overall thermal resistance to heat flow through this wall is :

$$R = \frac{1}{k} = \frac{1}{\alpha_1} + \frac{\delta_1}{\lambda_1} + \frac{\delta_2}{\lambda_2} + \frac{\delta_3}{\lambda_3} + \frac{1}{\alpha_2}$$

We first plot sections O_1A_1, A_1A_2, A_2A_3, A_3A_4 and A_4O_2 along the horizontal, equal each respectively to the thermal resistances $1/\alpha_1$, δ_1/λ_1, δ_2/λ_2, δ_3/λ_3, and $1/\alpha_2$ (Fig. 3.4), then raise perpendiculars at points O_1, A_1, A_2, A_3, A_4 and O_2, and on O_1K_1 and O_2K_2 plot the temperatures of the moving media t_{f1} and t_{f2} at any convenient scale. These points C_1 and B_2 are joined by a straight line.

The sections A_1E_1, A_2E_2, A_3E_3 and A_4E_4 are equal to the unknown temperatures t_{w1}, t_{w2}, t_{w3} and t_{w4}. Since the triangles $C_1B_1B_2$ and $C_1C_2E_1$ are similar, it follows that

$$\frac{C_1C_2}{C_1B_1} = \frac{C_2E_1}{B_1B_2} \quad \text{or} \quad \frac{C_1C_2}{t_{f1} - t_{f2}} = \frac{\frac{1}{\alpha_1}}{\frac{1}{k}} \qquad \text{...(3.30)}$$

It also follows from eq. (3.30) that $C_1C_2 = t_{f1} - t_{w1}$, hence

$$A_1E_1 = O_1C_1 - C_1C_2 = t_{w1}$$

By analogy we prove that lines A_2E_2, A_3E_3 and A_4E_4 are equal respectively to temperatures t_{w2}, t_{w3} and t_{w4}.

Boundary Conditions of the Second and Third Kind

Let us now consider the case of heat transfer through a homogeneous isotropic wall with boundary conditions of the second kind expressed as q_w = const (at $x = 0$) for one of the wall surfaces. For the second surface we know the heat-transfer coefficient α_2 and ambient temperature t_{f2}, i.e. boundary conditions of the third kind (Fig. 3.5). There are no inner heat sources within the wall ($q_v = 0$).

Solution of this problem consists in determining the temperature distribution across the wall and the surface temperatures. The process of heat transfer being steady, we can write the following system of equations:

$$\left.\begin{array}{l} q_w = (t_{w1} - t_{w2})\dfrac{\lambda}{\delta} \\[2mm] q_w = \alpha_2 (t_{w2} - t_{f2}) \end{array}\right\} \qquad ...(3.31)$$

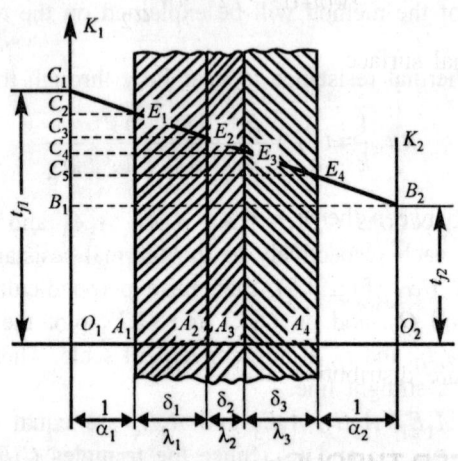

Fig. 3.4. Graphical method for determining temperature.

It follows from Eqs. (3.31) that for any given value of q_w

$$\left.\begin{array}{l} t_{w2} = t_{f2} + q_w \dfrac{1}{\alpha_2} \\[3mm] t_{w1} = t_{f2} + q_w \left(\dfrac{1}{\alpha_2} + \dfrac{\delta}{\lambda}\right) \end{array}\right\} \qquad ...(3.32)$$

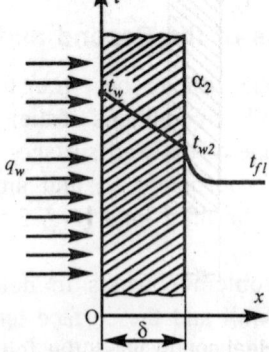

Fig. 3.5. Heat transfer through a plain wall (composite boundary conditions).

For a composite wall consisting of n homogeneous layers, the temperature of wall surface and interface temperatures can be determined by the following equations :
on the right external surface

$$t_{w(n+1)} = t_{f2} + q_w \frac{1}{\alpha_2}$$

on the left external surface

$$t_{w1} = t_{f2} + q_w \left(\frac{1}{\alpha_2} + \sum_{i=1}^{i=n} \frac{\delta_i}{\lambda_i} \right)$$

on the interface between the $m - 1$ and mth layer

$$t_{wm} = t_{f2} + q_w \left(\frac{1}{\alpha_2} + \sum_{i=m}^{i=n} \frac{\delta_i}{\lambda_i} \right) \qquad ...(3.33)$$

The temperature distribution across any layer may be found by Eqs. (3.7) or (3.14).

HEAT TRANSFER THROUGH A CYLINDRICAL WALL ($q_v = 0$)

Boundary Conditions of the First Kind

Consider the case of steady-state conduction through a cylindrical wall (tube) of inside diameter $d_1 = 2r_1$ and outside diameter $d_2 = 2r_2$ (Fig. 3.6).

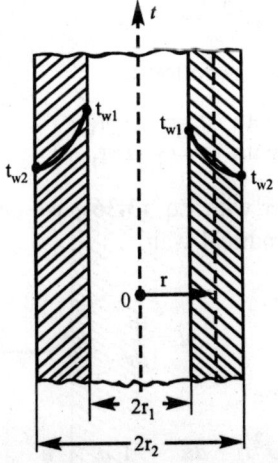

Fig. 3.6. Heat conduction through a cylindrical wall.

The surfaces of the wall are at constant temperatures t_{w1} and t_{w2}. The thermal conductivity τ of the wall material is constant in the given temperature range. It is necessary to find the temperature distribution in the cylindrical wall and the heat flow through it.

In this case it is convenient to write the differential equation of conduction in the cylindrical system of co-ordinates :

$$\nabla^2 t = \frac{\partial^2 t}{\partial r^2} + \frac{1}{r}\frac{\partial t}{\partial r} + \frac{1}{r^2}\frac{\partial^2 t}{\partial \varphi^2} + \frac{\partial^2 t}{\partial z^2} = 0 \qquad \ldots(3.34)$$

The Oz-axis coincides with the axis of the tube.

In the given conditions temperature changes only in radial direction, and the temperature field is one-dimensional. Consequently

$$\frac{\partial t}{\partial z} = 0 \quad \text{and} \quad \frac{\partial^2 t}{\partial z^2} = 0 \qquad \ldots(a)$$

In addition, since the temperatures of the outside and inside wall surfaces are constant, the isothermal surfaces are cylindrical and their axis is that of the tube. Temperature then must also not change along φ, i.e.

$$\frac{\partial t}{\partial \varphi} = 0 \quad \text{and} \quad \frac{\partial^2 t}{\partial \varphi^2} = 0 \qquad \ldots(b)$$

Taking (a) and (b) into account, Eq. (3.34) becomes :

$$\frac{d^2 t}{dr^2} + \frac{1}{r}\frac{dt}{dr} = 0 \qquad \ldots(3.35)$$

Boundary conditions are as follows :

$$\left.\begin{array}{ll} \text{with} & r=r_1, \quad t=t_{w1} \\ \text{with} & r=r_2, \quad t=t_{w2} \end{array}\right\} \qquad \ldots(3.36)$$

Solving Eq. (3.35) together with Eq. (3.36), we obtain the equation of the temperature field in a cylindrical wall.

If a new variable

$$u = \frac{dt}{dr} \qquad \ldots(c)$$

is introduced, then we have :

$$\frac{d^2 t}{dr^2} = \frac{du}{dr}; \quad \frac{1}{r}\frac{dt}{dr} = \frac{u}{r} \qquad \ldots(d)$$

Substituting (c) and (d) in Eq. (3.35), we get

$$\frac{du}{dr} + \frac{1}{r} u = 0 \qquad \qquad ...(3.37)$$

Integration of Eq. (3.37) gives

$$\ln u + \ln r = \ln C_1 \qquad \qquad ...(e)$$

Taking the antilogarithms of (e) and returning to the initial variables we get

$$dt = C_1 \frac{dr}{r} \qquad \qquad ...(f)$$

After integration we obtain :

$$t = C_1 \ln r + C_2 \qquad \qquad ...(3.38)$$

The constants C_1 and C_2 can be determined, if the given boundary conditions are substituted in Eq. (3.38)

$$\left. \begin{array}{ll} \text{with } r = r_1, & t = t_{w1}, \quad \text{hence } t_{w1} = C_1 \ln r_1 + C_2 \\ \text{with } r = r_2, & t = t_{w2}, \quad \text{hence } t_{w2} = C_1 \ln r_2 + C_2 \end{array} \right\} \qquad ...(g)$$

Solving Eqs. (g) with respect to C_1 and C_2, we get :

$$C_1 = \frac{t_{w1} - t_{w2}}{\ln \frac{r_1}{r_2}}$$

$$C_2 = t_{w1} - (t_{w1} - t_{w2}) \frac{\ln r_1}{\ln \frac{r_1}{r_2}}$$

Substituting these values of C_1 and C_2 in Eq. (3.38), we obtain :

$$t = t_{w1} - (t_{w1} - t_{w2}) \frac{\ln \frac{r}{r_1}}{\ln \frac{r_2}{r_1}}$$

or

$$t = t_{w1} - (t_{w1} - t_{w2}) \frac{\ln \frac{d}{d_1}}{\ln \frac{d_2}{d_1}} \qquad \qquad ...(3.39)$$

The derived expression is the equation of a logarithmic curve. A

The derived expression is the equation of a logarithmic curve. A curvilinear distribution of temperature in a cylindrical wall may be explained as follows.

In case of a plane wall the rate of heat flow q is the same for all isothermal surfaces, and the temperature gradient therefore remains constant for all isothermal surfaces. With a cylindrical wall the rate of heat flow through any isothermal surface depends on the radius.

The quantity of heat passing through a cylindrical surface F per unit time can be found by Fourier's law :

$$Q = -\lambda \frac{dt}{dr} F$$

Substituting the value of the temperature gradient given by equation (f) into Fourier's formula and taking into account that $F = 2\pi r l$, we obtain :

$$Q = \frac{2\pi\lambda l \,(t_{w1} - t_{w2})}{\ln \frac{d_2}{d_1}} \qquad \qquad ...(3.40)$$

Here Q is in W.

From Eq. (3.40) it follows that the quantity of heat passing through a cylindrical wall per unit time is completely defined by the given boundary conditions and does not depend on the radius.

Heat flow (3.40) can be related either to unit tube length or to unit area of inner or outer surface. We then obtain the following calculation formulae for the rate of heat flow, W/m^2:

$$\frac{Q}{\pi d_1 l} = q_1 = \frac{2\lambda(t_{w1} - t_{w2})}{d_1 \ln \frac{d_2}{d_1}} \qquad \qquad ...(3.41)$$

(heat flow through unit area of inner surface);

$$\frac{Q}{\pi d_2 l} = q_2 = \frac{2\lambda(t_{w1} - t_{w2})}{d_2 \ln \frac{d_2}{d_1}} \qquad \qquad ...(3.42)$$

(heat flow through unit area of outer surface);

$$\frac{Q}{l} = q_l = \frac{\pi(t_{w1} - t_{w2})}{\frac{1}{2\lambda} \ln \frac{d_2}{d_1}} \qquad \qquad ...(3.43)$$

(heat flow through unit tube length, W/m).

Heat flow reduced to unit tube length is measured in W/m, and is called the linear rate of heat flow. Equation (3.43) shows it to be independent of the surface area of the cylindrical wall at a constant ratio d_2/d_1. With heat transfer through a tube the rates of heat flow q_1 and q_2 via the inner and outer surfaces of the cylindrical wall are not equal, q_1 being always larger than q_2, as may be clearly seen from Eqs. (3.41) and (3.42).

The relation between q_l, q_1 and q_2 can be easily found from Eqs. (3.41), (3.42) and (3.43) :

$$q_l = \pi d_1 q_1 = \pi d_2 q_2 \qquad \qquad ...(3.44)$$

When thermal conductivity is a function of temperature of the kind $\lambda(t) = \lambda_0 (1 + bt)$ it can be shown that the rate of heat flow can be calculated by the formula used when $\lambda = $ const:

$$q_l = \frac{\pi(t_{w1} - t_{w2})}{\frac{1}{2\lambda_m} \ln \frac{d_2}{d_1}} \qquad (3.45)$$

bearing in mind that in formula (3.45) the term λ_m is the integral mean thermal conductivity :

$$\lambda_m = \frac{1}{(t_{w1} - t_{w2})} \int_{t_{w2}}^{t_{w1}} \lambda(t)\, dt$$

To find the temperature fields for the case when $\lambda = \lambda(t) = \lambda_0 (1 + bt)$, we can use the Fourier equation applied to a cylindrical wall :

$$q_l = -\lambda(t)\frac{dt}{dr} 2\pi r \qquad ...(3.46)$$

Separating the variables and integrating Eq. (3.46) within the limits $r = r_1$ and r, $t = t_{w1}$ and t and finding t from the integral obtained, we get the following expression for the temperature field :

$$t = \sqrt{\left(\frac{1}{b} + t_{w1}\right)^2 - \frac{q_l \ln \frac{d}{d_1}}{\pi b \lambda_0}} - \frac{1}{b} \qquad ...(3.47)$$

Boundary Conditions of the Third Kind (Heat Transfer)

Consider a homogeneous cylindrical wall (tube) of a material with constant thermal conductivity λ. The given quantities include the temperatures of the moving media, t_{f1} and t_{f2}, and the heat-transfer coefficients for the inner and outer surfaces of the tube α_1 and α_2 (Fig. 3.7), which are all constant.

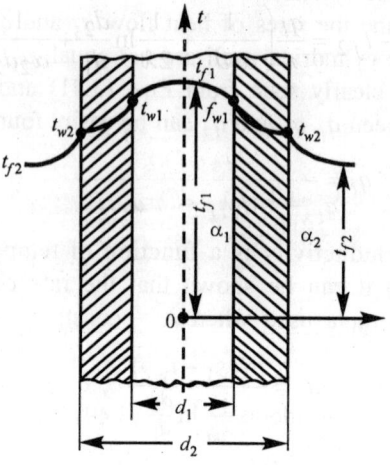

Fig. 3.7. Heat transfer through a cylindrical homogeneous wall.

It is necessary to find q_l and t_w. Let us assume the tube to be quite long as compared with its wall thickness. Heat loss through the tube ends can then be ignored, and in the steady-state conduction, the quantity of heat transferred from the hot fluid to the surface of the wall, transmitted across the wall, and given up to the cold fluid will be the same throughout. Thus, we can write :

$$\left.\begin{aligned} q_l &= \alpha_1 \pi d_1 (t_{f1} - t_{w1}) \\[2mm] q_l &= \frac{\pi(t_{w1} - t_{w2})}{\frac{1}{2\lambda} \ln \frac{d_2}{d_1}} \\[2mm] q_l &= \alpha_2 \pi d_2 (t_{w2} - t_{f2}) \end{aligned}\right\} \qquad ...(3.48)$$

Let us rearrange these equations as follows :

$$\left.\begin{aligned} t_{f1} - t_{w1} &= \frac{q_l}{\pi} \frac{1}{\alpha_1 d_1} \\[2mm] t_{w1} - t_{w2} &= \frac{q_l}{\pi} \frac{1}{2\lambda} \ln \frac{d_2}{d_1} \\[2mm] t_{w2} - t_{f2} &= \frac{q_l}{\pi} \frac{1}{\alpha_2 d_2} \end{aligned}\right\} \qquad ...(3.48')$$

Summing up the equations of (3.48'), we obtain the full temperature difference :

$$t_{f1} - t_{f2} = \frac{q_l}{\pi} \left(\frac{1}{\alpha_1 d_1} + \frac{1}{2\lambda} \ln \frac{d_2}{d_1} + \frac{1}{\alpha_2 d_2} \right)$$

Then

$$q_l = \frac{t_{f1} - t_{f2}}{\frac{1}{\alpha_1 d_1} + \frac{1}{2\alpha} \ln \frac{d_2}{d_1} + \frac{1}{\alpha_2 d_2}} \qquad ...(3.49)$$

Let us denote :

$$k_l = \frac{1}{\frac{1}{\alpha_1 d_1} + \frac{1}{2\lambda} \ln \frac{d_2}{d_1} + \frac{1}{\alpha_2 d_2}} \qquad ...(3.50)$$

Taking Eq. (3.50) into consideration, Eq. (3.49) may be written :

$$q_l = k_l \pi \, (t_{f1} - t_{f2}) \qquad ...(3.49')$$

The value k_l is called the linear overall heat-transfer coefficient, it is in W/m.K. It defines the intensity of heat transfer from one fluid to another through a wall separating them. Its value k_l numerically equal to the quantity of heat passing through the wall of a tube one metre long per unit time from one fluid to another at a temperature difference of one degree.

The reciprocal of the linear overall heat-transfer coefficient $R_l = \dfrac{1}{k_l}$ is

referred to as the linear thermal resistance of heat transfer :

$$R_l = \frac{1}{k_l} = \frac{1}{\alpha_1 d_1} + \frac{1}{2\lambda} \ln \frac{d_2}{d_1} + \frac{1}{\alpha_2 d_2} \qquad ...(3.51)$$

here R_l is in m.K/W.

The components of the full thermal resistance are :

$\dfrac{1}{\alpha_1 d_1}$ and $\dfrac{1}{\alpha_2 d_2}$ = thermal resistance of heat transfer of the wall surfaces, R_{l1} and R_{l2} respectively;

$\dfrac{1}{2\lambda} \ln \dfrac{d_2}{d_1}$ = thermal resistance of heat conduction of the wall R_{lw}.

It should be noted that the linear thermal resistances of a tube are determined not only by the heat-transfer coefficients α_1 and α_2, but also by the corresponding diameters.

If the heat flow through a cylindrical wall is reduced to its inner or outer surface, we obtain the rate of heat flow W/m^2 relative to unit area of a corresponding tube surface :

$$q_1 = \frac{Q}{\pi d_1 l} = \frac{k_l}{d_1}(t_{f1} - t_{f2})$$

or

$$q_2 = \frac{Q}{\pi d_2 l} = \frac{k_l}{d_2}(t_{f1} - t_{f2})$$

and

$$q_1 = k_1 \, (t_{f1} - t_{f2})$$
$$q_2 = k_2 \, (t_{f1} - t_{f2})$$

where

$$k_1 = \frac{k_l}{d_1} \quad \text{and} \quad k_2 = \frac{k_l}{d_2}$$

These last expressions establish the relationship for the heat-transfer coefficients where the heat flow is attributed to unit length and area of a cylindrical wall :

$$k_l = d_1 k_1 = d_2 k_2$$

here k_l is in W/m.K.

The formulae for k_1 and k_2, W/m^2.K, may be extended as follows :

$$\left. \begin{array}{l} k_1 = \dfrac{1}{\dfrac{1}{\alpha_1} + \dfrac{d_1}{2\lambda} \ln \dfrac{d_2}{d_1} + \dfrac{d_1}{\alpha_2 d_2}} \\[4ex] k_2 = \dfrac{1}{\dfrac{1}{\alpha_1} \dfrac{d_2}{d_1} + \dfrac{d_2}{2\lambda} \ln \dfrac{d_2}{d_1} + \dfrac{1}{\alpha_2}} \end{array} \right\} \qquad ...(3.52)$$

In practice the thickness of the walls of cylinders is usually small compared with their diameter. Calculations may therefore be made using simplified formulae which can be obtained in the following way.

Expand the value $\ln \dfrac{d_2}{d_1}$ in a series :

$$\ln \frac{d_2}{d_1} = \left(\frac{d_2}{d_1} - 1\right) - \frac{1}{2}\left(\frac{d_2}{d_1} - 1\right)^2 + ...$$

If the ratio $\dfrac{d_2}{d_1} \to 1$, the series rapidly becomes converging, and it allows us to obtain sufficiently accurate results using only its first term :

$$\ln\frac{d_2}{d_1} = \left(\frac{d_2}{d_1} - 1\right) = \frac{d_2 - d_1}{d_1} = \frac{2\delta}{d_1}$$

whereas δ is the thickness of cylindrical wall, m.

Substituting the obtained value $\ln\dfrac{d_2}{d_1}$ in Eq. (3.52), we get :

$$k_1 = \frac{1}{\dfrac{1}{\alpha_1} + \dfrac{\delta}{\lambda} + \dfrac{1}{\alpha_2}} \qquad \ldots(3.53)$$

Hence, for a thin-walled tube practical calculations may be based on the following formula :

$$Q = k\pi d_x l \ (t_{f1} - t_{f2}) \qquad \ldots(3.54)$$

where k, W/m².K, is to be taken according to formula (3.53), i.e. as for a plane wall. If $\dfrac{d_2}{d_1} < 2$, the calculation error does not exceed 4 per cent. A maximum error not exceeding 4 per cent is quite admissible for many engineering calculations; and Eq. (3.54) is commonly used when $\dfrac{d_2}{d_1} \leq 1.8$.

The error can be diminished, if for the reference surface in formula (3.54) is taken that with smaller α :

(1) if $\alpha_1 \gg \alpha_2$, then $d_x = d_2$

(2) if $\alpha_2 \gg \alpha_1$, then $d_x = d_1$

(3) if $\alpha_1 \simeq \alpha_2$, then $d_x = \dfrac{d_1 + d_2}{2}$

For heat transfer through a composite cylindrical wall the system of equations (3.48') should be replaced by another system to take into account the thermal resistance to conduction of all the layers, i.e.

$$t_{f1} - t_{w1} = \frac{q_l}{\pi} \frac{1}{\alpha_1 d_1}$$

$$t_{w1} - t_{w2} = \frac{q_l}{\pi} \frac{1}{2\lambda_1} \ln\frac{d_2}{d_1}$$

$$\cdots\cdots\cdots\cdots\cdots\cdots\cdots\cdots$$

$$\cdots\cdots\cdots\cdots\cdots\cdots\cdots\cdots$$

$$t_{wn} - t_{w(n+1)} = \frac{q_l}{\pi} \frac{1}{2\lambda_n} \ln\frac{d_{n+1}}{d_n}$$

$$t_{w(n+1)} - t_{f2} = \frac{q_l}{\pi} \frac{1}{\alpha_2 d_{n+1}}$$

...(3.55)

Summing up these equalities and solving for q_l, W/m, we get

$$q_l = \frac{\pi(t_{f1} - t_{f2})}{\dfrac{1}{\alpha_1 d_1} + \sum_{i=1}^{i=n} \dfrac{1}{2\lambda_i} \ln\dfrac{d_{i+1}}{d_i} + \dfrac{1}{\alpha_2 d_{n+1}}} \qquad ...(3.56)$$

or

$$q_l = k_l \pi (t_{f1} - t_{f2}) \qquad ...(3.56')$$

The value

$$\frac{1}{k_l} = R_l = \frac{1}{\alpha_1 d_1} + \sum_{i=1}^{i=n} \frac{1}{2\lambda_i} \ln\frac{d_{i+1}}{d_i} + \frac{1}{\alpha_2 d_{n+1}} \qquad ...(3.56'')$$

is called the total thermal resistance of a composite cylindrical wall and is measured in m.K/W.

It follows from Eq. (3.55) that

$$t_{w1} = t_{f1} - \frac{q_l}{\pi} \frac{1}{\alpha_1 d_1}$$

$$t_{w2} = t_{f1} - \frac{q_l}{\pi} \left(\frac{1}{\alpha_1 d_1} + \frac{1}{2\lambda_i} \ln\frac{d_2}{d_1} \right)$$

$$t_{w(i+1)} = t_{f1} - \frac{1}{\pi} \left(\frac{1}{\alpha_1 d_1} + \sum_{i=1}^{i} \frac{1}{2\lambda_i} \ln\frac{d_{i+1}}{d_i} \right)$$

...(3.57)

When boundary conditions of the first kind are given they can be considered as the limiting case of boundary conditions of the third kind, when the local heat-transfer coefficients α_1 and α_2 tend to infinity resulting in that fluid temperatures t_{f1} and t_{f2} become equal to wall temperatures t_{w1} and $t_{w(n+1)}$. Then, Eq. (3.56) takes the following appearance :

$$q = \frac{\pi(t_{w1} - t_{w(n+1)})}{\sum\limits_{i=1}^{i=n} \frac{1}{2\lambda_i} \ln \frac{d_{i+1}}{d_i}} \qquad \text{...(3.58)}$$

and the calculation formula for interface temperature :

$$t_{w(i+1)} = t_{w1} - \frac{q_l}{\pi} \sum\limits_{i=1}^{i} \frac{1}{2\lambda_1} \ln \frac{d_{i+1}}{d_i} \qquad \text{...(3.59)}$$

CRITICAL DIAMETER OF A CYLINDRICAL WALL

Let us consider how varying the outside diameter affects the thermal resistance of a homogeneous cylindrical wall (3.1). Equation (3.51) gives :

$$R_l = \frac{1}{\alpha_1 d_1} + \frac{1}{2\lambda} \ln \frac{d_2}{d_1} + \frac{1}{\alpha_2 d_2}$$

The values α_1, d_1, λ and α_2 being constant, the total thermal resistance of heat transfer of the cylindrical wall will depend only on the outside diameter. It follows then from Eq. (3.51) that in this case

$$\frac{1}{\alpha_1 d_1} \equiv R_{l1} = \text{constant}$$

The thermal resistance to conduction, $\frac{1}{2\lambda} \ln \frac{d_2}{d_1} \equiv R_{lw}$, increases with d_2, while the thermal resistance to heat transfer from the wall to the medium, $\frac{1}{\alpha_1 d_2} \equiv R_{l2}$, decreases with increasing d_2. The total thermal resistance will apparently be determined by variations of the components R_{lw} and R_{l2}. The variations of partial thermal resistances are illustrated in Fig. 3.8.

In order to find the variation of R_l with increasing thickness of a cylindrical wall, let us examine it as a function of d_2, differentiating R_l with respect to d_2 and setting the derivative to zero :

$$\frac{d(R_l)}{d(d_2)} = \frac{1}{2\lambda d_2} - \frac{1}{\alpha_2 d_2^2} = 0$$

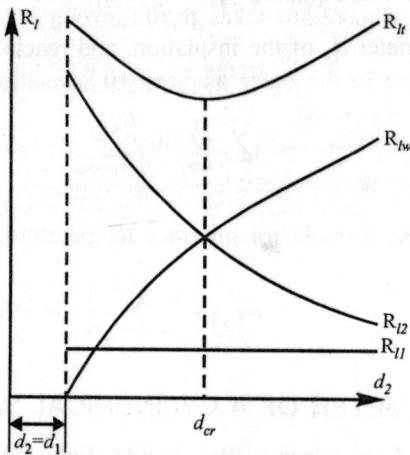

Fig. 3.8. Dependence of the thermal resistance of a cylindrical wall on d_2.

Here the value of d_2 corresponds to the extreme point of the curve $R_l = f(d_2)$. Investigating the curve for its maximum and minimum points by any of the known methods, we see that the minimum is located at the external point. Thus the thermal resistance is minimum at $d_2 = \dfrac{2\lambda}{\alpha_2}$.

The outside diameter of a tube corresponding to the minimum total thermal resistance is called the critical diameter and is denoted by d_{cr}. The critical diameter is calculated by the formula

$$d_{cr} = \frac{2\lambda}{\alpha_2} \qquad \qquad ...(3.60)$$

With $d_2 < d_{cr}$, the total thermal resistance drops with increasing d_2, because an increase of the outer surface affects the thermal resistance more than an increase in wall thickness.

With $d_2 > d_{cr}$, thermal resistance of heat transfer increases with d_2, which points out to dominating effect of the wall thickness.

These considerations should be taken into account in selecting heat insulation for various cylindrical apparatus and piping.

Consider the critical diameter of the insulation placed on a tube (Fig. 3.9). The thermal resistance of heat transfer of the tube is represented by the formula :

$$R_l = \frac{1}{\alpha_1 d_1} + \frac{1}{2\lambda_w} \ln \frac{d_2}{d_1} + \frac{1}{2\lambda_{ins}} \ln \frac{d_3}{d_2} + \frac{1}{\alpha_2 d_3}$$

As follows from the equation $q_l = \dfrac{\pi \Delta t}{R_l}$, q_l will first increase with increasing outer diameter d_3 of the insulation, and reach a maximum at $d_3 = d_{cr}$. Further increase in the outer diameter of insulation will then bring about a drop in q_l (Fig. 3.10).

Having chosen some heat-insulating material for a cylindrical surface, it is above all necessary to calculate the critical diameter using the given values of λ_{ins} and α_2 (Eq. (3.60)).

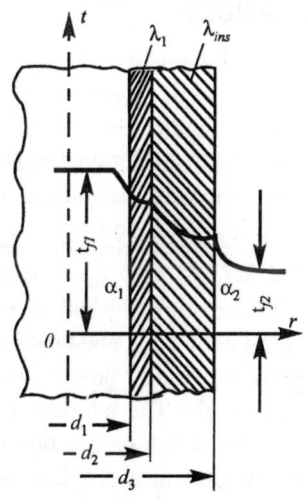

Fig. 3.9. To the notion of critical diameter of insulation.

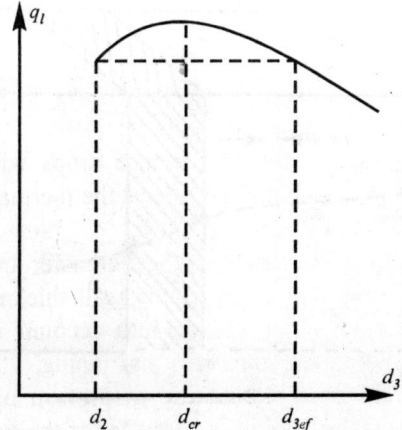

Fig. 3.10. Dependence of heat loss on thickness of insulation laid on a cylindrical wall.

If d_{cr} proves larger than the tube outside diameter d_2, it is not worth using the chosen material for heat insulation. An increase in heat loss will be observed with increasing insulation thickness in the zone characterised by $d_2 < d_3 < d_{cr\ ins}$. This will be clearly seen from Fig. 3.10. Only at $d_3 = d_{3ef}$ will the heat loss again equal the amount of heat lost from the initial non-insulated pipe. Consequently any layer of heat insulation will not justify its purpose.

This means that heat insulation, to be effective, must have a critical diameter $d_{cr\ ins}$ equal to, or smaller than, d_2 $(d_{cr\ ins} \leqslant d_2)$.

HEAT CONDUCTION THROUGH COMPOSITE WALLS: ADDITION OF RESISTANCES

In industrial heat transfer problems one is frequently concerned with conduction through walls made up of layers of various materials each with its own characteristic thermal conductivity. In this section we show how the various resistances to heat transfer are combined into a total resistance.

In Fig. 3.11 is shown a composite wall made up of three materials of different thickness, $x_1 - x_0$, $x_2 - x_1$, and $x_3 - x_2$ and different thermal conductivities k^{01}, k^{12}, and k^{23}. At $x = x_0$, substance "01" is in contact with a fluid with ambient temperature T_a, and at $x = x_3$ substance "23" is in contact with a fluid at temperature T_b. The heat transfer at the boundaries $x = x_0$ and $x = x_3$ is given by Newton's "law of cooling" with heat transfer coefficients h_0 and h_3, respectively. The temperature profile is sketched in Fig. 3.11; one can make such a sketch before one starts the problem—and a sketch is frequently very helpful in a problem setup.

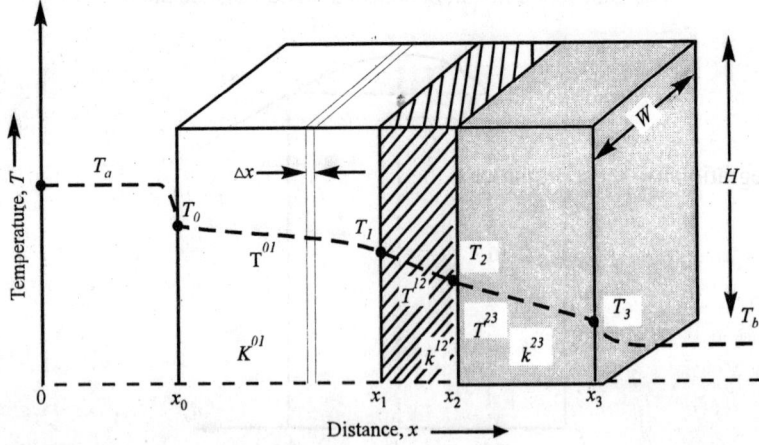

Fig. 3.11. Heat conduction through a composite wall, placed between two fluid streams of temperatures T_a and T_b.

First we derive the differential equation for heat conduction in region "01". A balance on the slab of volume $WH\,\Delta x$ gives

$$q_x{}^{01}\big|_x\,WH - q_x{}^{01}\big|_{x\,+\,\Delta x}\,WH = 0 \qquad\qquad ...(3.61)$$

or, after division by $WH\,\Delta x$ and taking the limit as Δx approaches zero,

$$\frac{dq_x{}^{01}}{dx} = 0 \qquad\qquad ...(3.62)$$

Integration of this equation gives

$$q_x{}^{01} = q_0 \qquad \text{(a constant)} \qquad\qquad ...(3.63)$$

The constant q_0 is the heat flux at the plane $x = x_0$. On physical grounds, we know that at steady-state the heat flux in all three regions will be the same. Hence

$$q_x{}^{01} = q_x{}^{12} = q_x{}^{23} = q_0 \qquad\qquad ...(3.64)$$

We also know that

$$q_x{}^{01} = -k^{01}\frac{dT^{01}}{dx} \qquad\qquad ...(3.65)$$

with similar relations for $q_x{}^{12}$ and $q_x{}^{23}$. Combination of these relations with Eq. (3.64) then gives

$$-k^{01}\frac{dT^{01}}{dx} = q_0 \qquad\qquad ...(3.66)$$

$$-k^{12}\frac{dT^{12}}{dx} = q_0 \qquad\qquad ...(3.67)$$

$$-k^{23}\frac{dT^{23}}{dx} = q_0 \qquad\qquad ...(3.68)$$

Integration of these equations then gives for constant k^{01}, k^{12}, and k^{23} :

$$T_0 - T_1 = -q_0\left(\frac{x_0 - x_1}{k^{01}}\right) \qquad\qquad ...(3.69)$$

$$T_1 - T_2 = -q_0\left(\frac{x_1 - x_2}{k^{12}}\right) \qquad\qquad ...(3.70)$$

$$T_2 - T_3 = -q_0\left(\frac{x_2 - x_3}{k^{23}}\right) \qquad\qquad ...(3.71)$$

In addition we have the two statements regarding the heat transfer at the surfaces :

$$T_a - T_0 = \frac{q_0}{h_0} \qquad ...(3.72)$$

$$T_3 - T_b = \frac{q_0}{h_3} \qquad ...(3.73)$$

Addition of all five of these equations gives

$$T_a - T_b = q_0 \left(\frac{1}{h_0} + \frac{x_1 - x_0}{k^{01}} + \frac{x_2 - x_1}{k^{12}} + \frac{x_3 - x_2}{k^{23}} + \frac{1}{h_3} \right) \qquad ...(3.74)$$

or

$$q_0 = \frac{T_a - T_b}{\left(\dfrac{1}{h_0} + \displaystyle\sum_{i=1}^{3} \dfrac{x_1 - x_{i-1}}{k^{i-1,i}} + \dfrac{1}{h_3} \right)} \qquad ...(3.75)$$

Sometimes this result is rewritten in a form reminiscent of Newton's law of cooling

$$q_0 = U(T_a - T_b) \qquad \text{or} \qquad Q_0 = U(WH)(T_a - T_b) \qquad ...(3.76)$$

and the quantity U (called the "over-all heat transfer coefficient") is given by

$$U = \left(\frac{1}{h_0} + \sum_{i=1}^{3} \frac{x_i - x_{i-1}}{k^{i-1,i}} + \frac{1}{h_3} \right)^{-1} \qquad ...(3.77)$$

This result can clearly be generalised to include more layers in the composite wall by replacing "3" in the upper limit of the sum by any integer "n" representing the number of layers in the wall and by replacing h_3 by h_n. Equation (3.75) is useful for calculating the heat transfer rate through a composite wall separating two fluid streams, when the heat transfer coefficients and the thermal conductivities are known.

One further comment needs to be made concerning the foregoing development. It has been tacitly assumed that the various layers are tightly fitted together with no intervening "air spaces". Clearly, if the layers touch each other only at several points, the resistance to heat transfer will be appreciably increased.

HEAT TRANSFER THROUGH A SPHERICAL WALL

Boundary Conditions of the First Kind

Consider a hollow sphere with radii r_1 and r_2, made of a material of constant thermal conductivity λ and given uniformly distributed surface temperatures t_{w1} and t_{w2}.

Since in the case considered temperature varies only in the radial direction of the sphere, the differential equation of conduction acquires in spherical coordinates the following appearance :

$$\nabla^2 t = \frac{d^2 t}{dr^2} + \frac{2}{r}\frac{dt}{dr} = 0 \qquad \qquad ...(3.78)$$

The boundary conditions are :

$$\left.\begin{array}{l} \text{at } r = r_1, \quad t = t_{w1} \\[2mm] \text{at } r = r_2, \quad t = t_{w2} \end{array}\right\} \qquad ...(3.79)$$

Integrating Eq. (3.61) first, we get :

$$\frac{dt}{dr} = \frac{C_1}{r^2} \qquad \qquad ...(a)$$

The second integration gives :

$$t = C_2 - \frac{C_1}{r} \qquad \qquad ...(3.80)$$

The integration constants of Eq. (3.63) are determined from the boundary conditions (3.62). In addition to this we get :

$$C_1 = -\frac{t_{w1} - t_{w2}}{\left(\frac{1}{r_1} - \frac{1}{r_2}\right)} \qquad \qquad ...(b)$$

$$C_2 = t_{w1} - \frac{t_{w1} - t_{w2}}{\frac{1}{r_1} - \frac{1}{r_2}}\frac{1}{r_1} \qquad \qquad ...(c)$$

Substituting the values of C_1 and C_2 in Eq. (3.63), we get the expression of the temperature field for a spherical wall :

$$t = t_{w1} - \frac{t_{w1} - t_{w2}}{\left(\frac{1}{r_1} - \frac{1}{r_2}\right)}\left(\frac{1}{r_1} - \frac{1}{r}\right) \qquad ...(3.81)$$

In order to find the quantity of heat transferred through an area F of the spherical wall per unit time, we use Fourier's law :

$$Q = -\lambda \frac{dt}{dr} f = -\lambda 4\pi r^2 \frac{dt}{dr}$$

here Q is in W.

Substituting the value of the temperature gradient dt/dr in this expression, we get :

$$Q = \frac{4\pi\lambda (t_{w1} - t_{w2})}{\left(\frac{1}{r_1} - \frac{1}{r_2}\right)} = \frac{2\pi\lambda\Delta t}{\frac{1}{d_1} - \frac{1}{d_2}} = \pi\lambda \frac{d_1 d_2}{\delta} \Delta t \qquad ...(3.82)$$

These are the calculation formulae for conduction heat transfer through a spherical wall. It follows from Eq. (3.64) that the temperature distribution across a spherical wall is represented by a hyperbola, provided thermal conductivity λ is constant.

Boundary Conditions of the Third Kind (Heat Transfer)

The boundary conditions given include, in addition to r_1 and r_2, the temperatures t_{f1} and t_{f2}, and the local coefficients of heat transfer between the surfaces of the spherical wall and surroundings α_1 and α_2. The temperatures t_{f1}, t_{f2}, and the heat-transfer coefficients α_1 and α_2 are assumed to be constant in time, α_1 and α_2 being constant over the surfaces as well.

Since conduction is steady and the full heat flow Q, W, is constant through all isothermal surfaces, we can write :

$$Q = \alpha_1 \pi d_1^2 (t_{f1} - t_{f1})$$

$$Q = \frac{2\pi\lambda}{\frac{1}{d_1} - \frac{1}{d_2}} (t_{w1} - t_{w2})$$

$$Q = \alpha_2 \pi d_2^2 (t_{f2} - t_{f2})$$

It follows from these equations that

$$Q = \frac{\pi(t_{f1} - t_{f2})}{\frac{1}{\alpha_1 d_1^2} + \frac{1}{2\lambda}\left(\frac{1}{d_1} - \frac{1}{d_2}\right) + \frac{1}{\alpha_2 d_2^2}} = k_{sph}\pi\Delta t \qquad ...(3.83)$$

The value

$$k_{sph} = \frac{1}{\frac{1}{\alpha_1 d_1^2} + \frac{1}{2\lambda}\left(\frac{1}{d_1} - \frac{1}{d_2}\right) + \frac{1}{\alpha_2 d_2^2}}$$

is the overall heat-transfer coefficient of a spherical wall measured in W/K, and its reciprocal

$$\frac{1}{k_{sph}} = R_{sph} = \frac{1}{\alpha_1 d_1^2} + \frac{1}{2\lambda}\left(\frac{1}{d_1} - \frac{1}{d_2}\right) + \frac{1}{\alpha_2 d_2^2}$$

is called the thermal resistance to heat flow of a spherical wall and measured in K/W.

GENERALISED SOLUTION METHOD FOR PROBLEMS OF HEAT TRANSFER THROUGH A PLAIN, CYLINDRICAL AND SPHERICAL WALL

The offered general solution can be used to calculate heat transfer through a plain, cylindrical and spherical wall, both at a constant thermal conductivity λ and with the latter depending on temperature.

Consider a unidimensional problem for the three different walls at a constant thermal conductivity of the wall material, with the temperature distribution through the wall being represented as $t = f_1(x)$ for a plain wall, $t = f_2(r)$ for a cylindrical wall, and $t = f_3(r)$ for a spherical wall.

If we admit that in the considered solids the isothermal surfaces are closed, temperature becomes a function only of the coordinate n, which is normal to the isothermal surfaces, and the amount of heat flow will be proportional to the temperature gradient $\partial t/\partial n$, and the size of the surface becomes a function of $F = F(n)$.

The isothermal surfaces of a cylinder and sphere are obviously closed, and the plate is to be considered as the extreme case of a closed system for which $n \rightarrow \infty$.

Since the isothermal surfaces are closed, the quantity of heat flowing through the wall of any of the solids considered can be expressed through

$$Q = -\lambda \frac{dt}{dn} F(n) \qquad \qquad ...(3.84)$$

Since Q is constant for any isothermal surface, then separating the variables in Eq. (3.84) and integrating between the limits $n = n_1$ and $n = n_2$, and the temperature interval from t_{w1} to t_{w2}, we get :

$$Q = \frac{\lambda(t_{w1} - t_{w2})}{\displaystyle\int_{n_1}^{n_2} \frac{dn}{F(n)}} \qquad \qquad ...(3.85)$$

It will be seen that formula (3.85) is similar to the expression previously derived for a plain wall :

$$q = \frac{\lambda(t_{w1} - t_{w2})}{\delta}$$

with Q being similar to the rate of heat flow q and the integral $\int_{n_1}^{n_2} \frac{dn}{F(n)}$,

to the wall thickness. From here on $\int_{n_1}^{n_2} \frac{dn}{F(n)} = I_{n_1}^{n_2}$ will be referred to as the

reduced wall thickness. Formula (3.85) can be used to describe the heat flow through the walls of all three geometrical forms.

The quantity $\int_{n_1}^{n_2} \frac{dn}{F(n)}$ depends only on the shape of the wall.

1. For a plain wall (plate) $n = x$, $n_1 = 0$ and $n_2 = \delta$; $F(n) = F = $ constant, then

$$I_{n_1}^{n_2} = \int_{n_1}^{n_2} \frac{dn}{F} = \frac{1}{F} \int_0^{\delta} dx = \frac{\delta}{F}$$

Substituting the obtained value of $I_{n_1}^{n_2}$ into Eq. (3.85), we obtain the expression for heat flow Q through a plain wall (plate):

$$Q = \frac{\lambda(t_{w1} - t_{w2})}{\delta} F \qquad \qquad ...(3.86)$$

2. For a cylindrical wall $n = r$, $n_1 = r_1$ and $n_2 = r_2$; $F(n) = F(r) = 2\pi r l$, then

$$I_{n_1}^{n_2} = \int_{n_1}^{n_2} \frac{d(n)}{F(n)} = \int_{r_1}^{r_2} \frac{dr}{2\pi r l} = \frac{1}{2\pi l} \ln \frac{r_2}{r_1}$$

Substituting the obtained value of $I_{n_1}^{n_2}$ into the equation (3.85), we get :

$$Q = \frac{2\pi l \lambda\, (t_{w1} - t_{w2})}{\ln \frac{r_2}{r_1}} \qquad \qquad ...(3.87)$$

3. For a spherical wall $n = r$, $n_1 = r_1$ and $n_2 = r_2$; $F(n) = F(r) = 4\pi r^2$, then

$$I_{n_1}^{n_2} = \int_{n_1}^{n_2} \frac{d(n)}{F(n)} = \int_{r_1}^{r_2} \frac{dr}{4\pi r^2} = \frac{1}{4\pi}\left(\frac{1}{r_1} - \frac{1}{r_2}\right)$$

and the formula (3.85) acquires the following appearance :

$$Q = \frac{2\pi\lambda(t_{w1} - t_{w2})}{\frac{1}{d_1} - \frac{1}{d_2}} \qquad \qquad ...(3.88)$$

Integrating the expression (3.84) between the limits n_1 and any coordinate n, and in the temperature interval from t_{w1} to t, we get the equation for the temperature field :

$$t = t_{w1} - \frac{Q}{\lambda}\int_{n_1}^{n} \frac{dn}{F(n)}$$

Denoting the integral $\int_{n_1}^{n} \frac{dn}{F(n)} = I_{n_1}^{n}$, the latter expression can be presented as :

$$t = t_{w1} - \frac{Q}{\lambda} I_{n_1}^{n}$$

Substituting into the obtained expression the value of Q from Eq. (3.85), we get :

$$t = t_{w1} - (t_{w1} - t_{w2})\frac{I_{n_1}^{n}}{I_{n_1}^{n_2}} \qquad \qquad ...(3.89)$$

The ratio $I_{n_1}^{n} / I_{n_1}^{n_2}$ of Eq. (3.89) can be considered as some reduced dimensionless coordinate X which depends on the geometrical shape of the wall. Eq. (3.89) can be presented in the dimensionless form :

$$\frac{t - t_{w2}}{t_{w1} - t_{w2}} = 1 - \frac{I_{n_1}^{n}}{I_{n_1}^{n_2}} \qquad ...(3.90)$$

If we denote the dimensionless temperature $\dfrac{t - t_{w2}}{t_{w1} - t_{w2}}$ by Θ and the

ratio $I_{n_1}^{n} / I_{n_1}^{n_2}$ by X, Eq. (3.90) acquires the appearance :

$$\Theta = 1 - X \qquad ...(3.90')$$

Equation (3.90) is the general expression of the temperature field of the three walls, in dimensionless coordinates.

The reduced dimensionless coordinate present in formula (3.90') is calculated with account taken of the shape of the wall :

for a plain wall

$$X = X_p = x/\delta \qquad ...(3.91)$$

for a cylindrical wall

$$X = X_c = \frac{\ln \frac{r}{r_1}}{\ln \frac{r_2}{r_1}} \qquad ...(3.92)$$

for a spherical wall

$$X = X_{sph} = \frac{\left(\frac{1}{d_1} - \frac{1}{d}\right)}{\left(\frac{1}{d_1} - \frac{1}{d_2}\right)} \qquad ...(3.93)$$

Eqs. (3.65) and (3.90') have been derived assuming the thermal conductivity of the wall to be constant. By analogy, generalised relationships can be derived for the case when the thermal conductivity λ is a function of temperature.

SHAPE FACTORS FROM VOLTAGE OR RESISTANCE MEASUREMENTS

It should be mentioned in connection with electrical analogs that it is possible to determine shape factors without making a flux plot. They can be determined by comparison of the voltage drop across or resistance of an analog of a system, for which the shape factor is desired, with a voltage drop across or resistance of a comparable analog of a system for which the shape factor is known. By way of illustration, consider two electrically conducting shapes connected in series, as shown in Fig. 3.12; both are made of the same material and are of the same thickness l normal to the diagram. The current flow would be

$$i = \frac{\Delta v_1}{R_1} = \frac{\Delta v_2}{R_2}$$

or

$$= G_1 \Delta v_1 = G_2 \Delta v_2 \qquad ...(3.94)$$

where R_1 and R_2 are the resistances and G_1 and G_2 the conductances of 1 and 2 respectively. If σ is the electrical conductivity of the material, the electrical conductance of shape 1 would be

$$G_1 = \sigma l \frac{W}{L}$$

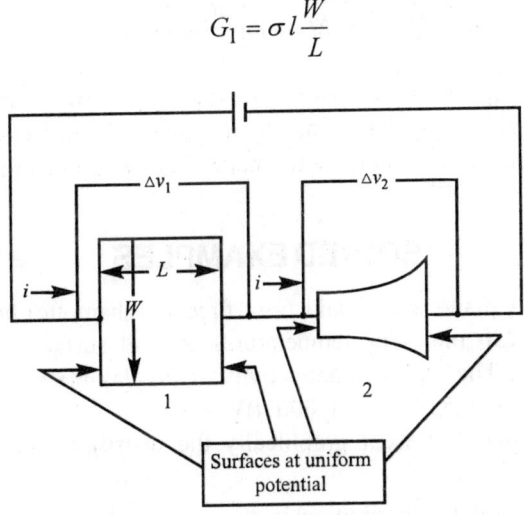

Fig. 3.12. Shape factor determination from voltage measurements.

Now note for a rectangular cross-section, a flow net (indicating either lines of constant temperature and heat flow lines or lines of constant voltage and current flow lines) would consist of a network of squares and the shape factor would be simply the width divided by the length, W/L. Thus

$$G_1 = \sigma l S_1$$

and

$$i = \sigma l S_1 \Delta v_1$$

It must also be apparent that

$$i = \sigma l S_2 \Delta v_2$$

where S_2 is the electrical or thermal shape factor for 2. Equating these last two expressions for i shows that

$$S_2 = \frac{\Delta v_1}{\Delta v_2} S_1 \qquad \qquad ...(3.95)$$

Substituting for Δv_1 and Δv_2 from equation (3.94) shows

$$S_2 = \frac{R_1}{R_2} S_1 \qquad \qquad ...(3.96)$$

The heat flow through a shape geometrically similar to 2 would then be

$$q = klS_2\Delta t_2 = kl\left(\frac{R_1}{R_2}S_1\right)\Delta t_2 \qquad \qquad ...(3.97)$$

Thus, the shape factor for heat flow through an irregular-shaped body can be quickly and easily determined by measuring the resistance of a model of the body and a rectangle both made from a conductor of uniform thickness.

SOLVED EXAMPLES

Example 3.1. A plane wall is laid from fireclay (chamotte) brick, and wall thickness $\delta = 250$ mm. The temperatures of wall surfaces $t_{w1} = 1350°C$ and $t_{w2} = 50°C$. The thermal conductivity of fireclay brick is a function of temperature, $\lambda = 0.838 (1 + 0.0007t)$W/m.°C.

Calculate and represent graphically the distribution of temperature through the wall.

Solution. Temperature distribution in the wall is shown in Fig. 3.13. With linear dependence of thermal conductivity on temperature the rate of heat flow

$$q = \frac{\lambda_m}{\delta}(t_{w1} - t_{w2}), \quad \text{W} / \text{m}^2,$$

where

$$\lambda_m = \lambda_0\left(1 + \beta_\lambda\frac{t_{w1} + t_{w2}}{2}\right), \quad \text{W} / \text{m.°C}$$

In the case considered

$$\lambda_m = 0.838\left(1 + 0.0007\frac{1350 + 50}{2}\right) = 1.25 \text{ W} / \text{m}^2.°C$$

and

$$q = \frac{1.25}{0.25}(1350 - 50) = 6500 \text{ W} / \text{m}^2.$$

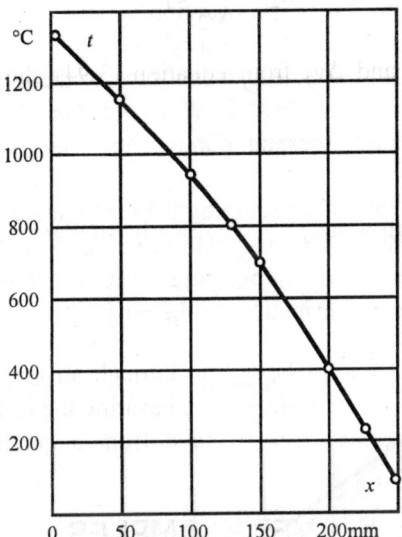

Fig. 3.13. Solution to Example 3.1.

At any distance x from the surface of the wall the temperature is determined from the following equation

$$t_x = \sqrt{\left(\frac{1}{\beta_\lambda} + t_{w1}\right)^2 - \frac{2qx}{\lambda_0 \beta_\lambda} - \frac{1}{\beta_\lambda}}, \ ^\circ\text{C}.$$

Substituting the known value of λ_0 and the calculated value of the rate of heat flow q into above equation, we find that

$$t_x = \sqrt{\left(\frac{1}{0.0007} + 1350\right)^2 - \frac{2 \times 6500 \, x}{0.838 \times 0.0007} - \frac{1}{0.0007}},$$

whence

$$t_x = \left(\sqrt{7.74 - 22.2\,x} - 1.43\right) \times 10^3, \ ^\circ\text{C}.$$

Substituting in the equation obtained the values of x expressed in metres, we find the corresponding values of the wall temperature.

Example 3.2. The surface temperatures of a 200 mm thick fireclay wall (δ = 200 mm) are: t_{w1} = 1000°C and t_{w2} = 200°C.

The thermal conductivity of fireclay (chamotte) varies depending on temperature, obeying the following equation

$$\lambda = 0.813 + 0.000582t, \quad W/m.°C$$

Show that in the event of linear dependence of thermal conductivity on temperature the rate of heat flow q, W/m^2, can be calculated from the formula for the constant thermal conductivity, taken at a mean wall temperature.

Find the error admitted in determining the temperature at points x = 57.5 mm; 110 mm; 157.5 mm, if the calculation is based on the mean thermal conductivity for the given temperature interval, and plot the graph representing temperature distribution in the wall.

Solution. q = 4650 W/m²

The temperature distribution through the wall is shown in Fig. 3.14.

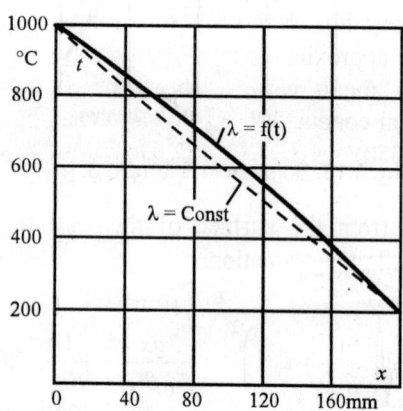

Fig. 3.14. Solution to Example 3.2.

$t x$,°C	x, mm	0	57.5	110	157.7	200
	$l = lm$ = const	1000	770	560	370	200
	$l = 0.813 + 5.82 \times 10^{-4}t$	1000	800	600	400	200

Example 3.3. The brickwork of a furnace is built up of a layer of fireclay brick with a thermal conductivity $\lambda = 0.84 (1 + 0.695 \times 10^{-3} t)$ W/m.°C; the brickwork is 250 mm thick.

Determine the loss of heat from one square metre of brickwork surface, q W/m², and the wall surface temperatures, if the temperature of the gas in the furnace t_{f1} = 1200°C and the room temperature t_{f2} = 30°C, and the local coefficient of heat transfer from the hot gas to the brickwork α_1 =

30 W/m².°C, and the local coefficient of heat transfer from the brickwork to surroundings $\alpha_2 = 10$ W/m².°C.

Solution. With the dependence of the thermal conductivity of fireclay brick on temperature given, the loss of heat can be calculated from the following equation :

$$q = \frac{\lambda}{\delta}(t_{w1} - t_{w2}) =$$

$$\frac{1}{\delta} = \left\{ \lambda_0 + \lambda_0\beta_\lambda \left[\frac{t_{f1} + t_{f2}}{2} - \frac{q}{2}\left(\frac{1}{\alpha_2} - \frac{1}{\alpha_1}\right) \right] \right\} \times$$

$$\left[t_{f1} - t_{f2} - q\left(\frac{1}{\alpha_1} + \frac{1}{\alpha_2}\right) \right]$$

or with the aid of the method of successive approximations (also known as the step-by-step or iteration method). The solution with the aid of the method of successive approximations is given below :

Let us assume a mean wall temperature $\bar{t}_w = 650°C$. At this temperature the thermal conductivity of fireclay (chamotte) brick $\lambda_m = 0.84$ $(1 + 0.695 \times 10^{-3} \times 650) = 1.12$ W/m.°C. Determine now the overall heat transfer coefficient

$$k = \frac{1}{\dfrac{1}{\alpha_1} + \dfrac{\delta}{\lambda} + \dfrac{1}{\alpha_2}} = \frac{1}{\dfrac{1}{30} + \dfrac{0.25}{1.12} + \dfrac{1}{10}} = 2.81 \text{ W}/\text{m}^2.°\text{C}$$

and the rate of heat flow

$$q = k(t_{f1} - t_{f2}) = 2.81(1200 - 30) = 3290 \text{ W/m}^2.$$

Making use of the determined rate of heat flow, calculate the surface temperatures for the brickwork:

$$t_{w1} = t_{f1} - q\frac{1}{\alpha_1} = 1200 - \frac{3290}{30} = 1091°C;$$

$$t_{w2} = t_{f2} - q\frac{1}{\alpha_2} = 30 + \frac{3290}{10} = 359°C.$$

Determine the mean temperature of the brickwork and specify the magnitude of the thermal conductivity:

$$t_w = 0.5(1091 + 359) = 725°C;$$

$$\lambda_m = 0.84 \ (1 + 0.695 \times 10^{-3} \times 725) = 1.265 \ \text{W/m.}°\text{C}$$

$$k = \cfrac{1}{\cfrac{1}{30} + \cfrac{0.25}{1.265} + \cfrac{1}{10}} = 3.02 \ \text{W / m}^2.°\text{C.}$$

The rate of heat flow will then be

$$q = 3.02 \ (1200 - 30) = 3530 \ \text{W/m}^2.$$

Using this new value of the rate of heat flow, calculate the wall surface temperatures t_{w1} and t_{w2}:

$$t_{w1} = 1200 - 3530 \frac{1}{30} = 1082 \ °\text{C};$$

$$t_{w2} = 30 + 3530 \frac{1}{10} = 383 \ °\text{C}.$$

Determine the mean surface temperature and thermal conductivity:

$$\overline{t_w} = 0.5 \ (1082 + 383) = 732°\text{C};$$

$$\lambda_m = 0.84 \ (1 + 0.695 \times 10^{-3} \times 732) = 1.267 \ \text{W/m.} °\text{C.}$$

Inasmuch as the obtained mean thermal value of conductivity coincides practically with the previously assumed value, no further calculations are necessary and it may be assumed that

$$q = 3530 \ \text{W/m}^2.$$

Example 3.4. A stainless steel tube with an inner diameter d_1 = 7.6. mm and an outer diameter d_2 = 8 mm is heated with electric current, connecting the tube directly into an electric circuit.

All the heat evolved in the wall of the tube is transferred from the tube through its inner surface.

Calculate the volumetric rate of heat liberation from the inner heat sources and the temperature drop across the wall of the tube, if the latter carries a current of 250 A intensity.

The specific resistance and the thermal conductivity of steel are respectively ρ = 0.85 Ohm · mm^2/m, λ = 18.6 W/m.°C.

Solution. The electric resistance per unit length of the tube

$$R_l = \frac{\rho}{\pi(r_2^2 - r_1^2)} = \frac{0.85}{3.14 \ (4^2 - 3.8^2)} = 0.174 \ \text{Ohm / m.}$$

The rate of heat flow per unit length of the tube

$$q_l = I^2 R_l = 250^2 \times 0.174 = 10\ 870\ \text{W/m}.$$

The volumetric rate of heat liberation from the inner heat sources

$$q_v = \frac{q_l}{\pi(r_2^2 - r_1^2)} = \frac{10\ 870}{3.14\ (4^2 - 3.8^2)10^{-6}} = 2.22 \times 10^9\ \text{W/m}^3.$$

The temperature drop across the wall of the tube

$$t_{w2} - t_{w1} = \frac{q_l r_2^2}{4\pi\lambda\ (r_2^2 - r_1^2)}\left[2\ln\frac{r_2}{r_1} + \left(\frac{r_1}{r_2}\right)^2 - 1\right]$$

$$= \frac{q_v r_2^2}{4\lambda}\left[2\ln\frac{r_2}{r_1} + \left(\frac{r_1}{r_2}\right)^2 - 1\right]$$

$$= \frac{2.22 \times 10^9 \times 0.004^2}{4 \times 18.6}\left[2 \times 2.3\ln\frac{4}{3.8} + \left(\frac{3.8}{4}\right)^2 - 1\right] \approx 2.4°\text{C}.$$

Example 3.5. Acting in a plate of thickness $s = 5$ mm are uniformly distributed heat sources with a volumetric rate of heat liberation $q_v = 2.7 \times 10^7$ W/m^3. The thermal conductivity of the plate material $\lambda = 25$ W/m°C. The local coefficients of heat transfer from the surfaces of the plate to the fluid to which it is exposed $\alpha_1 = 3000$ W/m^2.°C and $\alpha_2 = 1500$ W/m^2.°C and the temperatures of the fluid are respectively $t_{f1} = 130$°C and $t_{f2} = 140$°C.

Determine the coordinate and the magnitude of the maximum temperature in the plate, x_0 and t_0, and also the temperatures at the surfaces of the plate, t_{w1} and t_{w2}.

Solution. With $q_v = $ const and $\lambda = $ const, a non-symmetrical temperature field and boundary conditions of the third kind the relative coordinate or the maximum temperature in the plate

$$\frac{x_0}{s} = \frac{\dfrac{1}{2} + \dfrac{\lambda}{q_v s^2}\left(t_{f2} - t_{f1}\right) + \dfrac{\lambda}{\alpha_2 s}}{1 + \dfrac{\lambda}{s}\left(\dfrac{1}{\alpha_1} + \dfrac{1}{\alpha_2}\right)},$$

Where x_0 is measured from the surface which is exposed to the fluid at a temperature t_{f1}.

In the case considered

$$\frac{x_0}{s} = \frac{\frac{1}{2} + \frac{25}{2.7 \times 10^7 (5 \times 10^{-3})^2}(140 - 130) + \frac{25}{1.5 \times 10^3 \times 5 \times 10^{-3}}}{1 + \frac{25}{5 \times 10^{-3}}\left(\frac{1}{3 \times 10^3} + \frac{1}{1.5 \times 10^3}\right)} = 0.7;$$

$x_0 = 0.7 \times 5 = 3.5$ mm.

The surface temperatures of the plate

$$t_{w1} = t_{f1} + \frac{q_{w1}}{\alpha_1} = t_{f1} + \frac{q_v x_0}{\alpha_1}$$

$$= 130 + \frac{2.7 \times 10^7 \times 3.5 \times 10^{-3}}{3 \times 10^3} = 161.5°\text{C};$$

$$t_{w2} = t_{f2} + \frac{q_{w2}}{\alpha_2} = t_{f2} + \frac{q(s - x_0)}{\alpha_2}$$

$$= 140 + \frac{2.7 \times 10^7 \times (5 - 3.5) \times 10^{-3}}{1.5 \times 10^3} = 167°\text{C}.$$

The maximum temperature

$$t_0 = t_{w1} + q_{w1}\frac{x_0}{2\lambda} = t_{w1} + \frac{q_v x^2_0}{2\lambda}$$

$$= 161.5 + \frac{2.7 \times 10^7 \left(3.5 \times 10^{-3}\right)^2}{2 \times 25} = 168.1°\text{C}.$$

Example 3.6. Calculate the rate of heat flow through a plane homogeneous wall the thickness of which is considerably smaller than its width and height, assuming the wall to be made from: (a) steel ($\lambda = 40$ W/m.°C); (b) concrete ($\lambda = 1.1$ W/m.°C); (c) diatomite brick ($\lambda = 0.11$ W/m.°C).

In all the three cases wall thickness $\delta = 50$ mm. The temperatures of the wall surfaces are maintained constant :

$$t_{w1} = 100°\text{C and } t_{w2} = 90°\text{C}.$$

Solution. For a steel wall $q = 8000$ W/m^2;
for a concrete wall $q = 220$ W/m^2;
for a wall laid from diatomite brick $q = 22$ W/m^2.

PROBLEMS

Problem 3.1. The rate of heat flow through a plane wall of thickness δ = 50 mm., q = 70 W/m^2.

Determine the difference between the temperatures of wall surfaces and the numerical values of the temperature gradient through the wall, if it is made from: (a) brass (λ = 70 W/m.°C); (b) red brick (λ = 0.7 W/m.°C); (c) cork (λ = 0.07 W/m.°C).

Answer. For a brass wall Δt = 0.05°C and | grad t | = 1°C/m; for a brick wall Δt = 5°C and | grad t | = 1000°C; for a cork wall Δt = 50°C and | grad t | = 1000°C/m.

Problem 3.2. Determine the loss of heat Q,W, through a wall laid from red brick of length l = 5 m, height h = 4 m and thickness δ = 0.250 m, if the temperatures of wall surfaces are maintained equal to t_{w1} = 110°C and t_{w2} = 40°C. The thermal conductivity of red brick λ = 0.70 W/m.°C.
Answer. The loss of heat Q = 3920 W.

Problem 3.3. Determine the thermal conductivity of the material of a wall 40 mm thick, if at a temperature difference between surface temperatures Δt = 20°C the rate of heat flow q = 145 W/m^2.
Answer. The thermal conductivity λ = 0.29 W/m.°C.

Problem 3.4. It is necessary to insulate a flat surface so that the loss of heat from unit area of this surface per unit time will not exceed 450 W/m^2. The temperature of the surface underneath the layer of insulation t_{w1} = 450°C, and the temperature of the external surface of the insulating layer t_{w2} = 50°C.

Determine the thickness of insulation for two cases:
(a) the insulation is made from sovelite, for which
λ = 0.09 + 0.0000874t,   W/m.°C;
(b) the insulation is made from asbestos cement, for which
λ = 0.109 + 0.000146t, W/m.°C.

Answer.
(a) δ = 100 mm;
(b) δ = 130 mm.

Problem 3.5. The flat wall of a tank, of an area F = 5 m^2, is covered with two layers of a heat insulating material. The steel wall of the tank is 8 mm thick and its thermal conductivity λ_1 = 46.5 W/m.°C. The first layer of the insulation is made from an insulating material novoasbosurite of thickness δ_2 = 50 mm and the thermal conductivity of this insulating material is determined from the equation
λ_2 = 0.144 + 0.000174t, W/m.°C.

The second layer of insulation of thickness δ_3 = 10 mm is lime plaster whose thermal conductivity λ_3 = 0.698 W/m.°C.

The temperature of the inside surface of the tank wall t_{w1} = 250°C and the temperature of the outside surface of insulation t_{w4} = 50°C.

Calculate the amount of heat transferred through the wall, the boundary temperatures for the layers of insulation, and plot the graph of temperature distribution.

Answer. The heat flux through the wall Q = 3170 W; the boundary temperatures for the layers of insulation are t_{w2} = 249.9°C and t_{w3} = 59°C.

Problem 3.6. The brickwork of a furnace is built up of layers laid of fireclay and red brick and the space between the two layers of brickwork is filled with crushed diatomite brick (Fig. 3.15). The fireclay layer is 120 mm thick (δ_1); the thickness of the diatomite filling δ_2 = 50 mm, and the thickness of the red brick layer δ_3 = 250 mm. The thermal conductivity of the three materials are equal respectively:

$$\lambda_1 = 0.93; \quad \lambda_2 = 0.13; \quad \text{and} \quad \lambda_3 = 0.7 \text{ W/m.°C.}$$

What should be the thickness of the red brick layer δ_3, if the brickwork is to be laid without the diatomite filling between the two layers, so that the heat flux through the brickwork remains constant?

Answer. The layer of red brick must be 500 mm thick.

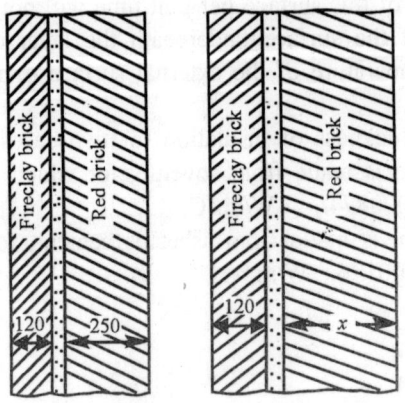

Fig. 3.15. Solution to Problem 3.6.

Problem 3.7. The wall of a steam boiler furnace which is not fitted with waterwalls is made of a layer of foamed fireclay of thickness δ_1 = 125 mm and a layer of red brick 500 mm thick (δ_2). The two layers fit each other tightly. The wall temperature inside the boiler furnace t_{w1} = 1100°C and the wall outside temperature t_{w3} = 50°C (Fig. 3.16). The thermal conductivity of foamed fireclay $\lambda_1 = 0.28 + 0.0023t$ and that of red brick, $\lambda_2 = 0.7$ W/m.°C.

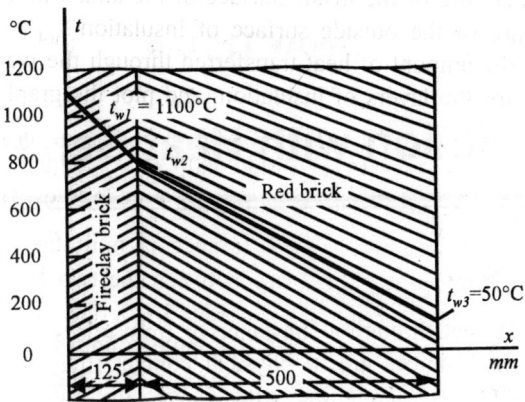

Fig. 3.16. Solution to Problem 3.7.

Calculate the amount of heat lost from 1 m² of the furnace wall by conduction and the temperature at the layer interface.

Answer. The loss of heat or the rate of heat flow through the wall $q = 1090$ W/m². The temperature at the layer interface $tw_2 = 828°C$.

CHAPTER 4

Conduction with Heat Generation

INTRODUCTION

Transference of heat through a substance or from one substance to another when the two substances are in physical contact (thermal conduction) is known as conduction. Crystalline solids (especially metals and alloys) are good thermal conductors because of their high density; liquids, such as water and glass, and high polymers, such as rubbers and cellulose, usually are not.

Transference of an electric current through a solid or liquid (electrical conduction). In metallic or electronic conductors, the current is carried by a flow of electrons from atom to atom, the atomic nuclei remaining stationary. This type of conduction is common to all metals and alloys, carbon and graphite, and certain solid compounds (manganese dioxide, lead sulphide). In electrolytic or ionic conductors, the current is carried by ions, as in solutions of acids, bases, and salts and in many fused compounds. In electrolytic conduction, as in metallic conduction, heat is generated, and a magnetic field is formed around the conductor; a transfer of matter also occurs. In a few materials, as solutions of alkali and alkaline earth metals in anhydrous liquid ammonia, both types of conduction take place simultaneously; such conductors are called mixed conductors.

CONDUCTION THROUGH A FLAT SLAB OR WALL

In this section Fourier's equation will be used to obtain equations for one-dimensional steady-state conduction of heat through some simple geometries. For a flat slab or wall where the cross-sectional area A and k are constant, then Fourier's equation can be rewritten as

$$\frac{q}{A} = \frac{k}{x_2 - x_1}(T_1 - T_2) = \frac{k}{\Delta x}(T_1 - T_2) \qquad \ldots (4.1)$$

This is shown in Fig. 4.1, where $\Delta x = x_2 - x_1$. Equation (4.1) indicates that if T is substituted for T_2 and x for x_2, the temperature varies linearly with distance as shown in Fig. 4.1(b).

If the thermal conductivity is not constant but varies linearly with temperature.

$$\frac{q}{A} = \frac{a + b\frac{T_1 + T_2}{2}}{\Delta x}(T_1 - T_2) = \frac{k_m}{\Delta x}(T_1 - T_2) \qquad \dots (4.2)$$

where

$$k_m = a + b\frac{T_1 + T_2}{2} \qquad \dots (4.3)$$

This means that the mean value of k (i.e., k_m) to use in Eq. (4.2) is the value of k evaluated at the linear average of T_1 and T_2.

As stated already the rate of a transfer process equals the driving force over the resistance. Equation (4.1) can be rewritten in that form.

$$q = \frac{T_1 - T_2}{\Delta x / kA} = \frac{T_1 - T_2}{R} = \frac{\text{driving force}}{\text{resistance}} \qquad \dots (4.4)$$

where $R = \Delta x/kA$ and is the resistance in K/W or h.°F/btu.

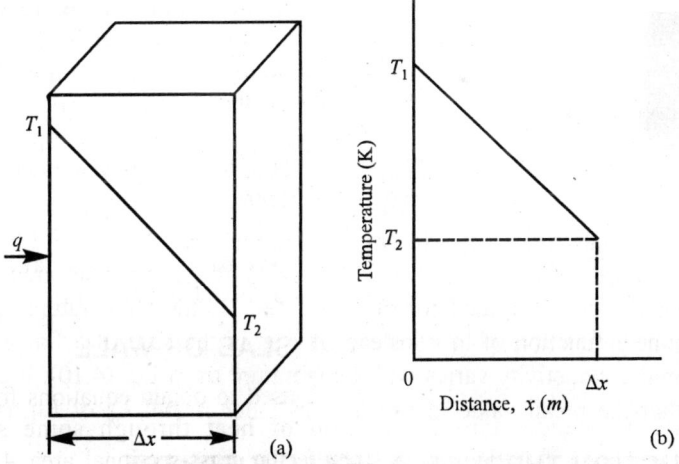

Fig. 4.1. Heat conduction in a flat wall: (a) geometry of wall, (b) temperature plot.

CONDUCTION THROUGH A HOLLOW CYLINDER

In many instances in the process industries, heat is being transferred through the walls of a thick-walled cylinder as in a pipe that may or may not be insulated. Consider the hollow cylinder in Fig. 4.2 with an inside radius of r_1, where the temperature is T_1, an outside radius of r_2 having a temperature of T_2, and a length of L m. Heat is flowing radially from the inside surface to the outside. Rewriting Fourier's law, with distance dr instead of dx,

$$\frac{q}{A} = -k\frac{dT}{dr} \qquad \ldots (4.5)$$

The cross-sectional area normal to the heat flow is

$$A = 2\pi r L \qquad \ldots (4.6)$$

Substituting Eq. (4.6) into (4.5), rearranging, and integrating,

$$\frac{q}{2\pi L}\int_{r_1}^{r_2}\frac{dr}{r} = -k\int_{T_1}^{T_2} dT \qquad \ldots (4.7)$$

$$q = k\frac{2\pi L}{\ln(r_2/r_1)}(T_1 - T_2) \qquad \ldots (4.8)$$

Multiplying numerator and denominator by $(r_2 - r_1)$,

$$q = kA_{\text{lm}}\frac{T_1 - T_2}{r_2 - r_1} = \frac{T_1 - T_2}{(r_2 - r_1)(kA_{\text{lm}})} = \frac{T_1 - T_2}{R} \qquad \ldots (4.9)$$

where

$$A_{\text{lm}} = \frac{(2\pi Lr_2) - (2\pi Lr_1)}{\ln(2\pi Lr_2/2\pi Lr_1)} = \frac{A_2 - A_1}{\ln(A_2/A_1)} \qquad \ldots (4.10)$$

$$R = \frac{r_2 - r_1}{kA_{\text{lm}}} = \frac{\ln(r_2/r_1)}{2\pi kL} \qquad \ldots (4.11)$$

The log mean area is A_{lm}. In engineering practice, if $A_2/A_1 < 1.5/1$, the linear mean area of $(A_1 + A_2)/2$ is within 1.5% of the log mean area. From Eq. (4.8), if r is substituted for r_2 and T for T_2, the temperature is seen to be a linear function of $\ln r$ instead of r as in the case of a flat wall. If the thermal conductivity varies with temperature as in Eq. (4.10), it can be shown that the mean value to use in a cylinder is still k_m of Eq. (4.3).

CONDUCTION THROUGH A HOLLOW SPHERE

Heat conduction through a hollow sphere is another case of one-dimensional conduction. Using Fourier's law for constant thermal conductivity with distance dr, where r is the radius of the sphere,

$$\frac{q}{A} = -k\frac{dT}{dr} \qquad \ldots(4.12)$$

The cross-sectional area normal to the heat flow is

$$A = 4\pi r^2 \qquad \text{...(4.13)}$$

Substituting Eq. (4.13) into (4.12), rearranging, and integrating,

$$\frac{q}{4\pi}\int_{r_1}^{r_2}\frac{dr}{r^2} = -k\int_{T_1}^{T_2} dt \qquad \text{...(4.14)}$$

$$q = \frac{4\pi k(T_1 - T_2)}{1/r_1 - 1/r_2} = \frac{T_1 - T_2}{(1/r_1 - 1/r_2)/4\pi k} \qquad \text{...(4.15)}$$

It can be easily shown that the temperature varies hyperbolically with the radius (Eq. 4.12).

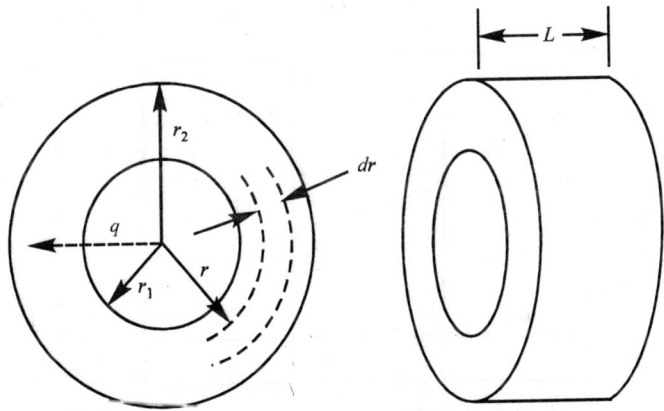

Fig. 4.2. Heat conduction in a cylinder.

CONDUCTION THROUGH SOLIDS IN SERIES

Plane Walls in Series

In the case where there is a multilayer wall of more than one material present as shown in Fig. (4.3), we proceed as follows. The temperature profiles in the three materials A, B, and C are shown. Since the heat flow q must be same in each layer, we can write Fourier's equation for each layer as

$$q = \frac{k_A A}{\Delta x_A}(T_1 - T_2) = \frac{k_B A}{\Delta x_B}(T_2 - T_3) = \frac{k_C A}{\Delta x_C}(T_3 - T_4) \qquad \text{...(4.16)}$$

Solving each equation for ΔT,

$$T_1 - T_2 = q \frac{\Delta x_A}{k_A A} \quad T_2 - T_3 = q \frac{\Delta x_B}{k_B A} \quad T_3 - T_4 = q \frac{\Delta x_C}{k_C A} \qquad ...(4.17)$$

Adding the equations for $T_1 - T_2$, $T_2 - T_3$, and $T_3 - T_4$, the internal temperatures T_2 and T_3 drop out and the final rearranged equation is

$$q = \frac{T_1 - T_4}{\Delta x_A / (k_A A) + \Delta x_B / (k_B A) + \Delta x_C / (k_C A)} = \frac{T_1 - T_4}{R_A + R_B + R_C} \qquad ...(4.18)$$

where the resistance $R_A = \Delta x_A / k_A A$, and so on.

Hence, the final equation is in terms of the overall temperature drop $T_1 - T_4$ and the total resistance, $R_A + R_B + R_C$.

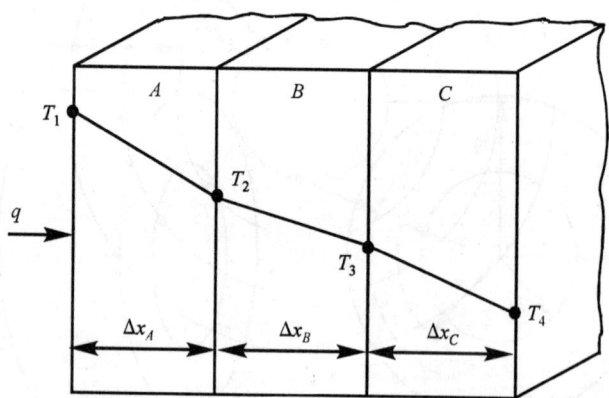

Fig. 4.3. Heat flow through a multilayer wall.

Multilayer Cylinders

In the process industries, heat transfer often occurs through multilayers of cylinders, as for example when heat is being transferred through the walls of an insulated pipe. Fig. 4.4 shows a pipe with two layers of insulation around it, i.e., a total of three concentric hollow cylinders. The temperature drop is $T_1 - T_2$ across material A, $T_2 - T_3$ across B, and $T_3 - T_4$ across C.

The heat-transfer rate q will, of course, be the same for each layer, since we are at steady-state. Writing an equation similar to Eq. (4.9) for each concentric cylinder,

$$q = \frac{T_1 - T_2}{(r_2 - r_1) / (k_A A_{A\mathrm{lm}})} = \frac{T_2 - T_3}{(r_3 - r_2) / (k_B A_{B\mathrm{lm}})}$$

$$= \frac{T_3 - T_4}{(r_4 - r_3)/(k_C A_{Clm})} \qquad ...(4.19)$$

where

$$A_{Alm} = \frac{A_2 - A_1}{\ln (A_2 / A_1)} \quad A_{Blm} = \frac{A_3 - A_2}{\ln (A_3 / A_2)} \quad A_{Clm} = \frac{A_4 - A_3}{\ln (A_4 / A_3)} \qquad ...(4.20)$$

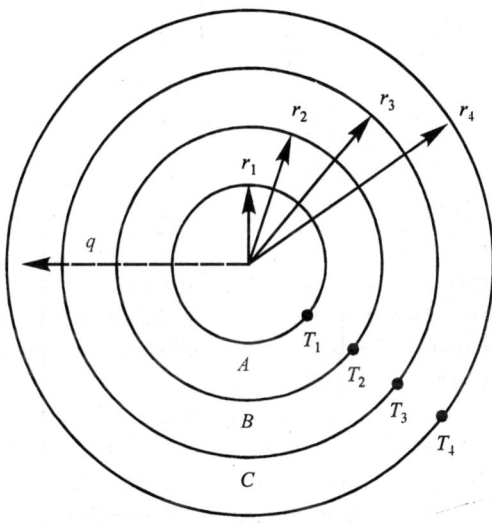

Fig. 4.4. Radial heat flow through multiple cylinders in series.

Using the same method to combine the equations to eliminate T_2 and T_3 as was done for the flat walls in series, the final equations are

$$q = \frac{T_1 - T_4}{(r_2 - r_1)/(k_A A_{Alm}) + (r_3 - r_2)/(k_B A_{Blm}) + (r_4 - r_3)/(k_C A_{Clm})}$$
$$...(4.21)$$

$$q = \frac{T_1 - T_4}{R_A + R_B + R_C} = \frac{T_1 - T_4}{\sum R} \qquad ...(4.22)$$

Hence, the overall resistance is again the sum of the individual resistances in series.

Conduction Through Materials in Parallel

Suppose that two plane solids A and B are placed side by side in parallel, and the direction of heat flow is perpendicular to the plane of the exposed surface of each solid. Then the total heat flow is the sum of the heat flow through solid A plus that through B. Writing Fourier's equation for each solid and summing,

$$q_T = q_A + q_B = \frac{k_A A_A}{\Delta x_A}(T_1 - T_2) + \frac{k_B A_B}{\Delta x_B}(T_3 - T_4) \quad ...(4.23)$$

where q_T is total heat flow, T_1 and T_2 are the front and rear surface temperatures of solid A; T_3 and T_4, for solid B.

If we assume that $T_1 = T_3$ (front temperatures the same for A and B) and $T_2 = T_4$ (equal rear temperatures),

$$q_T = \frac{T_1 - T_2}{\Delta x_A / k_A A_A} + \frac{T_1 - T_2}{\Delta x_B / k_B A_B} = \left(\frac{1}{R_A} + \frac{1}{R_B}\right)(T_1 - T_2) \quad ...(4.24)$$

An example would be an insulated wall (A) of a brick oven where steel reinforcing members (B) are in parallel and penetrate the wall. Even though the area A_B of the steel would be small compared to the insulated brick area A_A, the higher conductivity of the metal (which could be several hundred times larger than that of the brick) could allow a large portion of the heat lost to be conducted by the steel.

Another examples is a method of increasing heat conduction to accelerate the freeze drying of meat. Spikes of metal in the frozen meat conduct heat more rapidly into the insides of the meat.

It should be mentioned that in some cases some two-dimensional heat flow can occur if the thermal conductivities of the materials in parallel differ markedly. Then the results using Eq. (4.24) would be affected somewhat.

Combined Convection and Conduction and Overall Coefficients

In many practical situations the surface temperatures (or boundary conditions at the surface) are not known, but there is a fluid on both sides of the solid surfaces. Consider the plane wall in Fig. (4.5(a)) with a hot fluid at temperature T_1 on the inside surface and a cold fluid at T_4 on the outside surface. The outside convective coefficient is h_o W/m^2. K and h_i on the inside. (Methods to predict the convective h will be given later in this chapter).

The heat-transfer rate is given as

$$q = h_i A(T_1 - T_2) = \frac{k_A A}{\Delta x_A}(T_2 - T_3) = h_o A(T_3 - T_4) \quad ...(4.25)$$

Expressing $1/h_i A$, $\Delta x_A/k_A A$, and $1/h_o A$ as resistances and combining the equations as before,

$$q = \frac{T_1 - T_4}{1/h_i A + \Delta x_A / k_A A + 1/h_o A} = \frac{T_1 - T_4}{\sum R} \qquad ...(4.26)$$

The overall heat transfer by combined conduction and convection is often expressed in terms of an overall heat-transfer coefficient U defined by

$$q = UA\Delta T_{overall} \qquad ...(4.27)$$

where $\Delta T_{overall} = T_1 - T_4$ and U is

$$U = \frac{1}{1/h_i + \Delta x_A / k_A + 1/h_o} \frac{W}{m^2.K} \left(\frac{btu}{h.ft^2.°F} \right) \qquad ...(4.28)$$

A more important application is heat transfer from a fluid outside a cylinder, through a metal wall, and to a fluid inside the tube, as often occurs in heat exchangers. In Fig. 4.5(b) such a case is shown.

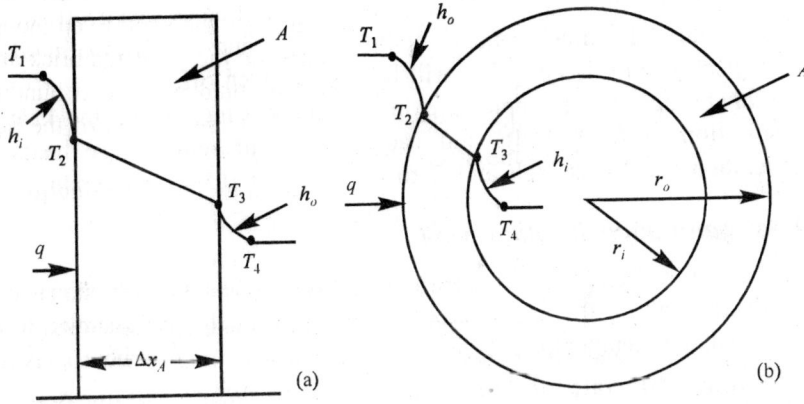

(a)   (b)

Fig. 4.5. Heat flow with convective boundaries: (a) plane wall (b) cylindrical wall.

Using the same procedure as before, the overall heat-transfer rate through the cylinder is

$$q = \frac{T_1 - T_4}{1/h_i A_i + (r_o - r_i)/k_A A_{A\,lm} + 1/h_o A_o} = \frac{T_1 - T_4}{\sum R} \qquad ...(4.29)$$

where A_i represents $2\pi L r_i$, the inside area of the metal tube; A_{Alm} the log mean area of the metal tube; and A_o the outside area.

The overall heat-transfer coefficient U for the cylinder may be based on the inside area A_i or the outside area A_o of the tube. Hence,

$$q = U_i A_i (T_1 - T_4) = U_o A_o (T_1 - T_4) = \frac{T_1 - T_4}{\sum R} \qquad ...(4.30)$$

$$U_i = \frac{1}{1/h_i + (r_o - r_i)A_i / k_A A_{A \text{ lm}} + A_i / A_o h_o} \qquad ...(4.31)$$

$$U_i = \frac{1}{A_o / A_i h_i + (r_o - r_i)A_o / k_A A_{A \text{ lm}} + 1/h_o} \qquad ...(4.32)$$

Conduction with Internal Heat Generation

In certain systems heat is generated inside the conducting medium; i.e., a uniformly distributed heat source is present. Examples of this are electric resistance heaters and nuclear fuel rods. Also, if a chemical reaction is occurring uniformly in a medium, a heat of reaction is given off. In the agricultural and sanitation fields, compost heaps and trash heaps in which biological activity is occurring will have heat given off.

Other important examples are in food processing, where the heat of respiration of fresh fruits and vegetables is present. These heats of generation can be as high as 0.3 to 0.6 W/kg or 0.5 to 1 btu/h.lb$_m$.

Heat generation in plane wall

In Fig. 4.6 a plane wall is shown with internal heat generation. Heat is conducted only in the one x direction. The other walls are assumed to be insulated. The temperature T_w in K at $x = L$ and $x = -L$ is held constant. The volumetric rate of heat generation is $\dot{q}$ W/m^3 and the thermal conductivity of the medium is k W/m.K.

To derive the equation for this case of heat generation at steady-state, we drop the accumulation term

$$q_x|_x + \dot{q}(\Delta x . A) = q_x|_{x + \Delta x} + 0 \qquad ...(4.33)$$

where A is the cross-sectional area of the plate. Rearranging, dividing by Δx, and letting Δx approach zero,

$$\frac{-dq_x}{dx} + q . A = 0 \qquad ...(4.34)$$

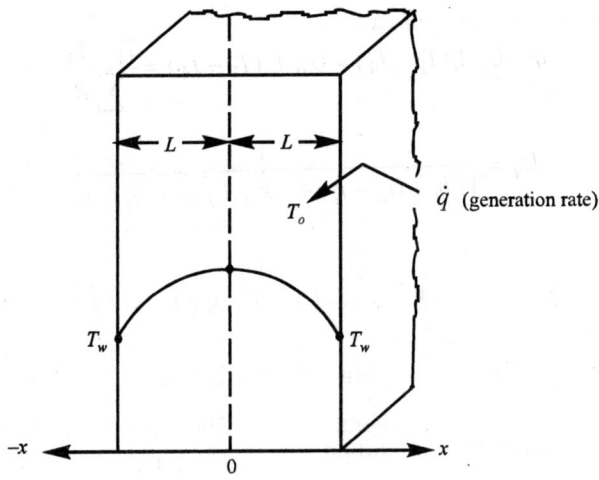

Fig. 4.6. Plane wall with internal heat generation at steady-state.

Substituting Eq. (4.12) for q_x,

$$\frac{d^2T}{dx^2} + \frac{\dot{q}}{k} = 0 \qquad \text{...(4.35)}$$

Integration gives the following for $\dot{q}$ constant :

$$T = -\frac{\dot{q}}{2k}x^2 + C_1 x + C_2 \qquad \text{...(4.36)}$$

where C_1, and C_2 are integration constants. The boundary conditions are at $x = L$ or $- L$, $T = T_w$, and at $x = 0$, $T = T_0$ (centre temperature). Then, the temperature profile is

$$T = -\frac{\dot{q}}{2k}x^2 + T_o \qquad \text{...(4.37)}$$

The centre temperature is

$$T_0 = \frac{\dot{q}L^2}{2k} + T_w \qquad \text{...(4.38)}$$

The total heat lost from the two faces at steady-state is equal to the total heat generated, $\dot{q}_r$, in W.

$$\dot{q}_T = \dot{q}\,(2LA) \qquad \text{...(4.39)}$$

where A is the cross-sectional area (surface area at T_w) of the plate.

Heat generation in cylinder

In a similar manner an equation can be derived for a cylinder of radius R with uniformly distributed heat sources and constant thermal conductivity. The heat is assumed to flow only radically; i.e., the ends are neglected or insulated. The final equation for the temperature profile is

$$T = \frac{\dot{q}}{4k}(R^2 - r^2) + T_w \qquad \qquad ...(4.40)$$

where r is distance from the centre. The centre temperature T_0 is

$$T_0 = \frac{\dot{q}R^2}{4k} + T_w \qquad \qquad ...(4.41)$$

Example 4.1. An electric current of 200 A is passed through a stainless steel wire having a radius R of 0.001268 m. The wire is $L = 0.91$ m long and has a resistance R of 0.126 Ω. The outer surface temperature T_w is held at 422.1 K. The average thermal conductivity is $k = 22.5$ W/m.K. Calculate the centre temperature.

Solution. First the value of $\dot{q}$ must be calculated. Since power = I^2R, where I is current in amps and R is resistance in ohms,

$$I^2R = \text{watts} = \dot{q}\,\pi R^2 L \qquad \qquad ...(4.42)$$

Substituting known values and solving,

$$(200)^2(0.126) = \dot{q}\,\pi(0.001268)^2(0.91)$$

$$\dot{q} = 1.096 \times 10^9 \text{ W/m}^3$$

Substituting into Eq. (4.41) and solving, $T_0 = 441.7$ K.

Critical Thickness of Insulation for a Cylinder

In Fig. 4.7 a layer of insulation is installed around the outside of a cylinder whose radius r_1 is fixed with a length L. The cylinder has a high thermal conductivity and the inner temperature T_1 at point r_1 outside the cylinder is fixed. An example is the case where the cylinder is a metal pipe with saturated steam inside. The outer surface of the insulation at T_2 is exposed to an environment at T_0 where convective heat transfer occurs. It is not obvious if adding more insulation with a thermal conductivity of k will decrease the heat transfer rate.

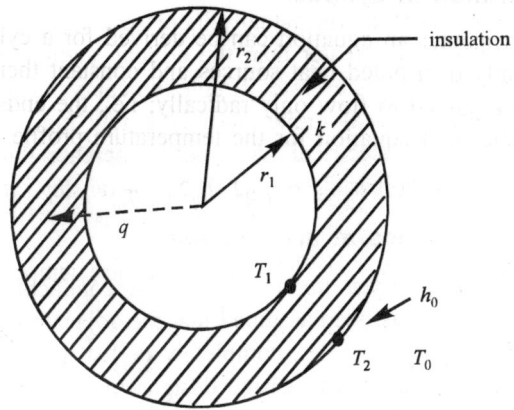

Fig. 4.7. Critical radius for insulation of cylinder or pipe.

At steady-state the heat-transfer rate q through the cylinder and the insulation equals the rate of convection from the surface.

$$q = h_o A(T_2 - T_0) \qquad \qquad ...(4.43)$$

As insulation is added, the outside area, which is $A = 2\pi r_2 L$, increases but T_2 decreases. However, it is not apparent whether q increases or decreases. To determine this, an equation similar to Eq. (4.29) with the resistance of the insulation represented by Eq. (4.25) is written using the two resistances.

$$q = \frac{2\pi L(T_1 - T_0)}{\dfrac{\ln\,(r_2\,/\,r_1)}{k} + \dfrac{1}{r_2 h_o}} \qquad \qquad ...(4.44)$$

To determine the effect of the thickness of insulation on q, we take the derivative of q with respect to r_2, equate this result to zero, and obtain the following for maximum heat flow.

$$\frac{dq}{dr_2} = \frac{-2\pi L(T_1 - T_0)(1\,/\,r_2 k - 1\,/\,r_2^2 h_o)}{\left[\dfrac{\ln\,(r_2\,/\,r_1)}{k} + \dfrac{1}{r_2 h_o}\right]^2} = 0 \qquad \qquad ...(4.45)$$

Solving,

$$(r_2)_{cr} = \frac{k}{h_o} \qquad \qquad ...(4.46)$$

where $(r_2)_{cr}$ is the value of the critical radius when the heat-transfer rate is a maximum. Hence, if the outer radius r_2 is less than the critical value, adding more insulation will actually increase the heat-transfer rate q. Also, if the outer radius is greater than the critical, adding more insulation will decrease the heat-transfer rate. Using typical values of k and h_o often encountered, the critical radius is only a few mm. As a result, adding insulation on small electrical wires could increase the heat loss. Adding insulation to large pipes decreases the heat-transfer rate.

Contact Resistance at an Interface

In the equations derived in this section for conduction through solids in series (Fig. 4.3) it has been assumed that the adjacent touching surfaces are at the same temperature; i.e., completely perfect contact is made between the surfaces. For many engineering designs in industry, this assumption is reasonably accurate. However, in cases such as in nuclear power plants where very heat fluxes are present, a significant drop in temperature may be present at the interface. This interface resistance, called contact resistance, occurs when the two solids do not fit tightly together and a thin layer of stagnant fluid is trapped between the two surfaces. At some points the solids touch at peaks in the surfaces and at other points the fluid occupies the open space.

This interface resistance is a complex function of the roughness of the two surfaces, the pressure applied to hold the surfaces in contact, the interface temperature, and the interface fluid. Heat transfer takes place by conduction, radiation, and convection across the trapped fluid and also by conduction through the points of contact of the solids. No completely reliable empirical correlations or theories are available to predict contact resistances for all types of materials.

The equation for the contact resistance is often given as follows :

$$q = h_c A \Delta T = \frac{\Delta T}{1/h_c A} = \frac{\Delta T}{R_c} \qquad ...(4.47)$$

where h_c is the contact resistance coefficient in W/m^2.K, ΔT the temperature drop across the contact resistance in K, and R_c the contact resistance. The contact resistance R_c can be added with the other resistances in Eq. (4.18) to include this effect for solids in series. For contact between two ground metal surfaces h_c values of the order of magnitude of about 0.2×10^4 to 1×10^4 W/m^2. K have been obtained.

An approximation of the maximum contact resistance can be obtained if the maximum gap Δx between the surfaces can be estimated. Then, assuming that the heat transfer across the gap is by conduction only through the stagnant fluid, h_c is estimated as

$$h_c = \frac{k}{\Delta x} \qquad \qquad ...(4.48)$$

If any actual convection, radiation, or point-to-point contact is present, this will reduce this assumed resistance.

HEAT CONDUCTION WITH A NUCLEAR HEAT SOURCE

We consider a nuclear fuel element of spherical form, as shown in Fig. 4.8. It consists of a sphere of fissionable material with radius $R^{(F)}$, surrounded by a spherical shell of aluminium "cladding" with outer radius $R^{(C)}$. Inside the fuel element fission fragments are produced which have very high kinetic energies. Collisions between these fragments and the atoms of the fissionable material provide the major source of thermal energy in the reactor. Such a volume source of thermal energy resulting from nuclear fission we call S_n (cal cm^{-3} sec^{-1}).

This source will not be uniform throughout the sphere of fissionable material; it will be the smallest at the centre of the sphere. For the purpose of this problem, we assume that the source can be approximated by a simple parabolic function :

$$S_n = S_{n0}\left[1 + b\left(\frac{r}{R^{(F)}}\right)^2\right] \qquad \qquad ...(4.49)$$

Here S_{n0} is the volume rate of heat production at the centre of the sphere, and b is a dimensionless constant between 0 and 1.

We begin by making a thermal energy balance over a spherical shell of thickness Δr within the sphere of fissionable material :

thermal energy in at r

$$q_r^{(F)}\Big|_r .4\pi r^2 \qquad \qquad ...(4.50)$$

thermal energy out at $r + \Delta r$

$$q_r^{(F)}\Big|_{r+\Delta r} .4\pi(r + \Delta r)^2 \qquad \qquad ...(4.51)$$

thermal energy produced

$$S_n.4\pi r^2\Delta r \qquad \qquad ...(4.52)$$

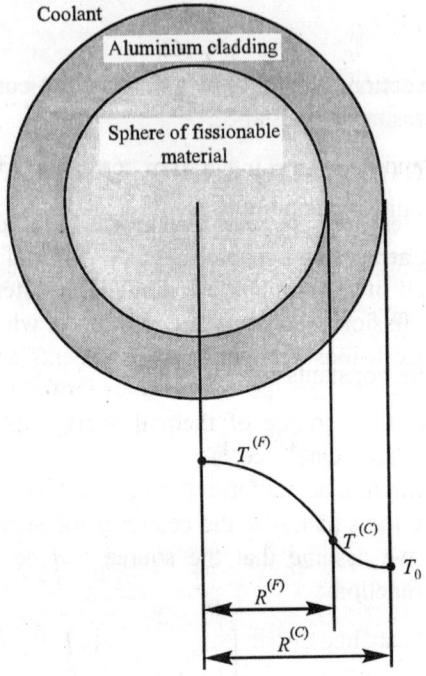

Fig. 4.8. Temperature distribution in a spherical nuclear fuel assembly.

Substitution of these terms into the thermal energy balance of Eq. (4.49) gives in the limit as $\Delta r \to 0$

$$\lim_{\Delta r \to 0} \frac{(r^2 q_r^{(F)})|_{r+\Delta r} - (r^2 q_r^{(F)})|_r}{\Delta r} = S_n r^2 \qquad ...(4.53)$$

whence

$$\frac{d}{dr}(r^2 q_r^{(F)}) = S_{n0} r^2 \left[1 + b \left(\frac{r}{R^{(F)}} \right)^2 \right] \qquad ...(4.54)$$

The differential equation for the heat flux $q_r^{(C)}$ in the cladding is of the same form as Eq. (4.54), except that there is no source term :

$$\frac{d}{dr}(r^2 q_r^{(C)}) = 0 \qquad ...(4.55)$$

Integration of these two equations gives

$$q_r^{(F)} = S_{n0}\left(\frac{r}{3} + \frac{b}{R^{(F)2}} \frac{r^3}{5}\right) + \frac{C_1^{(F)}}{r^2} \qquad ...(4.56)$$

$$q_r^{(C)} = \frac{C_1^{(C)}}{r^2} \qquad ...(4.57)$$

in which $C_1^{(F)}$ and $C_1^{(C)}$ are integration constants. These are evaluated by means of the boundary conditions :

B.C. 1 :   at $r = 0$, $q_r^{(F)}$ is not infinite   ...(4.58)

B.C. 2 :   at $r = R^{(F)}$ $q_r^{(F)} = q_r^{(C)}$ ...(4.59)

Evaluation of the constants then leads to

$$q_r^{(F)} = S_{n0}\left(\frac{r}{3} + \frac{b}{R^{(F)2}} \frac{r^3}{5}\right) \qquad ...(4.60)$$

$$q_r^{(C)} = S_{n0} R^{(F)3}\left(\frac{1}{3} + \frac{b}{5}\right)\frac{1}{r^2} \qquad ...(4.61)$$

These are the heat-flux distributions in the fissionable sphere and in the spherical shell cladding.

Into these distributions we now substitute Fourier's law of heat conduction :

$$-k^{(F)} \frac{dT^{(F)}}{dr} = S_{n0}\left(\frac{r}{3} + \frac{b}{R^{(F)2}} \frac{r^3}{5}\right) \qquad ...(4.62)$$

$$-k^{(C)} \frac{dT^{(C)}}{dr} = S_{n0} R^{(F)3}\left(\frac{1}{3} + \frac{b}{5}\right)\frac{1}{r^2} \qquad ...(4.63)$$

These equations may be integrated for constants $k^{(C)}$ and $k^{(F)}$ to give

$$T^{(F)} = -\frac{S_{n0}}{k^{(F)}}\left(\frac{r^2}{6} + \frac{b}{R^{(F)2}} \frac{r^4}{20}\right) + C_2^{(F)} \qquad ...(4.64)$$

$$T^{(C)} = +\frac{S_{n0}}{k^{(C)}} R^{(F)3}\left(\frac{1}{3} + \frac{b}{5}\right)\frac{1}{r} + C_2^{(C)} \qquad ...(4.65)$$

The integration constants $C_2^{(F)}$ and $C_2^{(C)}$ are determined from the following boundary conditions :

B.C. 3 : at $r = R^{(F)}$, $T^{(F)} = T^{(C)}$ ...(4.66)

B.C. 4 : at $r = R^{(C)}$, $T^{(C)} = T_0$ ...(4.67)

where T_0 is the known temperature at the outside of the cladding. The final expressions for the temperature profiles are

$$T^{(F)} - T_0 = \frac{S_{n0}R^{(F)2}}{6k^{(F)}}\left\{\left[1-\left(\frac{r}{R^{(F)}}\right)^2\right] + \frac{3}{10}b\left[1-\left(\frac{r}{R^{(F)}}\right)^4\right]\right\}$$
$$+ \frac{S_{n0}R^{(F)2}}{3k^{(C)}}\left(1+\frac{3}{5}b\right)\left(1-\frac{R^{(F)}}{R^{(C)}}\right) \qquad ...(4.68)$$

$$T^{(C)} - T_0 = \frac{S_{n0}R^{(F)2}}{3k^{(C)}}\left(1+\frac{3}{5}b\right)\left(\frac{R^{(F)}}{r} - \frac{R^{(F)}}{R^{(C)}}\right) \qquad ...(4.69)$$

Clearly one can find the maximum temperature in the sphere of fissionable material by setting $r = 0$ in Eq. (4.68). This is a quantity one might well want to know when making estimates of structural deterioration.

HEAT CONDUCTION WITH A VISCOUS HEAT SOURCE

We consider the flow of an incompressible Newtonian fluid between two coaxial cylinders, as shown in Fig. 4.9. As the outer cylinder rotates, each cylindrical shell of fluid rubs against an adjacent shell of fluid. This rubbing together of adjacent layers of fluid produces heat; that is, mechanical energy is steadily degraded into thermal energy. The volume heat source resulting from this "viscous dissipation" we designate by S_v. Its magnitude depends on the local velocity gradient; the more rapidly two adjacent layers move with respect to one another, the greater will be the viscous dissipation heating. The surfaces of the inner and outer cylinders are maintained at $T = T_0$ and $T = T_b$, respectively. Clearly T will be a function of r alone.

If the slit width b is small with respect to the radius R of the outer cylinder, then the problem may be solved approximately by using the somewhat simplified system shown in Fig. 4.10; that is, we ignore curvature effects and solve the problem in cartesian coordinates. For this modified problem, the viscous heat source is given by

Outer cylinder moves with angular velocity Ω

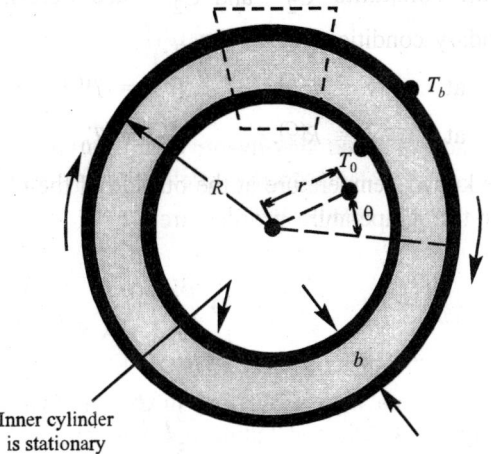

Fig. 4.9. Flow between cylinders with viscous heat generation. This part of the system enclosed within the dotted lines is shown in idealised form in Fig. 4.10.

$$S_v = -\tau_{xz}\left(\frac{dv_z}{dx}\right) = \mu\left(\frac{dv_z}{dx}\right)^2 \qquad \qquad ...(4.70)$$

Top surface moves with velocity $V = R\,\Omega$

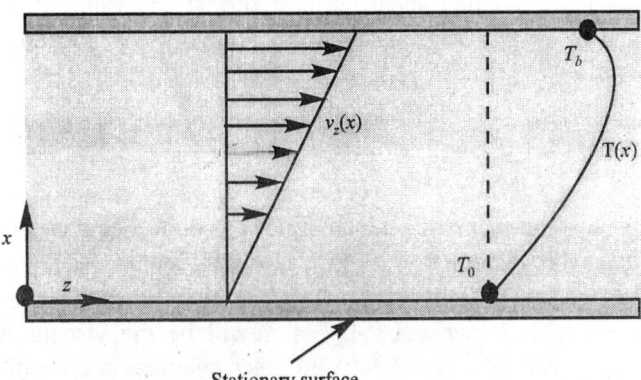

Fig. 4.10. Idealisation of portion of flow system in Fig. 4.9, in which curvature of the containing surfaces is neglected.

For the steady laminar flow of a fluid with constant viscosity in a slit, as shown in Fig. 4.10, the velocity profile is linear :

$$v_z = \left(\frac{x}{b}\right) V \qquad \qquad ...(4.71)$$

So that the rate of viscous heat production per unit volume is

$$S_v = \mu \left(\frac{V}{b} \right)^2 \qquad \qquad ...(4.72)$$

We are now ready to insert this quantity into an energy balance.

A thermal energy balance over a shell of thickness Δx, width W, and length L gives for the steady-state

$$W L q_x \big|_x - W L q_x \big|_{x+\Delta x} + W L \Delta x \mu \left(\frac{V}{b} \right)^2 = 0 \qquad ...(4.73)$$

Note that both "in" and "out" are taken to be in the $+x$ direction, even though in this problem heat is flowing in the $-x$ direction in part of the system. Division by $W L \Delta x$ and letting $\Delta x \to 0$ gives

$$\frac{dq_x}{dx} = \mu \left(\frac{V}{b} \right)^2 \qquad \qquad ...(4.74)$$

This may be integrated to give for constant μ

$$q_x = \mu \left(\frac{V}{b} \right)^2 x + C_1 \qquad \qquad ...(4.75)$$

Since we know nothing about the heat flux at any value of x, we cannot determine C_1 at this stage. Insertion of Fourier's law into Eq. (4.75) then gives

$$-k \frac{dT}{dx} = \mu \left(\frac{V}{b} \right)^2 x + C_1 \qquad \qquad ...(4.76)$$

in which k is the thermal conductivity of the fluid. Equation (4.76) may be integrated with respect to x to give (for constant k)

$$T = -\left(\frac{\mu}{k} \right) \left(\frac{V}{b} \right)^2 \frac{x^2}{2} - \frac{C_1}{k} x + C_2 \qquad \qquad ...(4.77)$$

The two integration constants C_1 and C_2 are determined from the boundary conditions :

B.C. 1 :   at $x = 0$, $T = T_0$ ...(4.78)

B.C. 2 :   at $x = b$, $T = T_b$ ...(4.79)

When the constants are thus determined and substituted back into Eq. (4.77), one obtains

$$\frac{T - T_0}{T_b - T_0} = \left(\frac{x}{b}\right) + \frac{1}{2}\mathrm{Br}\left(\frac{x}{b}\right)\left[1 - \left(\frac{x}{b}\right)\right] \qquad \text{...(4.80)}$$

Here $\mathrm{Br} = [\mu V^2/k(T_b - T_0)]$ is the "Brinkman number", which is a measure of the extent to which viscous heating is important relative to the heat flow resulting from the impressed temperature difference $(T_b - T_0)$. If $\mathrm{Br} > 2$, there is a maximum temperature at a position intermediate between the two walls.

In most flow problems viscous heating is not important. The viscous heating effect is important in several problems encountered in engineering work, however, in which large velocity changes occur over short distances: (i) flow of a lubricant between fast-moving parts; (ii) flow of plastics through dies in high-speed extrusion; and (iii) flow of air in the boundary layer near an earth satellite or a rocket (the re-entry problem). The first two of these problems are generally further complicated because many lubricants and molten plastics are non-Newtonian.

Heat Conduction in a Nuclear Fuel Rod Assembly

Consider a long nuclear fuel rod, which is surrounded by an annular layer of aluminium "cladding", as shown in Fig. 4.11. Within the fuel rod heat is produced by fission; this heat source is dependent on position, with a source strength varying approximately as

$$S_n = S_{n0}\left[1 + b\left(\frac{r}{R_F}\right)^2\right] \qquad \text{...(4.81)}$$

Here S_{n0} is the heat per unit volume per unit time produced at $r = 0$, and r is the distance from the axis of the fuel rod. If the outer surface of the cladding is in contact with a liquid coolant at temperature T_L, the heat transfer coefficient at the cladding-coolant interface being h_L. The thermal conductivities of fuel rod and cladding are k_F and k_C. The maximum temperature in the fuel rod can be then

$$(T_{F,\max} - T_L) = \frac{S_{n0}R_F^2}{4k_F}\left(1 + \frac{b}{4}\right) + \frac{S_{n0}R_F^2}{2k_C}\left(1 + \frac{b}{2}\right)\left(\frac{k_C}{R_C h_L} + \ln\frac{R_C}{R_F}\right)$$

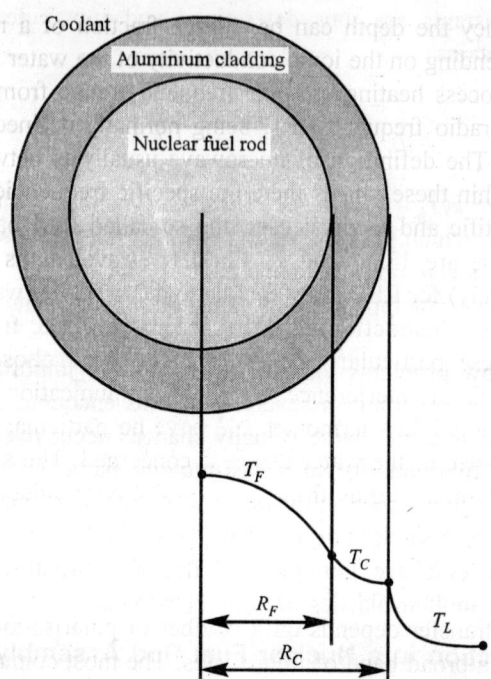

Fig. 4.11. Temperature distribution in a fuel rod assembly.

DIELECTRIC HEATING

Radio frequency and *microwave* are sometimes used as alternatives to convective, conductive or radiant heat transfer for the processing of "non-metals". Industries making use of these techniques include, textiles, paper, food, plastic and chemicals. The applications are many and varied including drying, baking, defrosting, welding and polymerisation. Known as *high frequency* or *dielectric heating*, both are forms of electromagnetic wave energy which share some characteristics but also have significant differences.

The perceived advantage of dielectric heating is based on the so-called "volumetric" effect arising from the fact that the energy is absorbed directly in the body of the material rather than being transferred to it via a surface. The concept of "volumetric" heating needs to be qualified since there is in reality a limiting penetration depth which depends on the properties of the material being heated as well as on the wavelength of the energy source. At the shorter wavelengths associated with microwave, penetration depth into a wet body is normally a few centimetres; at the longer wavelengths of radio frequency the depth can be a large fraction of a metre or a few centimetres depending on the ionic conductivity of the water in the material. Dielectric process heating uses the frequency range from about 5 MHz to 5 GHz with radio frequency, RF, being normally defined as being less than 100 MHz. The definition of microwave usually is between 500 MHz and 5 GHz. Within these ranges there are specific frequencies allocated for industrial, scientific and medical uses, the so-called ISM bands. The most common of these are 13.56 and 27.12 MHz (wavelengths 22.4 and 11.2 metres respectively) for RF with 900 MHz and 2.45 GHz (wavelengths 0.35 and 0.13 metres respectively) being the permitted frequencies for microwave. These particular frequencies have been chosen in order to minimise the risk of interference with telecommunications by either the fundamental or by higher harmonics and have no particular significance as far as the resonance of the water dipole is concerned. The actual frequency within the "900 band" varies from country to country, depending on local regulations.

Heat Transfer

Dielectric heat transfer depends on a number of polarisation effects which take place over a broad band of frequencies. The most commonly described one is dipolar orientational polarisation.

As can be seen from Fig. 4.10 which shows loss factor as a function of frequency, this is important at microwave frequency but of relatively little significance at the lower, radio frequencies. The dominant mode in the RF range is space charge orientation, which in turn is dependent on the ionic conductivity of the material being processed. If the dielectric properties of a particular material are known it is possible, in theory, to choose the most appropriate frequency from those available in the ISM bands. In reality unless the dielectric loss factor, ε_r'' is very low most products can be dried or processed by either RF or microwave. The choice can then be made on other considerations such as the engineering required to make a satisfactory heat transfer applicator compatible with the process line requirements, i.e., product width, height and shape.

Loss factor, i.e. the product of dielectric constant and loss tangent, varies with a number of parameters including frequency moisture content and temperature. The relationships are often quite complex as for example in drying where as the temperature increases the moisture content falls. These relationships can be such that preferential heating and drying of the wetter areas takes place, in the right circumstances leading to "moisture profile correction".

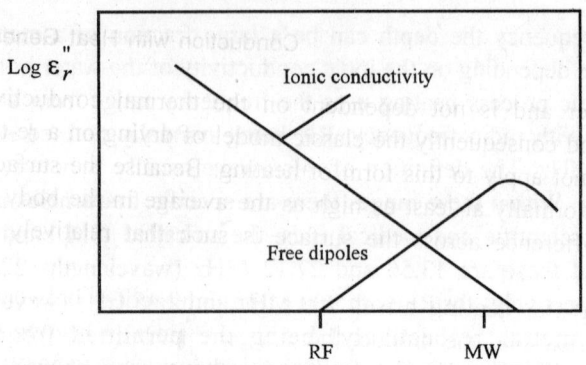

Fig. 4.12. Loss factor as function of frequency.

The heat transferred per unit volume of product is given by

$$P = 2\pi f \varepsilon_0 \varepsilon_r^{''} E^2 \qquad \text{(W/m)}$$

where f = frequency   (hertz)

E = electric field strength   (V/m)

$\varepsilon_r^{''}$ = loss factor or relative permittivity

ε_0 = permittivity of free space   (8.85 × 10^{12} farad/m)

When dealing with process heating, interpretation of published data, such as that by Von Hippel needs to be undertaken with care. For example the figures quoted for "water" often refer to deionised water, in this case the effect of having any level of ionic material present (as in most "real" water), will substantially increase the loss factor at RF but have relatively little effect at microwave frequency.

Mass Transfer Considerations

RF and microwave are alternatives to conventional heat transfer with several feature which need careful consideration. Internal mass transfer may be in liquid or vapour phase depending on the structure of the material, for example in certain capillary porous materials the internal generation of heat having first reduced the viscosity and surface tension of the bulk of the free water in the body will then cause a small quantity to be evaporated, raising the internal vapour pressure sufficiently for liquid phase flow to occur. In other materials where the pore structure is courser the more conventional mechanism of evaporation followed by vapour phase flow takes place. Perhaps the most significant aspect of this form of heating is that moisture gradients are avoided because the heat is transferred directly to the water and is not dependent on the thermal conductivity of the substrate and consequently the classic model of drying on a re-treating wet front does not apply to this form of heating. Because the surface moisture content is normally at least as high as the average in the body the vapour pressure difference across the surface is such that relatively modest air flows are required.

Because the heating arises from the interaction between the high frequency field and the product there is no incidental heating of the oven metal work or the air. In processes which involve the evaporation of water it is important that recondensation is avoided. The minimum requirement is a flow of heated air over the product to act as a means of mass transfer from the surrounding atmosphere; this same air flow is used to maintain the metal work above the dew point.

Radio Frequency Power Sources and Applicators

Radio frequency power supplies are usually a "class C" oscillator based on a triode valve which has been built into a cavity type tank circuit. With new regulations in place increasing importance is being attached to the need avoid electromagnetic interference with other equipment, and therefore alternative generator types are being considered for some applications, such as the use of crystal driven linear amplifiers in conjunction with 50 ohm transmission lines.

Radio frequency applicators are essentially capacitors which contain the product requiring heating as the whole or a part of its dielectric. The simplest and most widely used is through field or parallel plate electrode as shown in Fig. 4.13.

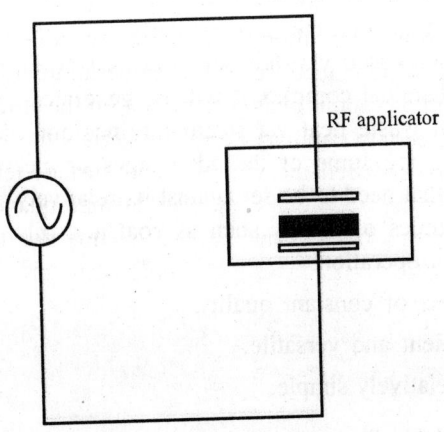

RF applicator

Fig. 4.13. Radio frequency applicators with parallel plate electrodes.

When used for drying an air space is required above the dielectric to allow for the movement of the product through the machine and for ventilation of the water vapour. This then means an increase in voltage between the plate in order to maintain an adequate field strength in the product.

It is therefore important to consider the relative dimensions of the dielectric and air space capacitors to give the desired heating effect without an electrical discharge taking place. For very thin materials such as paper it may be necessary to use an alternative electrode configuration.

Microwave Power Sources and Applicators

For industrial heating applications using microwaves the usual power source is the magnetron. The most common form of microwave heating applicator is the multimode cavity, similar in concept to the domestic oven. When used for continuous, conveyorised processing the design of the ports to allow the passage of product is critical in order to prevent the emission of microwave energy.

The limits on the dimensions of these apertures can be very restrictive, typically 20 to 30 mm for 2.45 GHz and perhaps 80 to 100 mm at 900 MHz.

Industrial microwave heating has been used extensively in the rubber industry for curing and preheating prior to moulding. In the food industry it has been used for tempering, melting, cooking and drying. Recently microwave vacuum dryers have been developed for drying expensive, high quality temperature sensitive pharmaceuticals.

JOULE HEATING

The use of electricity for heating purposes in the process industries is not widespread but it is increasing, particularly for special applications. The power may be purchased via the local electricity grid, or more often than not in a large chemical complex it will be generated by the recovery and transformation of waste heat via steam turbines and alternators.

The following are some of the advantages of electricity (a secondary form of energy) that need to be set against its relatively high cost compared with primary sources of energy such as coal and oil :

1. It is clean in operation.

2. The energy is of constant quality.

3. It is convenient and versatile.

4. Control is relatively simple.

There are a number of different ways that the energy can be utilised, providing an opportunity to optimise cost and convenience. Common forms of electrical heating include :

1. *Resistance heating* which involves passing an electric current through a resistance to generate heat. It is probably the most common method of using electrical power for process heating. The electric current may pass through an external resistance (indirect heating) or through the material to be heated (direct heating). It has been used for food processing.

2. *Induction heating* utilities the transfer of energy from a coil to the workpiece via an alternating magnetic field. Traditionally, the technique has been used for metal heating, but in more recent times it has been applied to chemical reactors.

3. *Dielectric heating* involves subjecting the material to be heated— provided it contains suitable molecules capable of excitation—to the effects of an electric field alternating at radio or microwave frequencies. It has been used in polymerisation and curing processes.

4. *Infra-red heating* depends on radiation effects. The principle is that an element (resistance-heated) radiates heat energy that may be focused or reflected in the same way as light energy, and can therefore be directed as determined by the process requirements. The principles of *radiative heat transfer* apply to this technique. Uses of infra-red heating include drying of sheet material and spray-painted articles.

CHAPTER 5

Extended Surfaces or Fins

INTRODUCTION

The heat conducted through solids or walls has to be continuously dissipated to the surroundings to maintain the steady state. In many cases large quantities of heat have to be dissipated from small areas. Heat transfer by convection between solid surface and the surrounding fluid can be increased by increasing heat transfer area by attaching to the surface thin strips of metals called fins.

The problems considered are encountered in practice when a solid of relatively small cross-sectional area protrudes from a large body into a fluid at a different temperature. Such extended surfaces have wide industrial application as fins attached to the walls of heat transfer equipment in order to increase the rate of heating or cooling.

FINS OF UNIFORM CROSS-SECTION

As a simple illustration, consider a pin fin having the shape of a rod whose base is attached to a wall at surface temperature T_s (Fig. 5.1). The fin is cooled along its surface by a fluid at temperature T_∞.

The fin has a uniform cross-sectional area A, is made of a material having uniform conductivity k, and the heat transfer coefficient between the surface of the fin and the fluid is $\bar{h}_c$.

We will assume that transverse temperature gradients are so small that the temperature at any cross-section of the rod is uniform, that is, $T = T(x)$ only. Even in a relatively thick fin the error in a one-dimensional solution is less than 1 per cent.

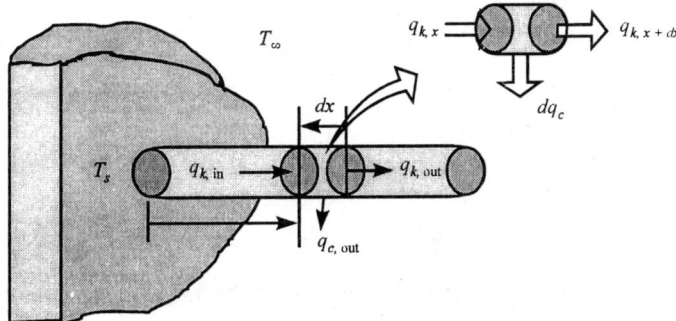

Fig. 5.1. Schematic diagram of a pin fin protruding from a wall.

To derive an equation for the temperature distribution, we make a heat balance for a small element of the fin. Heat flows by conduction into the left face of the element, while heat flows out of the element by conduction through the right face and by convection from the surface. Under steady-state conditions,

Rate of heat flow by conduction into element at x	=	rate of heat flow by conduction out of element at $x + dx$	+	rate of heat flow by convection from surface between x and $x + dx$

In symbolic form, this equation becomes

$$q_{k,x} = q_{k,x + dx} + dq_c$$

or

$$-kA\frac{dT}{dx}\bigg|_x = -kA\frac{dT}{dx}\bigg|_{x+dx} + \bar{h}_c P dx[T(x) - T_\infty] \qquad ...(5.1)$$

where P is the perimeter of the pin and Pdx is the pin surface area between x and $x + dx$.

If k and $\bar{h}_c$ are uniform, Eq. (5.1) simplifies to the form

$$\frac{d^2 T(x)}{dx^2} - \frac{\bar{h}_c P}{kA}[T(x) - T_\infty] = 0 \qquad ...(5.2)$$

It will be convenient to define an excess temperature of the fin above the environment, $\theta(x) = [T(x) - T_\infty]$, and transform Eq. (5.2) into the form

$$\frac{d^2\theta}{dx^2} - m^2\theta = 0 \qquad ...(5.3)$$

where $m^2 = \bar{h}_c P/kA$.

Equation (5.3) is a linear, homogeneous, second-order differential equation whose general solution is of the form

$$\theta(x) = C_1 e^{mx} + C_2 e^{-mx} \qquad ...(5.4)$$

To evaluate the constants C_1 and C_2 it is necessary to specify appropriate boundary conditions. One condition is that at the base ($x = 0$) the fin temperature is equal to the wall temperature or

$$\theta(0) = T_s - T_\infty = \theta_s$$

The other boundary condition depends on the physical condition at the end of the fin. We will treat the following four cases :

1. The fin is very long and the temperature at the end approaches the fluid temperature, or

$$\theta = 0 \quad \text{at} \quad x \to \infty$$

2. The end of the fin is insulated, or

$$\frac{d\theta}{dx} = 0 \quad \text{at } x = L$$

3. The temperature at the end of the fin is fixed, or

$$\theta = \theta_L \quad \text{at} \quad x = L$$

4. The tip loses heat by convection, or

$$-k \left. \frac{\partial\theta}{\partial x} \right|_{x=L} = \bar{h}_{c,L} \theta_L$$

Fig. 5.2 illustrates schematically the cases described by these conditions at the tip.

For case 1 the second boundary condition can be satisfied only if C_1 in Eq. (5.4) equals zero, or

$$\theta(x) = \theta_s e^{-mx} \qquad ...(5.5)$$

Usually we are interested not only in the temperature distribution, but also in the total rate of heat transfer to or from the fin. The rate of heat flow can be obtained by two different methods. Since the heat conducted across the root of the fin must equal the heat transferred by convection from the surface of the rod to the fluid.

$$q_{\text{fin}} = -kA \left. \frac{dT}{dx} \right|_{x=0} = \int_0^\infty \bar{h}_c P [T(x) - T_\infty] dx = \int_0^\infty \bar{h}_c P \theta(x) dx \qquad ...(5.6)$$

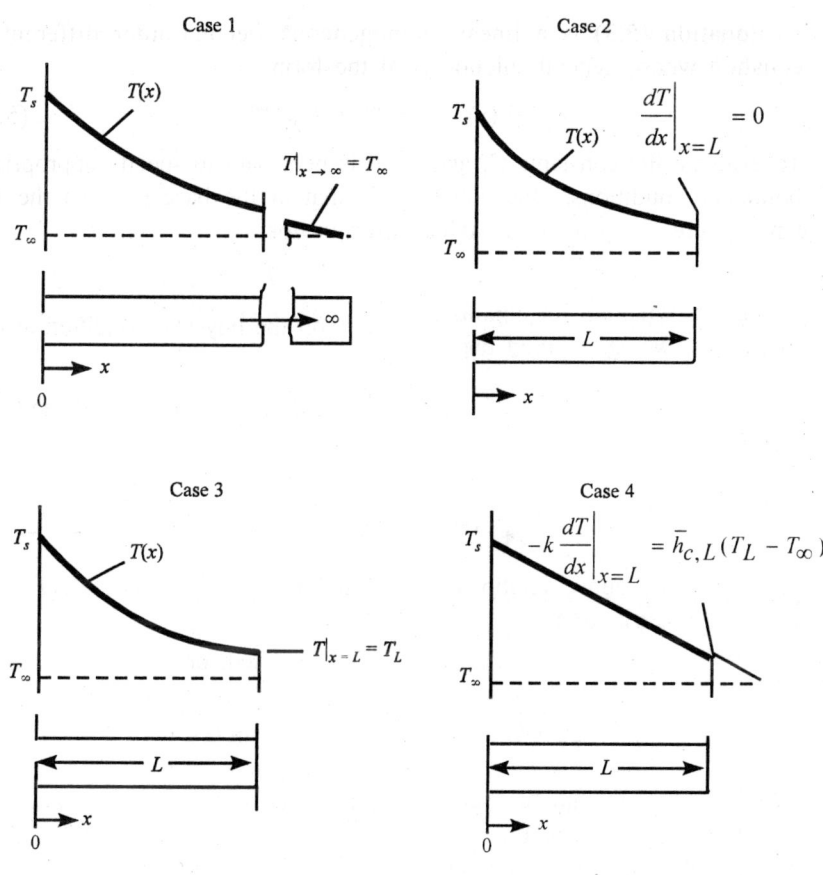

For all cases $T\big|_{x=0} = T_s$.

Fig. 5.2. Schematic representation of four boundary conditions at the tip of a fin.

Differentiating Eq. (5.5) and substituting the result for $x = 0$ into Eq. (5.6) yields

$$q_{fin} = -kA[-m\theta(0)e^{(-m)0}] = \sqrt{\bar{h}_c PAk} \; \theta_s \qquad ...(5.7)$$

The same result is obtained by evaluating the convective heat flow from the surface of the rod

$$q_{fin} = \int_0^\infty \bar{h}_c P\theta_s e^{-mx} dx = \frac{\bar{h}_c P}{m} \theta_s e^{-mx}\bigg|_0^\infty = \sqrt{\bar{h}_c PAk} \; \theta_s$$

Equations (5.5) and (5.6) are reasonable approximations of the temperature distribution and heat flow rate in a finite fin if its length is very large compared to its cross-sectional area. If the rod is of finite length but the heat loss from the end of the rod is neglected, or if the end of the rod is insulated, the second boundary condition requires that the temperature gradient at $x = L$ be zero, or $dT/dx = 0$ at $x = L$. These conditions require that

$$\left(\frac{d\theta}{dx}\right)_{x=L} = 0 = mC_1\, e^{mL} - mC_2\, e^{-mL}$$

Solving this equation for condition 2 simultaneously with the relation for condition 1, which required that

$$\theta\,(0) = \theta_s = C_1 + C_2$$

yields

$$C_1 = \frac{\theta_s}{1+e^{2mL}} \qquad C_2 = \frac{\theta_s}{1+e^{-2mL}}$$

Substituting the above relations for C_1 and C_2 into Eq. (5.4) gives the temperature distribution

$$\theta = \theta_s\left(\frac{e^{mx}}{1+e^{2mL}} + \frac{e^{-mx}}{1+e^{-2mL}}\right) = \theta_s\,\frac{\cosh m(L-x)}{\cosh (mL)} \qquad ...(5.8)$$

The heat loss from the fin can be found by substituting the temperature gradient at the root into Eq. (5.6). Noting that $\tanh(mL) = (e^{mL} - e^{-mL})/(e^{mL} + e^{-mL})$, we get

$$q_{\text{fin}} = \sqrt{\bar{h}_c PAk}\,\theta_s \tanh (mL) \qquad ...(5.9)$$

The results for the other two tip conditions can be obtained in a similar manner, but the algebra is more lengthy. For convenience, all four cases are summarised in Table 5.1.

FIN SELECTION AND DESIGN

In the preceding section, we developed equations for the temperature distribution and the rate of heat transfer of extended surfaces and fins. Fins are widely used to increase the rate of heat transfer from a wall. As an illustration of such an application, consider a surface exposed to a liquid at temperature T_∞ flowing over the surface. If the wall is bare and the surface temperature T_s is fixed, the rate of heat transfer per unit area from the plane wall is controlled entirely by the heat transfer coefficient $\bar{h}_c$. The coefficient at the plane wall may be increased by increasing the fluid velocity, but this also creates a larger pressure drop and requires increased pumping power.

Table 5.1. Equations for temperature distribution and rate of heat transfer for fins of uniform cross section[a].

Case	Tip condition (x = L)	Temperature of distribution, θ/θ_s	Fin heat transfer rate. q_f
1.	Infinite fin ($L \rightarrow \infty$): $\theta(L) = 0$	e^{-mx}	M
2.	Adiabatic: $\dfrac{d\theta}{dx}\bigg|_{x=L} = 0$	$\dfrac{\cosh m(L-x)}{\cosh mL}$	$M \tanh mL$
3.	Fixed temperature : $\theta(L) = \theta_L$	$\dfrac{(\theta_L/\theta_s)\sinh mx + \sinh m(L-x)}{\sinh mL}$	$M\,\dfrac{\cosh mL - \theta_L/\theta_s}{\sinh mL}$
4.	Convection heat transfer : $\bar{h}\,\theta(L) = \dfrac{-kd\theta}{dx}\bigg|_{x=L}$	$\dfrac{\cosh m(L-x) + (\bar{h}/mk)\sinh m(L-x)}{\cosh mL + (\bar{h}/mk)\sinh mL}$	$M\,\dfrac{\sinh mL + (\bar{h}/mk)\cosh mL}{\cosh mL + (h/mk)\sinh mL}$

[a]
$$\theta \equiv T - T_\infty$$
$$\theta_s \equiv \theta(0) = T_s - T_\infty$$
$$m^2 \equiv \frac{\bar{h}_c P}{kA}$$
$$M \equiv \sqrt{\bar{h}_c\, pkA\,\theta_x}$$

In many cases it is thus preferable to increase the rate of heat transfer from the wall by using fins that extend from the wall into the fluid and increase the contact area between the solid surface and the fluid. If the fin is made of a material with high thermal conductivity, the temperature gradient along the fin from base to tip will be small and the heat transfer characteristics of the wall will be greatly enhanced. Fins come in many shapes and forms, some of which are shown in Fig. 5.3. The selection of fins is made on the basis of thermal performance and cost. The selection of a suitable fin geometry requires a compromise among the cost, the weight, the available space, and the pressure drop of the heat transfer fluid, as well as the heat transfer characteristics of the extended surface. From the point of view of thermal performance, the most desirable size, shape, and length of the fin can be evaluated by an analysis such as that outlined below.

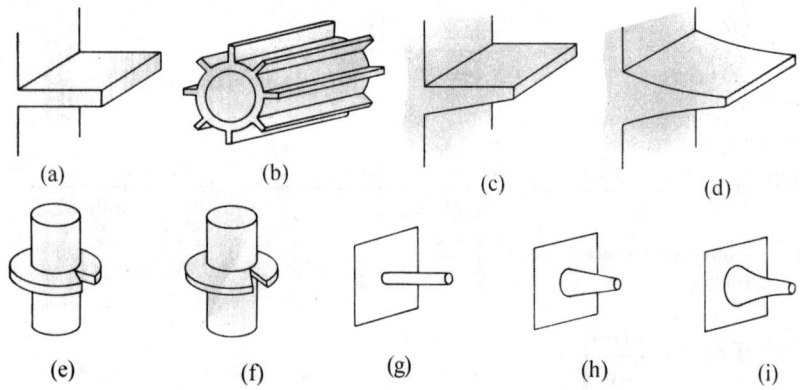

Fig. 5.3. Schematic diagrams of different types of fins: (a) longitudinal fin of rectangular profile; (b) cylindrical tube with fins of rectangular profile; (c) longitudinal fin of trapezoidal profile; (d) longitudinal fin of parabolic profile; (e) cylindrical tube with radial fin of rectangular profile; (f) cylindrical tube with radial fin of truncated conical profile; (g) cylindrical pin fin; (h) truncated conical spine; (i) parabolic spine.

The heat transfer effectiveness of a fin is measured by a parameter called the fin efficiency η_f, which is defined as

$$\eta_f = \frac{\text{actual heat transferred by fin}}{\substack{\text{heat that would have been transferred} \\ \text{if entire fin were at the base temperature}}}$$

Using Eq. (5.9), the fin efficiency for a circular pin fin of diameter D and length L with an insulated end is

$$\eta_f = \frac{\tanh \sqrt{4 L^2 \bar{h}_c / kD}}{\sqrt{4 L^2 \bar{h}_c / kD}} \qquad \qquad \dots(5.10)$$

whereas for a fin of rectangular cross-section (length L and thickness t) the efficiency of a fin with an insulated end is

$$\eta_f = \frac{\tanh \sqrt{\bar{h}PL^2 / kA}}{\sqrt{\bar{h}PL^2 / kA}} \qquad \dots(5.11)$$

If a rectangular fin is long, wide, and thin, $P/A \simeq 2/t$ and the heat loss from the end can be taken into account approximately by increasing L by $t/2$ and assuming that the end is insulated. This keeps the surface area from which heat is lost the same as in the real case, and the fin efficiency then becomes

$$\eta_f = \frac{\tanh \sqrt{2\bar{h}_c L_c^2 / kt}}{\sqrt{2\bar{h}_c L_c^2 / kt}} \qquad \dots(5.12)$$

where $L_c = (L + t/2)$

The error that results from this approximation will be less than 8 per cent when

$$\left(\frac{\bar{h}_c t}{2k} \right)^{1/2} \leq \frac{1}{2}$$

It is often convenient to use the profile area of a fin, A_m. For a rectangular shape A_m is Lt, whereas for a triangular cross-section A_m is $Lt/2$, where t is the base thickness. In Fig. 5.4 the fin efficiencies for rectangular and triangular fins are compared. Fig. 5.5 shows the fin efficiency for circumferential fins of rectangular cross-section.

For a plane surface of area A, the thermal resistance is $1/\bar{h}A$. Addition of fins increases the surface area, but at the same time it introduces a conductive resistance over that portion of the original surface to which the fins are attached.

Addition of fins will therefore not always increase the rate of heat transfer. In practice, addition of fins is hardly ever justified unless $\bar{h}A/Pk$ is considerably less than unity.

It is interesting to note that the fin efficiency reaches its maximum value for the trivial case of $L = 0$, or no fin at all. It is therefore not possible to maximise fin performance with respect to the quantity of fin material (mass, volume, or cost), because such an optimisation has obvious economic significance.

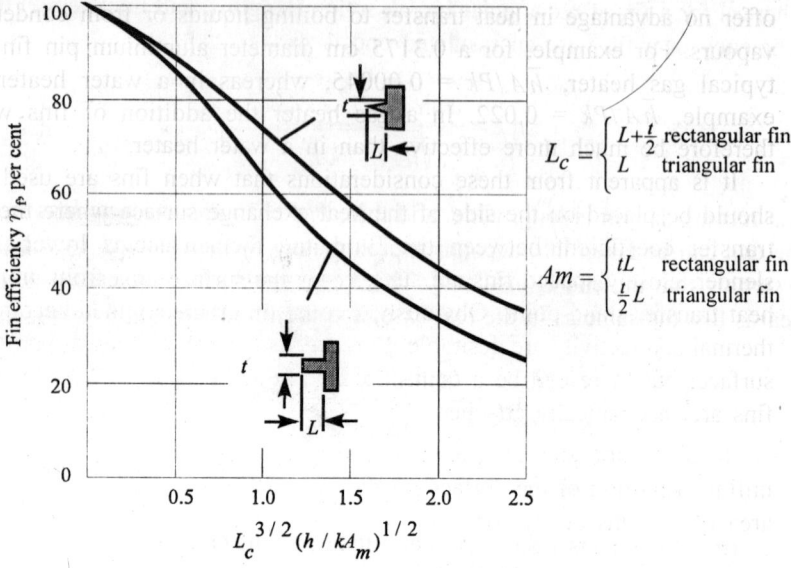

Fig. 5.4. Efficiency of rectangular and triangular fins.

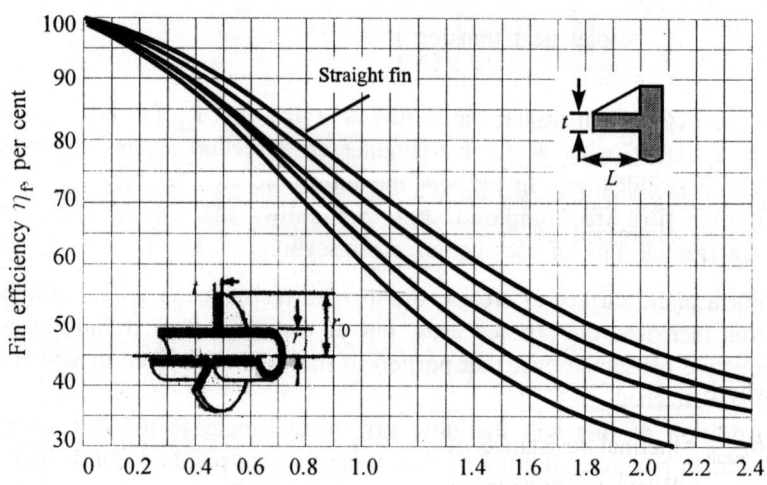

Fig. 5.5. Efficiency of circumferential rectangular fins.

$$(r_0 + \frac{t}{2} - r_i)^{3/2} \sqrt{2\bar{h} / kt(r_0 - r_i)} \text{ or } (L + \frac{t}{2})^{3/2} \sqrt{2\bar{h} / ktL}$$

Using the values of the average surface conductances as a guide, we can easily see that fins effectively increase the heat transfer to or from a gas, are less effective when the medium is a liquid in forced convection, but offer no advantage in heat transfer to boiling liquids or from condensing vapours. For example, for a 0.3175 cm diameter aluminium pin fin in a typical gas heater, $\bar{h}A/Pk$ = 0.00045, whereas in a water heater, for example, $\bar{h}A/Pk$ = 0.022. In a gas heater the addition of fins would therefore be much more effective than in a water heater.

It is apparent from these considerations that when fins are used they should be placed on the side of the heat exchange surface where the heat transfer coefficient between the fluid and the surface is lower. Thin, slender, closely spaced fins are superior to fewer and thicker fins from the heat transfer stand-point. Obviously, fins made of materials having a high thermal conductivity are desirable. Fins are sometimes an integral part of the surface, but there can be a contact resistance at the base of the fin if the fins are mechanically attached.

To obtain the total efficiency of a surface with fins η_t, we combine the unfinned portion of the surface at 100 per cent efficiency with the surface area of the fins at η_f, or

$$A_o\eta_t = (A_o - A_b) + A_f\eta_f \qquad ...(5.13)$$

where

A_o = total heat transfer area

A_b = base area of the fins

A_f = heat transfer area of the fins

The overall heat transfer coefficient U_o, based on the total outer surface area, for heat transfer between two fluids separated by a wall with fins can then be expressed as

$$U_o = \cfrac{1}{\cfrac{1}{\eta_{to}\bar{h}_o} + R_{k_{wall}} + \cfrac{A_o}{\eta_{ti}A_i\bar{h}_i}} \qquad ...(5.14)$$

where

$R_{k_{wall}}$ = thermal resistance of the wall to which the fins are attached, W/m^2 K (outside surface)

A_o = total outer surface area, m^2

A_i = total inner surface area, m^2

h_{to} = total efficiency for outer surface

η_{ti} = total efficiency for inner surface

$\bar{h}_c$ = average heat transfer coefficient for outer surface, W/m^2 K

$\bar{h}_i$ = average heat transfer coefficient for inner surface, W/m^2 K

For tubes with fins on the outside only, the usual case in practice, η_{ti} is unity and $A_i = \pi D_i L$.

In the analysis presented in this chapter, details of the convection heat flow between the fin surface and the surrounding fluid have been omitted. A complete engineering analysis not only requires an evaluation of the fin performance, but must also take the relation between the fin geometry and the convection heat transfer into account.

HEAT CONDUCTION THROUGH A ROD (FIN) OF CONSTANT CROSS-SECTION

Differential Equation and Its Solution

As already discussed fins may be of various cross-sections (rectangular, circular, triangular and so on, including irregular geometric forms). Consider the propagation of heat in a straight rod of constant cross-section. Let the cross-sectional area be denoted by f and its perimeter by u. The rod is placed in a medium of constant temperature t_f; the coefficient of heat transfer from the surface of the rod to the surroundings is assumed to be constant for the entire surface. Suppose also that the thermal conductivity of the rod material λ is quite large, and the cross-section of the rod extremely small in relation to its length. The latter assumption allows us to neglect temperature variations over the cross-section and to consider that the temperature changes only along the axis of the rod. To simplify calculation, t_f = constant will be considered as a reference temperature. The excess in temperature of the rod over this reference value will be denoted by ϑ. It is obvious that

$$\vartheta = t - t_f$$

where

$t_f =$ temperature of the medium;

$t =$ rod temperature.

For a given temperature t_1 of the base of the rod, the excess temperature of the base (Fig. 5.6) is :

$$\vartheta_1 = t_1 - t_f$$

Consider a rod element dx long at a distance x from the base. The heat balance equation for this rod element can be written as follows :

$$Q_x - Q_{x + dx} = dQ \qquad \text{..(a)}$$

where

Q_x =   quantity of heat entering the left face of the element per unit time;

Q_{x+dx}=   quantity of heat flowing from the opposite face of the element in the same time;

dQ =   quantity of heat lost per unit time to the surroundings by the outer surface of the rod.

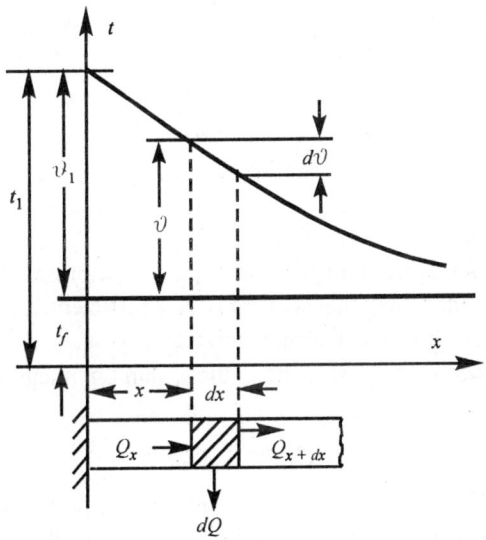

Fig. 5.6. Heat transfer through a rod.

In accordance with Fourier's law

$$Q_x = -\lambda \frac{d\vartheta}{dx} f$$

and

$$Q_{x+dx} = -\lambda \frac{d}{dx}\left(\vartheta + \frac{d\vartheta}{dx} dx\right) f$$

whence

$$Q_{x+dx} = -\lambda \frac{d}{dx} - \lambda f \frac{d^2\vartheta}{dx^2} dx$$

Consequently,

$$Q_x - Q_{x+dx} = \lambda f \frac{d^2\vartheta}{dx^2} dx \qquad \ldots(b)$$

On the other hand, according to Newton's law

$$dQ = \alpha_f \vartheta u \, dx \qquad \qquad ...(c)$$

Equating (b) and (c), we get the following differential equation for the temperature distribution in the rod :

$$\frac{d^2 \vartheta}{dx^2} = \frac{\alpha_f u}{\lambda f} \vartheta = m^2 \vartheta \qquad \qquad ...(5.15)$$

where

$$m = +\sqrt{\frac{\alpha_f u}{\lambda f}} \qquad \qquad ...(d)$$

here m is in 1/m.

It is clear from expression (d) that for a fin of preset shape and size m is constant, provided that the heat-transfer coefficient α_f is constant over the entire surface, and the thermal conductivity λ is constant within the considered temperature range. Then, the common integral for Eq. (5.15) will be

$$\vartheta = C_1 e^{mx} + C_2 e^{-mx} \qquad \qquad ...(5.16)$$

The constants C_1 and C_2 are determined from the boundary conditions, which may be given in several ways, depending on the length of the rod and other factors.

Infinite Rod

A constant temperature is maintained at the base or in the initial cross-section of the rod, i.e. at $x = 0$, $\vartheta = \vartheta_1$. If the rod is of infinite length ($l \to \infty$), all the heat imparted to it will be lost to the surroundings, and at $x \to \infty$, $\vartheta = 0$.

Substituting the boundary conditions in Eq. (5.16) gives

at $x = 0$, $\vartheta_1 = C_1 + C_2$

at $x \to \infty$, $C_1 e^{\infty} = 0$

The second equality is valid only when $C_1 = 0$. Thus, $C_2 = \vartheta_1$. Substituting the values of C_1 and C_2 in Eq. (5.16), we get :

$$\vartheta = \vartheta_1 e^{-mx} \qquad \qquad ...(5.17)$$

Equality (5.16) may also be written as follows :

$$\Theta = \frac{\vartheta}{\vartheta_1} = e^{-mx} \qquad \text{...(5.17')}$$

where Θ is the dimensionless temperature expressed as a fraction of the temperature of the rod base ϑ_1.

Fig. 5.7 shows the dependence of the dimensionless temperature Θ on the length of a rod for different values of m ($m_1 < m_2 < m_3$).

Fig. 5.7 shows that the dimensionless temperature falls more with increasing factor m. With $x \to \infty$ all curves approach $\Theta = 0$ asymptotically.

From the equation $m = \sqrt{\dfrac{\alpha_f u}{\lambda f}}$ it follows that m is proportional to the

rate of heat removal from the side surface and inversely proportional to $\sqrt{\lambda f}$, i.e. to the factor determining heat transfer by conduction along the rod. Consequently, fins should be made of a material of high thermal conductivity, which will give a lower m and preserve large excess temperatures along the rod.

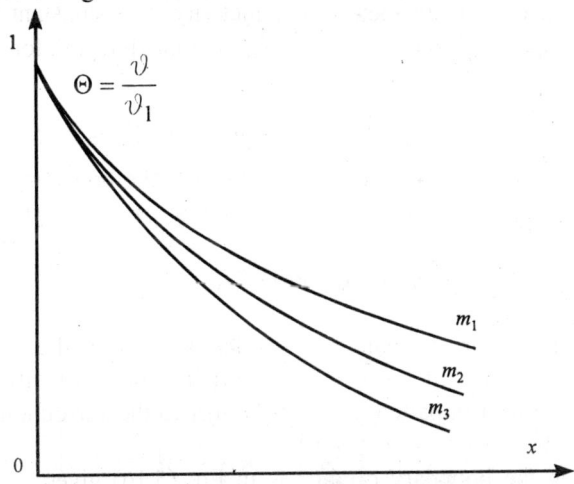

Fig. 5.7. Temperature variation in a rod length.

When $\dfrac{\alpha_f}{\lambda}$ is constant, m increases with increasing $\dfrac{u}{f}$. This indicates

that the effective fins are those with a small ratio $\dfrac{u}{f}$ at a given cross-

sectional area.

The quantity of heat removed from the rod to the surroundings will seemingly be equal to that passing through its base.

The heat flow is passed through the base of the rod

$$Q = -\lambda f \left(\frac{d\vartheta}{dx} \right)_{x=0}$$

Here Q is in W.

From Eq. (5.17) we find :

$$\left(\frac{d\vartheta}{dx} \right)_{x=0} = -me^{-mx} \vartheta_1 \big|_{x=0} = -m\vartheta_1$$

Substituting the temperature gradient at $x = 0$ in the preceding equation of the heat flow, we obtain the quantity of heat (heat flow) imparted to, or removed by, the rod to its surroundings :

$$Q = \lambda fm\vartheta_1 = \vartheta_1 \sqrt{\alpha_f u \lambda f} \qquad \qquad ...(5.18)$$

Finite Rod

The differential equation (5.15) and its solution (5.16) are also valid for a finite rod, but the boundary conditions are different :

$$
\left.
\begin{array}{l}
\text{at } x = 0, \quad \vartheta = \vartheta_1 \\[2ex]
\text{at } x = l, \quad -\lambda \left(\dfrac{d\vartheta}{dx} \right)_{x=l} = \alpha_l \vartheta_l \\[3ex]
\left(\dfrac{d\vartheta}{dx} \right)_{x=l} = -\dfrac{\alpha_l}{\lambda} \vartheta_l
\end{array}
\right\} \qquad ...(5.19)
$$

or

where ϑ_l = temperature at the end of the rod;

α_l = coefficient of heat transfer from the end of the rod.

When $x = l$, the quantity of heat imparted to the end of the rod by conduction equals the quantity lost from the end of the rod to the surroundings.

The constants C_1 and C_2 in Eq. (5.16) are determined by using boundary conditions (5.19) :

$$
\left.
\begin{aligned}
&\text{at } x = 0, \quad \vartheta_1 = C_1 + C_2 \\[2mm]
&\text{at } x = l, \quad \left(\frac{d\vartheta}{dx}\right)_{x=l} = C_1 m e^{ml} - C_2 m e^{-ml} = -\frac{\alpha_l}{\lambda}\lambda_l
\end{aligned}
\right\} \qquad ...(5.19')
$$

and

$$
\vartheta_l = C_1 e^{ml} + C_2 e^{-ml}
$$

From Eqs. (5.19), C_1 and C_2 are determined as follows :

$$
C_1 = \frac{\vartheta_1 \left(m - \dfrac{\alpha_l}{\lambda}\right)}{e^{2ml}\left(m + \dfrac{\alpha_l}{\lambda}\right) + m - \dfrac{\alpha_l}{\lambda}}
$$

$$
C_2 = \vartheta_1 \frac{e^{2ml}\left(m + \dfrac{\alpha_l}{\lambda}\right)}{e^{2ml}\left(m + \dfrac{\alpha_l}{\lambda}\right) + m - \dfrac{\alpha_l}{\lambda}}
$$

Substituting C_1 and C_2 in Eq. (5.16), we obtain :

$$
\vartheta = \vartheta_1 \left[\frac{e^{mx}\left(m - \dfrac{\alpha_l}{\lambda}\right)}{e^{2ml}\left(m + \dfrac{\alpha_l}{\lambda}\right) + m - \dfrac{\alpha_l}{\lambda}} + \frac{e^{-mx} e^{2ml}\left(m + \dfrac{\alpha_l}{\lambda}\right)}{e^{2ml}\left(m + \dfrac{\alpha_l}{\lambda}\right) + m - \dfrac{\alpha_l}{\lambda}} \right] \qquad ...(5.20)
$$

Multiplying and dividing the right side of Eq. (5.20) by e^{-ml} and making simple algebraic transformations, we get :

$$
\vartheta = \vartheta_1 \left[\frac{m\left[e^{m(l-x)} + e^{-m(l-x)}\right] + \dfrac{\alpha_l}{\lambda}\left[e^{m(l-x)} - e^{-m(l-x)}\right]}{m\left(e^{ml} + e^{-ml}\right) + \dfrac{\alpha_l}{\lambda}\left(e^{ml} - e^{-ml}\right)} \right]
$$

Recalling that

$$\frac{e^x + e^{-x}}{2} = \cosh(x) \quad \text{and} \quad \frac{e^x - e^{-x}}{2} = \sinh(x)$$

equation (5.20) may be written in the following form :

$$\vartheta = \vartheta_1 \frac{\cosh[m(l-x)] + \dfrac{\alpha_1}{\lambda m} \sinh[m(l-x)]}{\cosh(ml) + \dfrac{\alpha_l}{\lambda m} \sinh(ml)} \qquad \text{...(5.20')}$$

When the loss of heat from the end of the rod can be neglected, the boundary conditions (5.19) can be presented as follows :

at $\quad x = 0, \ \vartheta = \vartheta_1;$

at $\quad x = l, \ \left(\dfrac{d\vartheta}{dx}\right)_{x=l} = 0$

The latter can be admitted for the case where α_l is small at the end of the rod, and the thermal conductivity λ of the material is large and the ratio

$\dfrac{\alpha_l}{\lambda} \to 0$, i.e., the loss of heat from the end of the rod can be neglected.

With these boundary conditions the second terms of the numerator and denominator of Eq. (5.20') turn into zero and the equation acquires the appearance :

$$\vartheta = \vartheta_1 \frac{\cosh[m(l-x)]}{\cosh(ml)} \qquad \text{...(5.21)}$$

here ϑ is in °C.

Eqs. (5.20) and (5.21) can be used to calculate temperature in any cross-section of the rod. The fraction of heat lost from the end of the rod is usually small compared to the quantity of heat lost from the surface of the fins, and formula (5.21) is usually used for practical engineering calculations.

In the extreme case, when $x = l$, formula (5.21) acquires the following form :

$$\vartheta_{x=l} = \frac{\vartheta_1}{\cosh(ml)}$$

The amount of heat Q_f lost to the surroundings from the surface of a fin is equal to the amount of heat imparted to its base :

$$Q_f = -\lambda f \left(\frac{d\vartheta}{dx}\right)_{x=0}$$

From Eq. (5.21) we find :

$$\left(\frac{d\vartheta}{dx}\right)_{x=0} = -\vartheta_1 m \frac{\sinh\ (ml)}{\cosh\ (ml)} = -\vartheta_1 m \tanh\ (ml)$$

Then

$$Q_f = \lambda f \vartheta_1 m \tanh\ (ml) \qquad ...(5.22)$$

Substituting $m = \sqrt{\dfrac{\alpha_f u}{\lambda f}}$ in Eq. (5.22) we get :

$$Q_f = \vartheta_1 \sqrt{\alpha_f u \lambda f}\ \tanh\ (ml) \qquad ...(5.22')$$

If the rod is very long, then $\cosh\ (ml) \to \infty$, and $\tanh\ (ml) \cong 1$. Consequently, and Eq. (5.22) is transformed into Eq. (5.18).

HEAT TRANSFER THROUGH A FINNED PLANE WALL

The problem is to find the rate of heat flow through a finned wall of infinite area, with fins located on the side of the smaller heat-transfer coefficient (Fig. 5.8).

The constant values of the heat transfer α_1 are given on the finless side of the wall, the flat finned surface α_w and the surface of the fins themselves, α_f. The size of the fins is also given (Fig. 5.8), and the temperatures of the heat-carrying agents, t_{f1} and t_{f2}.

Since the width of a fin is larger than its thickness ($b \gg \delta$), we assume the perimeter of a fin cross-section to be $u = 2b$. The cross-sectional area of a fin is therefore $f = b\delta$.

Consequently,

$$m = \sqrt{\frac{\alpha_f u}{\lambda_f}} = \sqrt{\frac{2\alpha_f}{\lambda\delta}},\ m^{-1}$$

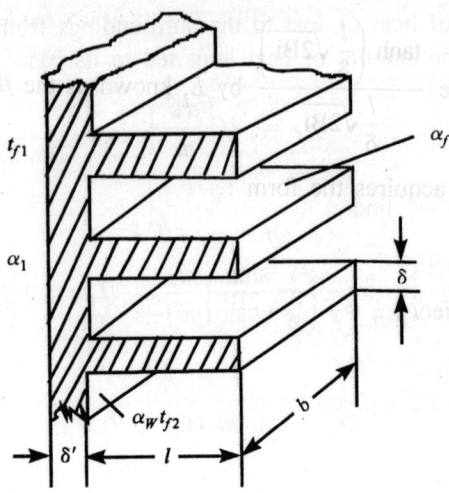

Fig. 5.8. Heat transfer through a finned wall.

Substituting this expression in Eq. (5.22), and multiplying and dividing by $2l$, we get :

$$Q_f = \vartheta_1 \sqrt{\alpha_f 2b\lambda b\delta}\, \frac{2l}{\lambda\delta} \tanh\left(\frac{l}{\delta}\sqrt{\frac{2\alpha_f\delta}{\lambda}}\right) = \alpha_f \vartheta_1 F_f \frac{\tanh\left(\dfrac{l}{\delta}\sqrt{\dfrac{2\alpha_f\delta}{\lambda}}\right)}{\dfrac{l}{\delta}\sqrt{\dfrac{2\alpha_f\delta}{\lambda}}}$$

where $\dfrac{\alpha_f\delta}{\lambda}$ = Bi is a dimensionless group, known as the Biot number, which is an important characteristic of the process of heat conduction. The Biot number represents the ratio between internal thermal resistance to conduction and external thermal resistance to heat transfer :

$$\text{Bi} = \frac{\dfrac{\delta}{\lambda}}{\dfrac{1}{\alpha_f}}$$

The equation for heat flow from the surface of a fin can be written in final form as follows :

$$Q_f = \alpha_f \vartheta_1 F_f \frac{\tanh\left(\dfrac{l}{\delta}\sqrt{2\text{Bi}}\right)}{\dfrac{l}{\delta}\sqrt{2\text{Bi}}} \qquad\qquad ...(5.23)$$

Let us denote $\dfrac{\tanh\left(\dfrac{l}{\delta}\sqrt{2\mathrm{Bi}}\right)}{\dfrac{l}{\delta}\sqrt{2\mathrm{Bi}}}$ by E, known as the *fin efficiency*. Then

equation (5.23) acquires the form :

$$Qf = \alpha_f \vartheta_1 F_f E$$

The fin efficiency $E = f\left(\dfrac{l}{\delta}\sqrt{2\mathrm{Bi}}\right)$ tends to its maximum of unity when

$\dfrac{l}{\delta}\sqrt{2\mathrm{Bi}} \to 0$ (for the given fin dimensions, this is possible when $\lambda \to \infty$,

i.e. $\mathrm{Bi} \to 0$).

The amount of heat Q_w, W, removed from the flat surface between fins is :

$$Q_w = \alpha_w \vartheta_1 F_w$$

and the total quantity of heat is

$$Q = Q_f + Q_w = \alpha_f \vartheta_1 F_f E + \alpha_w \vartheta_1 F_w \qquad ...(a)$$

or

$$\left.\begin{array}{l} Q = \alpha_{red} \vartheta_1 F_{fw} \\[2mm] F_{fw} = F_f + F_w \end{array}\right\} \qquad ...(b)$$

Correlation of (a) and (b) shows that

$$\alpha_{red} = \alpha_f E \frac{F_f}{F_{fw}} + \alpha_w \frac{F_w}{F_{fw}} \qquad ...(5.24)$$

The value α_{red} in Eq. (5.24) is called the reduced heat-transfer coefficient, and is a mean heat-transfer coefficient for a finned wall, accounting for the removal of heat from the fin surfaces and the plain section of the wall between fins, and for fin efficiency.

The process of heat transfer through a finned wall may then be described by the following set of equations :

$$Q = \alpha_1 F_1 (t_{f1} - t_{f1})$$

$$Q = \frac{\lambda}{\delta}(t_{w1} - t_{w2})F_1$$

$$Q = \alpha_{red} (t_{w2} - t_{f2}) F_{fw}$$

From these we obtain :

$$Q = \frac{t_{w1} - t_{f2}}{\dfrac{1}{\alpha_1 F_1} + \dfrac{\delta'}{\lambda F_1} + \dfrac{1}{\alpha_{red} F_{fw}}} \qquad \text{...(5.25)}$$

If we relate the heat flow to unit area of the finned wall surface, then

$$\frac{Q}{F_{fw}} = q_{fw} = \frac{t_{f1} - t_{f2}}{\dfrac{1}{\alpha_1 F_1} + \dfrac{\delta'}{\lambda} \dfrac{F_{fw}}{F_1} + \dfrac{1}{\alpha_{red}}} = k_{fw}(t_{f1} - t_{f2}) \qquad \text{...(5.26)}$$

where

$$k_{fw} = \frac{1}{\dfrac{1}{\alpha_1} \dfrac{F_{fw}}{F_1} + \dfrac{\delta'}{\lambda} \dfrac{F_{fw}}{F_1} + \dfrac{1}{\alpha_{red}}}$$

is the overall heat-transfer coefficient of a finned wall, with heat flow being reduced to the finned surface of the wall, $W/m^2.K$.

If we reduce the heat flow to the area of non-finned surface of the wall, we get :

$$\frac{Q}{F_1} = q_1 = \frac{t_{f1} - t_{f2}}{\dfrac{1}{\alpha_1} + \dfrac{\delta'}{\lambda} + \dfrac{1}{\alpha_{red}} \dfrac{F_1}{F_{fw}}} = k_1(t_{f1} - t_{f2}) \qquad \text{...(5.27)}$$

where

$$k_1 = \frac{t_{f1} - t_{f2}}{\dfrac{1}{\alpha_1} + \dfrac{\delta'}{\lambda} + \dfrac{1}{\alpha_{red}} \dfrac{F_1}{F_{fw}}}$$

is the overall heat-transfer coefficient for heat flow reduced to the non-finned surface of the wall.

The ratio of the finned surface, F_{fw}, to the plane one, F_1, is called the *finning factor*.

The effect of finning on the heat-transfer coefficient can be illustrated by an example. Let $\alpha_1 = 1000$ and $\alpha_2 = 20$ $W/m^2.K$. Assuming the ratio δ'/λ to be negligible, we have

$$k'_1 = \frac{1}{\dfrac{1}{\alpha_1} + \dfrac{1}{\alpha_{red}} \dfrac{F_1}{F_{fw}}}$$

For a plane surface (with the finning factor F_{fw}/F_1 equal to unity) we get:

$$k'_1 = \frac{1}{\dfrac{1}{1000} + \dfrac{1}{20}} \cong 20 \ \text{W} / \text{m}^2.\text{K}$$

If a wall has fins on one side, and the finning factor is $F_{fw}/F_1 = 2$, then

$$k'_1 = \frac{1}{\dfrac{1}{1000} + \dfrac{1}{20}} \cong 40 \ \text{W} / \text{m}^2.\text{K}$$

Consequently, with any given values of the heat-transfer coefficients, the rate of heat flow is about doubled by putting fins on the side of a plane wall with the smaller α, and observing a finning factor $\dfrac{F_{fw}}{F_1} = 2$.

HEAT TRANSFER THROUGH A CIRCULAR FIN OF CONSTANT THICKNESS

The calculation of fins with a variable cross-section is more complicated than that of straight fins of constant thickness. Consider the case of heat transfer through a circular fin of constant thickness (Fig. 5.9), such as those employed for extending cylindrical surfaces (tubes).

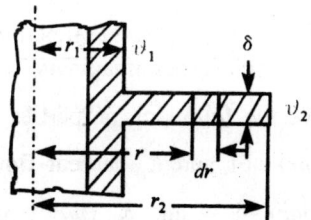

Fig. 5.9. Heat transfer through a circular fin of constant thickness.

The data given for the fin include its inner radius r_1, outer radius r_2, thickness δ, and thermal conductivity λ. The ambient temperature t_f is constant. The excess temperature of the fin is :

$$\vartheta = t - t_f$$

The constant heat-transfer coefficient α for the entire surface of the fin, and the temperature ϑ_1 at its base are given.

The process is a steady-state one with temperature varying only along the height of the fin. From these conditions, we find the differential equation that describes conduction of heat through the fin, first working out the energy balance equation for an annular element of the fin dr thick :

$$Q_r - Q_{r+dr} = dQ \qquad \qquad ...(5.28)$$

Finding the terms of Eq. (5.28), we obtain a differential equation of the following form :

$$\frac{d^2\vartheta}{dr^2} + \frac{1}{r}\frac{d\vartheta}{dr} - \frac{2\alpha}{\lambda\delta}\vartheta = 0 \qquad \qquad ...(5.29)$$

Denoting $\dfrac{2\alpha}{\lambda\delta} = m^2$, $mr = z$ and $\dfrac{1}{r} = \dfrac{m}{z}$, and substituting $\dfrac{d\vartheta}{dr} = \dfrac{d\vartheta}{dz}m$

and $\dfrac{d^2\vartheta}{dr^2} = m^2 \dfrac{d^2\vartheta}{dz^2}$, Eq. (5.28) becomes :

$$\frac{d^2\vartheta}{dz^2} + \frac{1}{z}\frac{d\vartheta}{dz} - \vartheta = |0 \qquad \qquad ...(5.30)$$

Equation (5.30) is the Bessel equation whose general solution is :

$$\vartheta = C_1 I_0(z) + C_2 K_0(z) \qquad \qquad ...(5.31)$$

where $I_0(z) = I_0(mr) =$ modified Bessel function of the first kind and zero-order imaginary argument

$K_0(z) = K_0(mr) =$ modified Bessel function of the second kind and zero-order imaginary argument.

These functions have the following properties :

at $r = 0$, $I_0(mr) = 1$ and $K_0(mr) \to \infty$

at $r = \infty$, $I_0(mr) \to \infty$ and $K_0(mr) = 0$

The constants C_1 and C_2 are determined from the boundary conditions.

If the loss of heat from the tip of the circular fin is assumed to be negligible, the calculation formulae acquire the following form :

for the current temperature along the fin

$$\vartheta = \vartheta_1 \frac{I_0(mr)K_1(mr_2) + I_1(mr_2)K_0(mr)}{I_0(mr_1)K_1(mr_2) + I_1(mr_2)K_0(mr_1)} \qquad \qquad ...(5.32)$$

for the temperature at the tip of the fin

$$\vartheta_2 = \vartheta_1 \frac{I_0(mr_2)K_1(mr_2) + I_1(mr_2)K_0(mr_2)}{I_0(mr_1)K_1(mr_2) + I_1(mr_2)K_0(mr_1)} \qquad ...(5.33)$$

for the quantity of heat

$$Q = -\lambda 2\pi r_1 \delta \left(\frac{d\vartheta}{dr}\right)_{r=r_1} = 2\pi r_1 \alpha \delta m \vartheta_1 \psi \qquad ...(5.34)$$

where

$$\psi = \frac{I_1(mr_2)K_1(mr_1) - I_1(mr_1)K_1(mr_2)}{I_0(mr_1)K_1(mr_2) + I_1(mr_2)K_0(mr_1)}$$

In using these formulae, the heat flow from the top of the fin can be accounted for by conditionally increasing its height (r_2) by half its thickness.

The formulae obtained, (5.32), (5.33), and (5.34) are cumbersome and inconvenient for engineering calculations. Circular fins of constant thickness and various straight fins of variable cross-section may, therefore, be treated and calculated as straight fins of constant cross-section. Then the quantity of heat lost by the surface of a circular fin of constant thickness is

$$Q' = \varepsilon' F' q \qquad ...(5.35)$$

where Q' = quantity of heat lost by the circular fin, W;

F' = area of the circular fin, m^2;

$Q = \dfrac{Q}{F}$ = quantity of heat lost per unit time from unit area of a straight fin of thickness equal to that of the circular fin and one metre long;

$\varepsilon' = f\left(\dfrac{\vartheta_2}{\vartheta_1}, \dfrac{r_2}{r_1}\right)$ = correction factor determined from the graph in

Fig. 5.10.

Thus, by calculating the temperature at the tip of the fin and the rate of heat flow for a straight fin, and substituting q and ε' in Eq. (5.35), we obtain the heat flow for a circular fin.

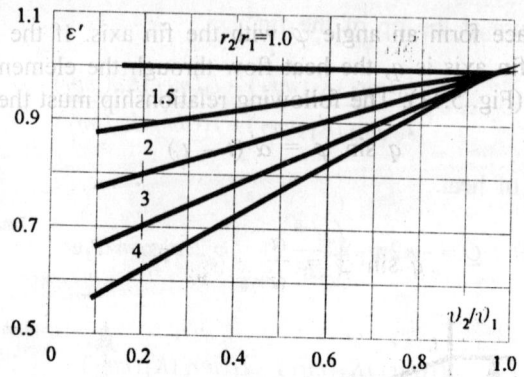

Fig. 5.10. $\varepsilon' = f(\vartheta_2/\vartheta_1; r_2/r_1)$—auxiliary calculation graph for circular fins of constant thickness.

CONDUCTION THROUGH A STRAIGHT FIN OF VARIABLE THICKNESS

In the design of cooling systems for a variety of machines and equipment, particularly for aircraft, solution of the problem of maximum heat transfer at minimum weight of the heat-exchanger is of special importance. The question arises: what is the optimal profile for a fin with the lowest weight at a given rate of heat transfer.

Fin of Minimum Weight

The essence of the problem resolves itself into the requirement that all sections of the fin should be of equal efficiency, i.e. the specific rate of heat flow must remain constant over the entire cross-section of the fin. This means that the lines of heat flow must be parallel to the axis of the fin. Under such conditions, temperature distribution along the lines of heat flow will be linear (Fig. 5.11).

With the temperature at the base of the fin t_1 given, and with the temperature at the fin tip close to the ambient temperature t_f, the problem becomes one-dimensional; therefore we may write for any cross-section of the fin :

$$t - t_f = \frac{x}{h}(t_1 - t_f) \qquad \qquad ...(5.36)$$

where x = distance measured along the axis of the fin from its tip;

h = total height of the fin.

Consider an elementary surface of the fin at a distance x. Let this elementary surface form an angle φ with the fin axis. If the rate of heat flow along the fin axis is q, the heat flow through the elementary surface will be $q \sin \varphi$ (Fig. 5.11). The following relationship must therefore hold:

$$q \sin \varphi = \alpha \, (t - t_f)$$

or

$$q \sin \varphi = \frac{\alpha}{h} \times (t_1 - t_f) \qquad \qquad ...(5.37)$$

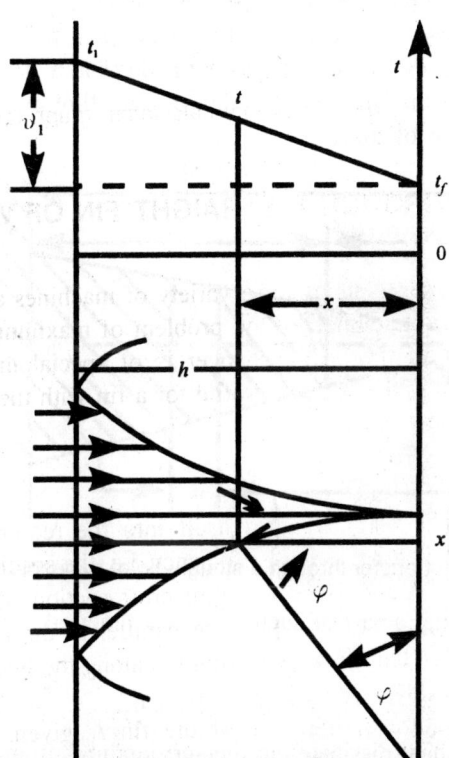

Fig. 5.11. Selection through a fin of minimum weight.

It follows from Eq. (5.37) that the angle φ is a function only of x, i.e.

$$\sin \varphi = \frac{\alpha \vartheta_1}{qh} x \qquad \qquad ...(5.37')$$

The contour of the fin found in this way is a circular arc of radius r,

since, $\sin \varphi = \dfrac{x}{r}$. It follows from Eq. (5.37') that $r = \dfrac{qh}{\alpha \vartheta_1}$. It has been shown that a fin with a profile formed by circular arcs possesses the least weight. Such a fin differs little in weight from a straight triangular fin; technologically, however, the latter is easier to manufacture which is why fins of this kind are more common in practice than circular-profile fins.

Triangular and Trapezoidal Fins

Straight fins of both triangular (pointed) and trapezoidal (truncated) profile are widely used in engineering practice.

Let us assume that the size of a trapezoidal fin (Fig. 5.12) and the excess temperature at its base ϑ_1 are given. It is convenient to take the origin of coordinates at the apex of the triangle, with the x-axis directed along the axis of symmetry of the fin. The direction of heat flow is then opposite to the positive direction of the x-axis.

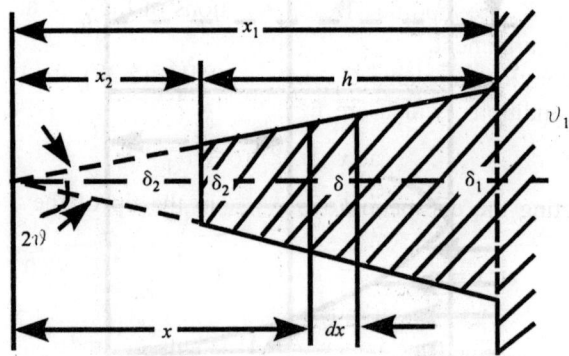

Fig. 5.12. Heat transfer through a straight fin of trapezoidal section.

The cross-sectional area f of such a fin is a function only of coordinate x, i.e.

$$f = l\delta = 2lx\,\tan\varphi \qquad \text{...(a)}$$

The quantity of heat dissipated to the surroundings from the element dx is equal to :

$$d\left(\lambda f\frac{d\vartheta}{dx}\right) = \alpha u\vartheta\,dx' \qquad \text{...(b)}$$

where α = heat-transfer coefficient of the fin surface;

u = perimeter of the fin profile at a distance x, which can be expressed as $u = 2l$; $dx' = dx/\cos\varphi$.

Differentiating (b) and accounting for (a), we obtain :

$$\frac{d^2\vartheta}{dx^2} + \frac{1}{x}\frac{d\vartheta}{dx} - \frac{1}{x}\frac{\alpha}{\lambda\sin\varphi} = 0 \qquad \qquad ...(c)$$

Upon introduction of a new variable $z = \dfrac{\alpha}{\lambda\sin\varphi}x$ it becomes :

$$\frac{d^2\vartheta}{dz^2} + \frac{1}{z}\frac{d\vartheta}{dz} - \frac{1}{z}\vartheta = 0 \qquad \qquad ...(5.38)$$

The differential equation (5.38) is a modified Bessel equation whose solution is

$$\vartheta = C_1 I_0(2\sqrt{z}) + C_2 K_0(2\sqrt{z}) \qquad \qquad ...(5.39)$$

where I_0 and K_0 are modified Bessel functions of the first and second kind of imaginary argument.

The integration constants C_1 and C_2 of Eq. (5.38) are found from the boundary conditions, which are here :

$$\text{at } x = x_1, \quad \vartheta = \vartheta_1$$

and, neglecting the dissipation of heat from the tip of the fin, we have :

$$\text{at } x = x_2, \quad \vartheta = \vartheta_2, \quad \text{and} \quad \left(\frac{d\vartheta}{dx}\right)_{x=x_2} = 0$$

Having determined the integration constants C_1 and C_2, we obtain: for the current temperature along the fin

$$\vartheta = \vartheta_1\frac{I_0(2\sqrt{z})K_1(2\sqrt{z_2}) + I_1(2\sqrt{z_2})K_0(2\sqrt{z})}{I_0(2\sqrt{z_1})K_1(2\sqrt{z_2}) + I_1(2\sqrt{z_2})K_0(2\sqrt{z_1})} \qquad ...(5.40)$$

for the temperature at the tip of the fin

$$\vartheta_2 = \vartheta_1\frac{I_0(2\sqrt{z_2})K_1(2\sqrt{z_2}) + I_1(2\sqrt{z_2})K_0(2\sqrt{z_2})}{I_0(2\sqrt{z_1})K_1(2\sqrt{z_2}) + I_1(2\sqrt{z_2})K_0(2\sqrt{z_1})} \qquad ...(5.41)$$

The heat flow can be determined by Fourier's law

$$Q = -\lambda f_1\left(\frac{d\vartheta}{dx}\right)_{x=x_1} = \frac{\alpha\vartheta_1\delta_1 l}{\sqrt{z_1}\sin\varphi} \times$$

$$\times \left[\frac{I_1(2\sqrt{z_1})K_1(2\sqrt{z_2}) - I_1(2\sqrt{z_2})K_0(2\sqrt{z_1})}{I_0(2\sqrt{z_1})K_1(2\sqrt{z_2}) + I_1(2\sqrt{z_2})K_0(2\sqrt{z_1})}\right] \qquad ...(5.42)$$

When these formulae are used, the dissipation of heat from the tip can be accounted for by fictitiously increasing the height of the fin h by half of its thickness $\delta_2/2$.

For a fin of triangular profile, $x_2 = 0$; consequently, $z_2 = 0$, $I_1(0) = 0$, and formulae (5.39), (5.40), and (5.41) become :

$$\vartheta = \vartheta_1 \frac{I_0(2\sqrt{z})}{I_0(2\sqrt{z_1})} \qquad \ldots(5.43)$$

$$\vartheta_2 = \vartheta_1 \frac{1}{I_0(2\sqrt{z_1})} \qquad \ldots(5.44)$$

$$Q = \frac{\alpha\delta_1\vartheta_1 l}{\sqrt{z_1}\sin\varphi}\left[\frac{I_1(2\sqrt{z_1})}{I_0(2\sqrt{z_1})}\right] \qquad \ldots(5.45)$$

Maximum heat flow through a triangular fin of given weight will take place provided that the equality

$$\frac{h}{\delta_1/2} = 1.309 \sqrt[3]{\frac{2\lambda}{\alpha\delta_1}} \qquad \ldots(5.46)$$

is fulfilled.

Formulae (5.40), (5.41) and (5.42) are cumbersome and inconvenient for practical use. Therefore variable-section fins are calculated as fins of constant cross-section or profile.

Then

$$Q'' = \varepsilon''qF'' \qquad \ldots(5.47)$$

where

$\qquad Q'' = $ quantity of heat flowing per unit time;

$\qquad F'' = $ cooling surface of the fin;

$\qquad q = \dfrac{Q}{F} = $ rate of heat flow for a straight fin whose length, height and thickness are respectively equal to the length, height, and mean thickness of a tapered fin;

$\varepsilon'' = f\left(\dfrac{\vartheta_2}{\vartheta_1}, \dfrac{\delta_2}{\delta_1}\right) = $ correction factor on a tapered fin; ε'' is determined from the chart in Fig. 5.13.

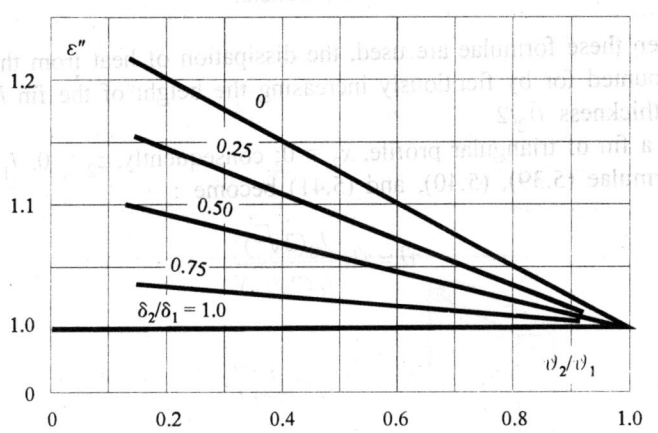

Fig. 5.13. $\varepsilon'' = f(\vartheta_2/\vartheta_1, \delta_2/\delta_1)$—auxiliary calculation graph for fins of trapezoidal and triangular section.

The lower curve $\left(\text{at } \dfrac{\delta_2}{\delta_1} = 1 \right)$ corresponds to a triangular fin of constant

section, and the upper curve $\dfrac{\delta_2}{\delta_1} = 0$ corresponds to a triangular fin.

The ratio $\dfrac{\vartheta_2}{\vartheta_1}$ is calculated by formula (5.21), the heat transfer from the

tip of the fin being taken into account by increasing the height of the fin h by half of its thickness.

HEAT CONDUCTION IN A COOLING FIN

Another simple, but practical, application of heat conduction is in the calculation of the efficiency of a cooling fin. Such fins are used to increase the area available for heat transfer between metal walls and poorly conducting fluids such as gases. A simple rectangular fin is sketched in Fig. 5.14.

A reasonably good description of the system may be obtained by approximating the true physical situation by a simplified model :

True situation	Model
T is a function of both x and z, but the dependence on z is more important.	T is a function of z alone.

(Table contd...)

True situation	Model
A small quantity of heat is lost from the fin at the end (area $2BW$) and the edges (area $2BL + 2BL$).	No heat is lost from the end or from the edges.
The heat transfer coefficient is a function of position.	The heat flux at the surface is given by $q = h(T - T_a)$, in which h is constant and $T = T(z)$.

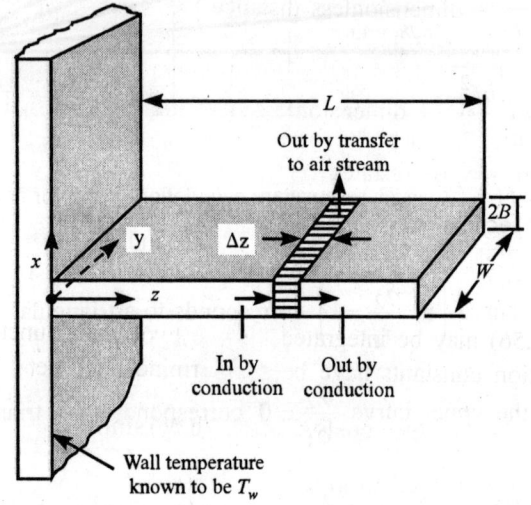

Fig. 5.14. A simple cooling fin with $B \ll L$.

A thermal energy balance on a segment Δz of the bar gives

$$q_z\big|_z \cdot 2BW - q_z\big|_{z+\Delta z} \cdot 2BW - h(2W\,\Delta z)(T - T_a) = 0 \qquad ...(5.48)$$

Division by $2BW\,\Delta z$ and taking the limit as Δz approaches zero gives

$$-\frac{dq_z}{dz} = \frac{h}{B}(T - T_a) \qquad ...(5.49)$$

Insertion of Fourier's law ($q_z = -k\,dT/dz$) in which k is the thermal conductivity of the metal gives the constant k

$$\frac{d^2T}{dz^2} = \frac{k}{kB}(T - T_a) \qquad ...(5.50)$$

This equation is to be solved with the boundary conditions

B.C. 1: at $z = 0$ $\qquad T = T_w$ $\qquad ...(5.51)$

B.C. 2:   at $z = L$ $\dfrac{dT}{dz} = 0$ ...(5.52)

We can now introduce the following dimensionless quantities :

$$\Theta = \frac{T - T_a}{T_w - T_a} = \text{dimensionless temperature}$$ ...(5.53)

$$\zeta = \frac{z}{L} = \text{dimensionless distance}$$ ...(5.54)

$$N = \sqrt{\frac{hL^2}{kB}} = \text{dimensionless heat transfer coefficient}$$ ...(5.55)

This problem may be restated :

$$\frac{d^2\Theta}{d\zeta^2} = N^2\Theta \quad \text{with} \quad \Theta\big|_{\zeta=0} = 1 \quad \text{and} \quad \frac{d\Theta}{d\zeta}\bigg|_{\zeta=1} = 0 \qquad \text{...(5.56)}$$

Equation (5.56) may be integrated to give hyperbolic functions. When the two integration constants have been determined, we get

$$\Theta = \cosh N\zeta - (\tanh N)\sinh N\zeta \qquad \text{...(5.57)}$$

This may be rearranged to give

$$\Theta = \frac{\cosh N(1 - \zeta)}{\cosh N} \qquad \text{...(5.58)}$$

It should be emphasised that this expression is reasonable only if the heat lost at the edges is negligible.

The "effectiveness of a fin" is defined by

$$\eta = \frac{\text{(heat which is actually dissipated by the fin surface)}}{\begin{bmatrix} \text{heat which would be dissipated if (without change} \\ \text{in } h) \text{ the fin surface were held at } T_w \end{bmatrix}} \qquad \text{...(5.59)}$$

The theoretical value of η for the problem considered here is then

$$\eta = \frac{\displaystyle\int_0^w \int_0^L h(T - T_a)\,dz\,dy}{\displaystyle\int_0^W \int_0^L h(T_w - T_a)\,dz\,dy} \qquad \text{...(5.60)}$$

or

$$\eta = \frac{\int_0^1 \Theta d\zeta}{\int_0^1 d\zeta}$$

$$= \frac{1}{\cosh N}\left(-\frac{1}{N}\sinh N(1-\zeta)\right)\Bigg|_0^1 = \frac{\tanh N}{N} \quad ...(5.61)$$

in which N is the dimensionless quantity defined in Eq. (5.55).

SOLVED EXAMPLES

Example 5.1. Calculate the mean heat-transfer coefficient for transformer oil flowing in a tube with a diameter d = 8 mm and length l = 1 m, if the lengthwise mean oil temperature t_f= 80°C; the mean temperature of the wall of the tube t_w = 20°C, and the velocity of the oil w = 0.6 m/s (Fig. 5.15.)

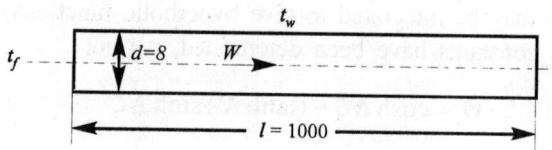

Fig. 5.15. Solution to Example 5.1.

Solution. To determine the pattern of oil flow, we calculate the value of the Reynolds number.

At t_f = 80°C the kinematic viscosity of oil v_f = 3.66 × 10⁻⁶ m²/s and the Reynolds number

$$\mathrm{Re}_f = \frac{wd}{vf} = \frac{0.6 \times 8 \times 10^{-3}}{3.66 \times 10^{-6}} = 1310.$$

Inasmuch as Re_f < 2300 the pattern of flow is laminar.

In order to determine the effect of free convection on heat transfer, we must calculate the value of the product $(GrPr)_h$, where as the reference temperature is used

$$t_h = 0.5 \ (t_f + t_w) \quad \text{and} \quad t_f = 0.5(t_{f1} + t_{f2}).$$

In the case considered

$$t_h = 0.5 \ (80 + 20) = 50°C.$$

At this temperature

$$v_h = 7.58 \times 10^{-6} \text{ m}^2/\text{s}; \quad \beta_h = 7.05 \times 10^{-4} \text{ K}^{-1}; \quad Pr_h = 111$$

and

$$(GrPr)_h = g\beta_h \frac{(t_f - t_w)d^3}{v_h^2} Pr_h =$$

$$= 9.81 \times 7.05 \times 10^{-4} \frac{(80 - 20)(8 \times 10^{-3})^3}{(7.58 \times 10^{-6})^2} 111 = 36 \times 10^5.$$

Since $(GrPr)_h < 8 \times 10^5$, the free convection does not affect materially the intensity of heat transfer and the pattern of flow is viscous.

With fluid in viscous flow in tubes at a constant wall temperature (t_w = const) the mean rate of heat transfer can be calculated from :

$$Nu_h = 1.55 \left(Pe_h \frac{d}{l} \right)^{1/3} \left(\frac{\mu_f}{\mu_w} \right)^{0.14} \varepsilon, \qquad \qquad ...(5.1)$$

where

$$Nu_h = \frac{\alpha d}{\lambda h}; \quad Pe_h \frac{d}{l} = \frac{4Gc_{ph}}{\pi l \lambda_h}; \quad \alpha = \frac{q}{t_f - t_w};$$

the subscripts "w" and "h" indicate that the physical properties of the fluid (gas or liquid) are selected respectively at the wall temperature t_w and temperature $t_h = 0.5 (t_f + t_w)$;

$$\varepsilon = 0.6 \left(\frac{1}{Re_f} \frac{l}{d} \right)^{-1/7} \left(1 + 2.5 \frac{1}{Re_f} \frac{l}{d} \right)$$

is the correction for the section of hydrodynamic stabilisation. This correction is introduced when there is no section of hydrodynamic stabilisation in front of the heated section of the tube and

$$\frac{1}{Re_f} \frac{l}{d} < 0.1.$$

Equation (5.1) is valid at $Re_f < 2300$;

$$\frac{1}{Pe_h} \frac{l}{d} \le 0.05; \quad (GrPr)_h \le 8 \times 10^5; \quad 0.07 \le \frac{\mu_w}{\mu_f} \le 1500^*.$$

In the case considered $t_f = 80°C$, $t_w = 20°C$, and $t_h = 50°C$.

The physical properties of oil:

$$\rho_f = 844 \text{ kg/m}^3; \quad \mu_f = 30.8 \times 10^4 \text{ Pa.s}; \quad \lambda_h = 0.108 \text{ W/m.°C};$$

$$c_{ph} = 1.846 \ \text{kJ/kg. °C}; \quad \mu_w = 198.2 \times 10^{-4} \ \text{Pa.s.}$$

The rate of oil flow

$$G = \rho_f w \frac{\pi d^2}{4} = 844 \times 0.6 \frac{\pi (8 \times 10^{-3})^2}{4} = 2.53 \times 10^{-2} \ \text{kg / s.}$$

The Peclet number at the reference temperature

$$Pe_h \frac{d}{l} = \frac{4G}{\pi l} \frac{c_{ph}}{\lambda_h} = \frac{4 \times 2.53 \times 10^{-2}}{3.14 \times 1.0} \frac{1.846 \times 10^3}{0.108} = 550;$$

$$\frac{1}{pe_h} \frac{1}{d} < 0.05 \ \text{and, consequently, equation (5.1) is applicable.}$$

The correction for the initial hydrodynamic section

$$\frac{1}{Re_f} \frac{l}{d} = \frac{1}{1310} \frac{1}{8 \times 10^{-3}} = 0.0955 < 0.1 \ \text{and}$$

$$\varepsilon = 0.6 \ (0.0955)^{-1/7} \ (1 + 2.5 \times 0.0955) = 1.05.$$

The Nusselt number at the reference temperature

$$Nu_h = 1.55(550)^{1/3} \left(\frac{30.8}{198.2} \right)^{0.14} 1.05 = 10.2.$$

The heat-transfer coefficient

$$\alpha = Nu_h \frac{\lambda_h}{d} = 10.2 \frac{0.108}{8 \times 10^{-3}} = 138 \ \text{W / m}^2.\text{°C.}$$

Example 5.2. Determine the temperature of oil at the tube inlet and outlet and the pressure drop along the tube under the conditions assumed in example (5.1).

Solution. The solution of example (5.1) gives:

$$\alpha = 138 \ \text{W/m}^2.\text{°C}; \quad t_f = 80\text{°C}; \quad t_w = 20\text{°C}; \quad G = 2.53 \times 10^{-2} \ \text{kg/s.}$$

The quantity of the heat transferred

$$Q = \alpha \ (t_f - t_w) \ \pi d l = 138(80 - 20)3.14 \times 8 \times 10^{-3} \times 1.0 = 207 \ \text{W.}$$

At the temperature $t_f = 80\text{°C}$ the heat capacity of oil $c_{pf} = 2.03$ kJ/kg.°C and the variation of oil temperature along the tube

$$t_{f1} - t_{f2} = \frac{Q}{Gc_{pf}} = \frac{207}{2.53 \times 10^{-2} \times 2.03 \times 10^{3}} = 4°C.$$

and the arithmetic mean oil temperature $t_f = 0.5(t_{f1} + t_{f2}) = 80°C$, whence $t_{f1} = 82°C$ and $t_{f2} = 78°C$.

With liquid (fluid) in viscous non-isothermal flow in tubes the friction factor can be determined from the following equation :

$$\xi = \xi_i \left(\frac{\mu_w}{\mu_{f1}} \right)^n, \qquad \qquad ...(5.2)$$

where ξ_i is the friction factor for isothermal flow,

$$\xi_i = \frac{64}{Re}; \quad n = C \left(Pe_1 \frac{d}{l} \right)^m \left(\frac{\mu_w}{\mu_{f1}} \right)^{-0.062};$$

with $Pe_1 d/l \leq 1500$ $C = 2.3$, $m = -0.3$;

with $Pe_1 d/l > 1500$ $C = 0.535$, $m = 0.1$.

In the case considered the inlet temperature of oil $t_{f1} = 82°C$ and at this temperature the heat capacity of oil $c_{pf\,1} = 2.04$ kJ/kg.°C; $\lambda_{f1} = 0.105$ W/m.°C; $\mu_{f1} = 29.7$ Pa.s. from the solution of example (5.1) we have: $Re_f = 1310$ and $\mu_w = 198.2$ Pa.s, $\rho_f = 844$ kg/m^3, $w = 0.6$ m/s. Then

$$Pe_1 \frac{d}{l} = \frac{4G}{\pi l} \frac{c_{pf1}}{\lambda_{f1}} = \frac{4 \times 2.53 \times 10^{-2}}{3.14 \times 1.0} \frac{2.04 \times 10^3}{0.105} = 625,$$

Since $Pe_1 \frac{d}{l} < 1500$, then $C = 2.3$ and $m = -0.3$. The exponent n in equation (5.2)

$$n = 2.3(625)^{-0.3} \left(\frac{198.2}{29.7} \right)^{-0.062} = 0.3$$

The friction factor

$$\xi = \frac{64}{Re_f} \left(\frac{\mu_w}{\mu_{f1}} \right)^n = \frac{64}{1310} \left(\frac{198.2}{29.7} \right)^{0.3} = 0.0865.$$

The drop in oil pressure across the tube

$$\Delta p = \xi \frac{\rho_{fw^2}}{2} \frac{l}{d} = 0.0865 \frac{844 \times 0.6^2}{2} \frac{1}{8 \times 10^{-3}} = 1640 \text{ Pa}$$

Example 5.3. Water flows along a tube of diameter $d = 6$ mm with a velocity $w = 0.4$ m/s. The temperature of the tube wall $t_w = 50°C$. What should the length of the tube be to ensure a water outlet temperature $t_{f2} = 20°C$ at a water inlet temperature $t_{w1} = 10°C$?

Solution. The lengthwise mean temperature of water

$$t_f = 0.5 \ (t_{f1} + t_{f2}) = 0.5 \ (10 + 20) = 15°C$$

the kinematic viscosity of water $v_f = 1.16 \times 10^{-6}$ m²/s and the Reynolds number

$$\text{Re}_f = \frac{wd}{vf} = \frac{0.4 \times 6 \times 10^{-3}}{1.16 \times 10^{-6}} = 2065.$$

The pattern of water flow is laminar. At a reference temperature

$$t_h = 0.5 \ (t_f + t_w) = 0.5 \ (15 + 50) = 32.5°C;$$

$$v_h = 0.769 \times 10^{-6} \text{ m}^2/\text{s}; \quad \beta_h = 3.37 \times 10^{-4} \text{ K}^{-1};$$

$$\text{Pr}_h = 5.14;$$

$$(\text{Gr Pr})_h = g\beta_h \frac{(t_w - t_f)d^3}{v_h^2} \text{Pr}_h = 9.81 \times 3.37$$

$$\times 10^{-4} \frac{(50 - 15)(6 \times 10^{-3})^3}{(0.769 \times 10^{-6})^2} 5.14 = 2.17 \times 10^5 < 8 \times 10^5.$$

Consequently, the water is in viscous flow and to determine the local coefficient of heat transfer let us use equation (5.1). Since the relative length of the tube is not known, the problem must be solved by the method of successive approximations. Let us assume a relative length of the tube $l/d = 100$ and, consequently, $l = 100 \times 6 \times 10^{-3} = 0.6$ m.

The physical properties of water are as follows:

at $t_f = 15°C$ $\mu_f = 1155 \times 10^{-6}$ Pa.s, $\rho_f = 999$ kg/m³;

at $t_h = 32.5°C$ $\lambda_h = 0.631$ W/m.°C, c_{pr} 4.174 kJ/kg.°C; at $t_w = 50°C$ $\mu_w = 549.4 \times 10^{-6}$ Pa.s.

The rate of water flow

$$G = \rho_f{}^w \frac{\pi d^2}{4} = 999 \times 0.4\frac{3.14\,(6 \times 10^{-3})^2}{4} = 0.0113\ \text{kg}\,/\,\text{s}.$$

The Peclet number at the reference temperature

$$Pe_h \frac{d}{l} = \frac{4G}{\pi_l}\frac{c_{ph}}{\lambda_h} = \frac{4.0 \times 0.0113}{3.14 \times 0.6}\frac{4174}{0.631} = 159.$$

The correction for the section of hydrodynamic stabilisation

$$\varepsilon = 0.6\left(\frac{1}{Re}\frac{l}{d}\right)^{-1/7}\left(1 + 2.5\frac{1}{Re_f}\frac{l}{d}\right)$$

$$= 0.6\left(\frac{100}{2065}\right)^{-1/7}\left(1 + 2.5\frac{100}{2065}\right) \approx 1.04.$$

The Nusselt number at the reference temperature

$$Nu_h = 1.55\left(Pe_h \frac{d}{l}\right)^{1/3}\left(\frac{\mu_f}{\mu_w}\right)^{0.14}\varepsilon$$

$$= 1.55\,(159)^{1/3}\left(\frac{1155}{549.4}\right)^{0.14} \times 1.04 = 9.7.$$

The heat-transfer coefficient

$$\alpha = Nu_h \frac{\lambda_h}{d} = 9.7\frac{0.631}{6 \times 10^{-3}} = 1020\ \text{W}\,/\,\text{m}^2.{}^\circ\text{C}.$$

The amount of heat transferred through the wall

$$Q = Gc_{pf}\,(t_{f2} - t_{f1}) = 0.0113 \times 4187 \times 10 = 473\ \text{W}$$

where c_{pf} is taken based on the mean temperature of the fluid $t_f = 15^\circ\text{C}$. On the other hand, the amount of heat transferred

$$Q = \alpha\,(t_w - t_f)\,\pi dl.$$

Thus, after the first approximation we find

$$l = \frac{Q}{\alpha(t_{w2} - t_{f1})\pi d} = \frac{473}{1020\,(50 - 15)\,3.14 \times 6 \times 10^{-3}} = 0.705\ \text{m}.$$

For the second approximation we select l = 0.75 m, repeat the calculation and obtain: $Pe_h \dfrac{d}{l} = 183$, $\varepsilon = 1.03$, $Nu_h = 8.94$, $\alpha = 940$ W/m^2.°C. After the second approximation we get

$$l = \frac{473}{940 \times 35 \times 3.14 \times 6 \times 10^{-3}} = 0.765 \text{ m.}$$

Since the assumed length of the tube coincides with a sufficient accuracy with the result obtained after the second approximation, there is no need in the third approximation, and we may assume a length of the tube l = 0.76 m.

Example 5.4. Water flows in a tube of diameter d = 8 mm. The tube is heated in a way that the rate of heat flow through the wall of the tube is constant over the perimeter and length and amounts to $q_w = 4 \times 10^4$ W/m^2.

Determine the value of the local heat-transfer coefficient and the tube wall temperature at a distance x = 20d from the inlet into the heated section of the tube.

The water inlet temperature $t_{f1} = 10$°C. The mean water flow velocity w = 0.15 m/s. Upstream of the heated section of the tube there is a section of hydrodynamic stabilisation (Fig. 5.16).

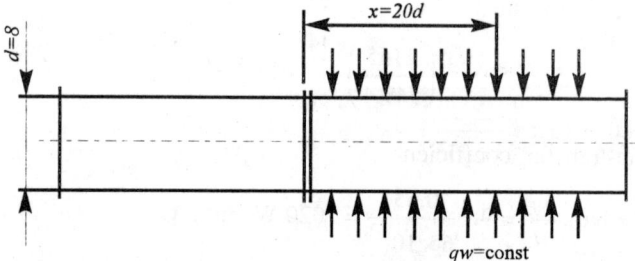

Fig. 5.16. Solution to Example 5.4.

Solution. To determine the pattern of water flow, we calculate the Reynolds number. The rate of water flow

$$G = \rho w \frac{\pi d^2}{4} = 999 \times 0.15 \frac{\pi (8 \times 10^{-3})^2}{4} = 7.54 \times 10^{-3} \text{ kg / s.}$$

The bulk mean temperature of water in the section x = 20d = 20 × 8 × 10^{-3} = 0.16 m long

$$t_f = t_{f1} + \frac{q_w \pi d}{G c_p} x = 10 + \frac{4 \times 10^4 \times 3.14 \times 8 \times 10^{-3}}{7.54 \times 10^{-3} \times 4187} 0.16$$

$$= 10 + 5.1 = 15.1 \ °C,$$

where the heat capacity and density of the water are selected at a temperature of 15°C:

$$\rho = 999 \ \text{kg/m}^3 \text{ and } c_{pf} = 4187 \ \text{J/kg.°C}.$$

At $t_f = 15°C$ the kinematic viscosity of water $v_f = 1.153 \times 10^{-6}$ m²/s

and the Reynolds number $\text{Re}_f = \dfrac{wd}{v_f} = \dfrac{0.15 \times 8 \times 10^{-3}}{1.153 \times 10^{-6}} = 1040.$

Since the Reynolds number $\text{Re}_f < 2300$, water is in a laminar flow. In order to determine whether or not heat transfer is affected by free convection, it is necessary to calculate the magnitude of the product $(GrPr)_h$, but for this purpose the wall temperature should be known. Therefore, this check will be carried out at the end of the calculations, after the wall temperature t_w is determined. Calculations are done, assuming that natural, or free, convection does not affect heat transfer.

To calculate the local heat transfer when fluid is in viscous flow in tubes at a constant rate of heat flow, q_w = const, use can be made of the following equation :

$$\text{Nu}_h = 1.31 \varepsilon \left(\frac{1}{\text{Pe}_h} \frac{x}{d} \right)^{-1/3} \left(1 + 2 \frac{1}{\text{Pe}_h} \frac{x}{d} \right) \left(\frac{\mu_w}{\mu_f} \right)^{-1/6} \qquad ...(5.3)$$

where

$$\text{Nu}_h = \frac{\alpha d}{\lambda_h}; \quad \frac{1}{\text{Pe}_h} \frac{x}{d} = \frac{\pi x \lambda_h}{4 G c_{ph}}; \quad \alpha = \frac{q_w}{t_w - t_f};$$

the subscripts "w" and "h" indicate that the physical properties of the fluid are selected at the wall temperature t_w and a reference temperature $t_h = 0.5$ $(t_f + t_w)$ respectively.

The correction for the section of hydrodynamic stabilisation

$$\varepsilon = 0.35 \left(\frac{1}{\text{Re}_f} \frac{x}{d} \right)^{-1/6} \left[1 + 2.85 \left(\frac{1}{\text{Re}_f} \frac{x}{d} \right)^{0.42} \right].$$

This correction is introduced when there is no section of hydrodynamic stabilisation upstream of the heated section of the tube and $\dfrac{1}{\mathrm{Re}_f}\dfrac{x}{d} \le 0.064$.

Equation (5.3) is valid at $\mathrm{Re}_f < 2300$;

$$\frac{1}{\mathrm{Pe}_h}\frac{x}{d} \le 0.04 \text{ and } 0.04 \le \frac{\mu_w}{\mu_f} \le 1 .$$

In order to calculate with the aid of equation (5.3), the temperature of the wall, t_w, must be known. The calculation is, therefore, carried out by the method of successive approximations.

Evaluating the local coefficient of heat transfer from the wall of the tube to the flowing water to be $\alpha = 1000$ W/m^2.°C, we get:

$$t_w = t_f + \frac{q_w}{\alpha} \approx 15.1 + \frac{4 \times 10^4}{1 \times 10^3} \approx 55°C,$$

then, the reference temperature $t_h = 0.5\,(t_f + t_w) = 0.5\,(15.1 + 55) \approx 35°C$.

At this temperature

$$c_{ph} = 4174 \text{ J/kg.°C}; \quad \lambda_h = 0.624 \text{ W/m.°C};$$

with $t_f = 15.1°C$ $\mu_f = 1152 \times 10^{-6}$ Pa.s;

with $t_w = 55°C$ $\mu_w = 509.6 \times 10^{-6}$ Pa.s

and

$$\frac{1}{\mathrm{Pe}_h}\frac{x}{d} = \frac{\pi_x \lambda_h}{4Gc_{ph}} = \frac{3.14 \times 0.16 \times 0.624}{4 \times 7.54 \times 10^{-3} \times 4174} = 2.5 \times 10^{-3};$$

$$\frac{1}{\mathrm{Pe}_h}\frac{x}{d} < 0.04; \quad \frac{\mu_w}{\mu_f} = \frac{509.6}{1152} = 0.442 > 0.04;$$

consequently, equation (5.3) is applicable.

In accordance with conditions assumed , $\varepsilon = 1$ and the number

$$\mathrm{Nu}_h = 1.31\,(2.5 \times 10^{-3})^{-1/3}\,(1 + 2 \times 2.5 \times 10^{-3})\,(0.442)^{-1/6} = 11.1$$

The local heat-transfer coefficient

$$\alpha = \mathrm{Nu}_h\frac{\lambda_h}{d} = 11.1\frac{0.624}{8 \times 10^{-3}} = 866 \text{ W / m}^2.°C.$$

As the second approximation, the temperature of the wall

$$t_w = 15.1 \times \frac{4 \times 10^4}{866} = 15.1 + 46.2 = 61.3 \ ^\circ C.$$

Assuming the wall temperature $t_w = 60^\circ C$, we repeat the calculation and obtain:

$$t_h = 37.5 \ ^\circ C; \quad \frac{1}{Pe_h} \frac{x}{d} = 2.54 \times 10^{-3};$$

$$\frac{\mu_w}{\mu_f} = 0.4; \quad Nu_h = 11.2; \quad \alpha = 885 \ W / m^2.^\circ C;$$

as the third approximation, the temperature of the wall

$$t_w = 15.1 + \frac{4 \times 10^4}{885} = 15.1 + 45.2 = 60.3 \ ^\circ C.$$

Since the calculated wall temperature coincides with sufficient accuracy with the temperature assumed, the local heat-transfer coefficient can be assumed equal to $\alpha = 885 \ W/m^2.^\circ C$ and $t_w = 60^\circ C$.

Having determined the wall temperature t_w, let us check now whether heat transfer is affected by free convection.

At a reference temperature $t_h = 37.5^\circ C$ $v_h = 0.695 \times 10^{-6} \ m^2/s$, $\beta_h = 3.71 \times 10^{-4} \ K^{-1}$, $Pr = 4.59$, then

$$(Gr \ Pr)_h - g\beta_h \frac{(t_w - t_f)d^3}{v_h^2} Pr_h =$$

$$= 9.81 \times 3.71 \times 10^{-4} \frac{(60 - 15.1)(8 \times 10^{-3})^3}{(0.695 \times 10^{-6})^2} 4.59 =$$

$$= 7.9 \times 10^5 < 8 \times 10^5$$

and it may be assumed that free convection will not affect materially the process of heat transfer.

Example 5.5. Determine the relative length of the section of thermal stabilisation, $l_{in.t}/d$, for a tube having a diameter $d = 14$ mm and carrying water in a laminar flow with a wall temperature maintained constant along the tube ($t_w = \text{const}$), if the mean temperature of water $t_f = 50^\circ C$ and $Re_f = 1500$. Also calculate the local heat-transfer coefficient on the section of the tube where $l > l_{in.t}$.

Solution. With water in a liminar flow and provided t_w = const, the relative length of the section of thermal stabilisation may be assumed:

$$l_{in.t} / d = 0.5 Pe_f;.$$

In the case considered at t_f = 50°C Pr_f = 3.55 and Pe_f = $Re_f Pr_f$ = 1500 × 3.55 = 5320, consequently, $l_{in.t}/d$ = 0.05 × 5320 = 266.

With $l > l_{in.t}$ the maximum value of the Nusselt number Nu_∞ = 3.66, consequently,

$$\alpha = Nu\infty \frac{\lambda_f}{d} = 3.66 \frac{0.648}{14 \times 10^{-3}} = 170 \text{ W} / \text{m}^2.°\text{C},$$

where at t_f = 50°C λ_f = 0.648 W/m.°C.

Example 5.6. Solve example 5.5, if heat transfer takes place at a constant rate of heat transfer along the tube (q_w = const).

Solution. With water in a laminar flow under the condition q_w = const the relative length of the section of thermal stabilisation may be assumed equal to $l_{in.t}/d \approx 0.07 Re_f$, and Nu_∞ = 4.36.

In this case considered (see example 5.5)

$$\frac{l_{in.t}}{d} = 0.07 \times 5320 \approx 372;$$

$$\alpha = 4.36 \frac{0.648}{14 \times 10^{-3}} = 203 \text{ W} / \text{m}^2.°\text{C}.$$

Example 5.7. Calculate the loss of heat per unit time from one square metre of a horizontal cylindrical heat exchanger which is cooled by the ambient air. The outside diameter of the heat-exchanger shell d = 400 mm, the shell surface temperature t_w = 200°C and the air temperature in the room t_f = 30°C (Fig. 5.3).

Solution. The rate of heat flow on the outside surface of the cylindrical heat exchanger $q = \alpha (t_w - t_f)$, W/m^2.

At the given surface temperature and the ambient air temperature at a distance from the wall the solution of the problem comes to determining the heat-transfer coefficient. For cases of free convection the mean heat-transfer coefficient can be calculated with the aid of following equation:

$$Nu_f = C(GrPr)_f^n \left(\frac{Pr_f}{Pr_w}\right)^{0.25} \qquad ...(5.3)$$

where the constants C and n depend on the pattern of free flow and the conditions of flow past a surface. They are functions of GrPr and are determined from the following Table (5.17).

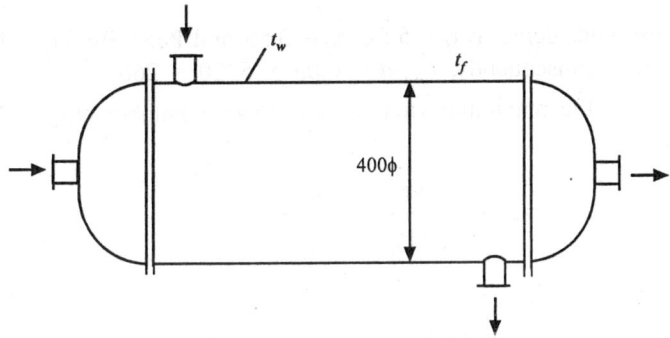

Fig. 5.17. Solution to Example 5.7.

Table 5.1. Function of $(GrPr)_f$ and conditions of flow.

$(GrPr)_f$	C	n	Conditions of flow
$1 \times 10^3 - 1 \times 10^9$	0.75	0.25	Along a vertical wall
$\geqslant 6 \times 10^{10}$	0.15	1.3	Along a vertical wall
$1 \times 10^3 - 1 \times 10^9$	0.50	0.25	On a horizontal tube

The subscripts "f" and "w" in equation (5.3) indicate that the physical properties of the fluid are selected respectively at the fluid temperature t_f, at a distance from the heat-transfer surface, and the wall temperature t_w. In the case of flow along a vertical wall taken as the reference dimension is the height of the heat-transfer surface and for a horizontal cylinder—its outside diameter.

In the case considered the reference temperature $t_f = 30°C$. At that temperature air has the following physical properties :

$$v_f = 16.0 \times 10^{-6} \text{ m}^2/\text{s}; \quad \lambda_f = 2.67 \times 10^{-2} \text{ W/m.°C};$$

$$\beta_f = \frac{1}{t_f + 273} = \frac{1}{300} \text{K}^{-1}; \quad \text{Pr}_f = 0.701.$$

Calculate the magnitude of the dimensionless term

$$(\text{GrPr})_f = \frac{g\beta_f \Delta t d^3}{v_f^2} \text{Pr}_f = \frac{9.81(200-30) \, 0.4^3}{303(16 \times 10^{-6})^2} 0.701 = 9.75 \times 10^8.$$

From the Table (5.1) we find that for the calculated values of the dimensionless term $(GrPr)_f$ the constants $C = 0.5$ and $n = 0.25$.

The Nusselt number

$$Nu_f = 0.50 \ (9.75 \times 10^8)^{0.25} = 88.2,$$

whence

$$\alpha = Nu_f = \frac{\lambda_f}{d} = 88.2 \frac{2.67 \times 10^{-2}}{0.4} = 5.9 \ W/m^2 \cdot{}^\circ C.$$

The loss of heat per unit time from unit surface area of the heat exchanger is

$$q = 5.9 \ (200 - 30) = 1000 \ W/m^2.$$

Example 5.8. Determine the coefficient of heat transfer from a vertical plate of height $H = 2$ m to the surrounding still air, if it is known that the surface temperature of the plate $t_w = 100^\circ C$, the ambient air temperature at a distance from the plate $t_f = 20^\circ C$.

Solution. The transfer of heat by free convection from a vertical plate can be determined by equation (5.3) :

$$Nu_f = C(GrPr)_f^n \left(\frac{Pr_f}{Pr_w} \right)^{0.25}.$$

where taken as the reference dimension is the height of plate H.

At $t_f = 20^\circ C$ the physical properties of air:

$$\lambda_f = 2.59 \times 10^{-2} \ W/m.^\circ C; \quad v_f = 15.06 \times 10^{-6} \ m^2/s;$$

$$Pr_f = 0.703; \quad \beta_f = \frac{1}{t_f + 273} = \frac{1}{293} \ K^{-1}.$$

Under these conditions the dimensionless term

$$(GrPr)_f = g\beta_f \frac{\Delta t H^3}{v_f^2} Pr_f =$$

$$= 9.81 \frac{1}{293} \frac{80 \times 2^3 \times 10^{12}}{(15.06)^2} 0.703 = 6.64 \times 10^{10}.$$

Using the obtained value of $(GrPr)_f$, find from the Table, compiled for equation (5.3), $C = 0.15$ and $n = 1.3$. Then,

$$\text{Nu}_f = 0.15 \ (6.64 \times 10^{10})^{1/3} = 610$$

and

$$\alpha = \text{Nu}_f \frac{\lambda_f}{H} = 610 \frac{2.59 \times 10^{-2}}{2} = 7.92 \ \text{W}/\text{m}^2.°\text{C}.$$

Example 5.9. Determine the coefficient of heat transfer from a plate, arranged with its heat-transfer surface facing upward and measuring $a \times b = 2 \times 3$ m^2, to the surrounding still air, if the plate surface temperature $t_w = 100°$C and the ambient air temperature, at a distance from the plate, $t_f = 20°$C.

Solution. The rate of heat transfer from horizontal plates can be calculated approximately with the aid of equation (5.3). Then, the smaller side of the plate is taken as the reference dimension. If the heat-transfer surface faces upward, the value of the heat-transfer, calculated with the air of equation (5.3) is increased by 30 per cent and, on the contrary, if the heat-transfer surface faced downward, the heat-transfer coefficient is reduced by 30 per cent.

In the case considered $t_f = 20°$C and at that temperature, for air, $v_f = 15.06 \times 10^{-6}$ m^2/s;

$$\lambda_f = 2.59 \times 10^{-2} \ \text{W/m·°C}; \quad \text{Pr}_f = 0.703.$$

The reference dimension is the smaller side of the plate, i.e. $a = 2$ m. Then, the dimensionless term

$$(\text{GrPr})_f = g\beta_f \frac{\Delta t a^3}{v_f^2} \ \text{Pr}_f = 9.81 \frac{80 \times 2^3 \times 0.703}{293(15.06 \times 10^{-6})^2} = 6.64 \times 10^{10}.$$

Using the obtained value of (GrPr), find from the Table, compiled for equation (5.3), $C = 0.15$ and $n = 1/3$. Then,

$$\text{Nu}_f = 0.15 \ (6.64 \times 10^{10})^{1/3} = 610,$$

whence

$$\alpha' = \text{Nu}_f \frac{\lambda_f}{a} = 610 \frac{2.59 \times 10^{-2}}{2} = 7.9 \ \text{W}/\text{m}^2.°\text{C}$$

and

$$\alpha = 1.3\alpha' = 1.3 \times 7.9 = 10.3 \ \text{W/m}^2.°\text{C}.$$

Example 5.10. In the service tank of a circuit, used to study hydrodynamics and heat transfer in liquid-metal coolants, a metal coolant is heated with the aid of a horizontal electric heater, of a cylindrical shape and 50 mm in diameter.

Calculate the coefficient of heat transfer from the surface of the heater to the liquid metal, for the case when the circuit is filled with sodium at a temperature $t_f = 200°C$, and the heater surface temperature $t_w = 400°C$.

Solution. With liquid metals in free flow the rate of heat transfer can be calculated by the following equation :

$$\text{Nu}_h = C \text{Gr}_h^n \text{Pr}_h^{0.4}. \qquad \qquad ...(5.4)$$

In equation (5.4) the constants C and n depend on the magnitude of the Grashof number :

with $\text{Gr}_h = 10^2$ to 10^9; $C = 0.52$ and $n = 0.25$;

with $\text{Gr}_h = 10^9$ to 10^{13}; $C = 0.106$ and $n = 0.33$.

The physical properties are chosen at the temperature $t_h = 0.5\,(t_w + t_f)$.

For the case considered $t_h = 0.5\,(200 + 400) = 300°C$ and at that temperature the physical properties of sodium are :

$$v_h = 39.4 \times 10^{-8} \text{ m}^2/\text{s}; \quad \lambda_h = 71 \text{ W/m·°C}; \quad \text{Pr}_h = 0.63 \times 10^{-2}$$

$$\beta_h \approx \frac{\rho_f - \rho_w}{\rho_f (t_w - t_f)} = \frac{903 - 854}{903\,(400 - 200)} = 2.71 \times 10^{-4} \text{ K}^{-1};$$

$$\text{Gr}_h = g\beta_h \frac{\Delta t d^3}{v_h^2} = 9.81 \times 2.71 \times 10^{-4} \frac{200(5 \times 10^{-2})^3}{(39.4 \times 10^{-8})^2} = 4.28 \times 10^8$$

With this value of the Grashof number

$$C = 0.52 \text{ and } n = 0.25.$$

Then

$$\text{Nu}_h = 0.52\,(4.28 \times 10^8)^{0.25}\,(6.3 \times 10^{-3})^{0.4} = 11.1,$$

whence

$$\alpha = \text{Nu}_h \frac{\lambda_h}{d} = 11.1 \frac{71}{2 \times 10^{-2}} = 15750 \text{ W / m}^2.°C.$$

Example 5.11. A copper bus bar of round cross-section, $d = 15$ mm, is cooled with a cross flow of dry air (Fig. 5.18). The velocity and temperature of the free-stream air flow are respectively $w = 1$ m/s and $t_f = 20°C$.

Calculate the coefficient of heat transfer from the surface of the copper bus bar to the cooling air and the admissible current intensity for the bus bar on condition that its surface temperature should not exceed $t_w = 80°C$.

Solution. The resistivity of copper $\rho = 0.0175$ ohm $\cdot$ mm²/m.

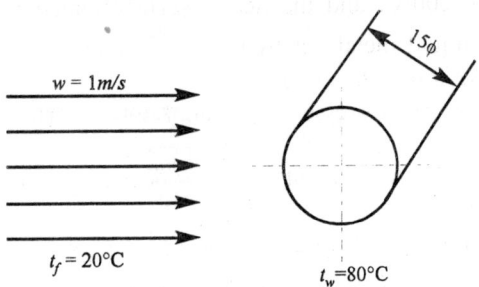

$w = 1 m/s$

$t_f = 20°C$

$t_w = 80°C$

Fig. 5.18. Solution to Example 5.11.

At a temperature $t_f = 20°C$ the physical properties of air are: $\nu_f = 15.06 \times 10^{-6}$ m²/s, $\lambda_f = 2.59 \times 10^{-2}$ W/m.°C.

The Reynolds number

$$\text{Re}_f = \frac{wd}{\nu_f} = \frac{1 \times 0.015}{15.06 \times 10^{-6}} = 995.$$

For a single cylinder placed in cross air flow heat transfer can be calculated with the aid of the following :

at $10 \le \text{Re}_f \le 1 \times 10^3$

$$\text{Nu}_f = 0.44 \text{ Re}_f^{0.5}$$

at $1 \times 10^3 \le \text{Re}_f \le 2 \times 10^5$ ...(5.11)

$$\text{Nu}_f = 0.22 \text{ Re}_f^{0.6}$$

where taken as the reference dimension is the diameter of the cylinder and as the reference temperature that of the free-stream air flow, t_f.

In the case considered

$$\text{Nu}_f = 0.44 \ (995)^{0.5} = 13.8$$

and the heat-transfer coefficient

$$\alpha = Nu \frac{\lambda_f}{d} = 13.8 \frac{2.59 \times 10^{-2}}{15 \times 10^{-2}} = 23.8 \ W/m^2/°C.$$

The admissible current intensity is determined from the energy balance equation

$$\alpha \, (t_w - t_f) \, \pi dl = I^2 R,$$

where

$$R = \frac{\rho l}{\dfrac{\pi d^2}{4}}, \text{ ohm,} \quad \bullet$$

whence the following expression for admissible current intensity

$$I = 10^3 \pi d \sqrt{\frac{\alpha \Delta t d}{4\rho}}.$$

Substituting known quantities, we obtain:

$$I = 10^3 \times 3.14 \times 1.5 \times 10^{-2} \sqrt{\frac{23.8 \times (80 - 20)1.5 \times 10^{-2}}{4 \times 0.0175}} = 825\text{A}.$$

Example 5.12. A tubular water calorimeter with an outside diameter $d = 15$ mm is placed in a cross flow of air. The velocity of the air, $w = 2$ m/s, is directed at an angle of $90°$ to the axis of the calorimeter and the mean air temperature $t_f = 20°C$. Under steady-state conditions the outer surface of the calorimeter acquires a constant mean temperature $t_w = 80°C$.

Calculate the coefficient of heat transfer from the tube to the air and the heat transfer rate from unit length of the calorimeter.

Solution. At a temperature $t_f = 20°C$ the physical properties of air are: $v_f = 15.06 \times 10^{-6}$ m²/s, $\lambda_f = 2.59 \times 10^{-2}$ W/m.°C.

The Reynolds number

$$\text{Re}_f = \frac{wd}{v_f} = \frac{2 \times 15 \times 10^{-3}}{15.06 \times 10^{-6}} = 1990.$$

Since $1 \times 10^3 < \text{Re} < 2 \times 10^5$, according to equation (5.11).

$$\text{Nu}_f = 0.22\text{Re}_f^{0.6}$$

Then

$$\text{Nu}_f = 0.22 \, (1.99 \times 10^3)^{0.6} = 21$$

and the heat-transfer coefficient

$$\alpha = \text{Nu}_f \frac{\lambda_f}{d} = \frac{21 \times 2.59 \times 10^{-2}}{1.5 \times 10^{-2}} = 36.3 \text{ W} / \text{m}^2.°C.$$

The heat transfer rate from unit length

$$q_1 = \alpha \, (t_w - t_f) \, \pi d = 36.3 \,\, (80 - 20) \,\, 3.14 \times 1.5 \times 10^{-2} = 102 \,\, \text{W/m.}$$

Example 5.13. A cylindrical tube having a diameter $d = 20$ mm is cooled in a cross flow of water. Flow velocity $w = 1$ m/s.

The mean temperature of the cooling water $t_f = 10°\text{C}$ and the surface of the tube is at a temperature $t_w = 50°\text{C}$.

Determine the coefficient of heat transfer from the surface of the tube to the cooling water.

Solution. At a water temperature $t_f = 10°\text{C}$

$$v_f = 1.3 \times 10^{-6} \,\, \text{m}^2/\text{s.}$$

The Reynolds number

$$\text{Re}_f = \frac{wd}{v_f} = \frac{1 \times 0.02}{1.3 \times 10^{-6}} = 1.54 \times 10^4.$$

For a single cylinder placed in cross flow of a liquid heat transfer can be calculated with the aid of the following equation :

at $8 < \text{Re}_f < 1 \times 10^3$

$$\text{Nu}_f = 0.50 \text{Re}_f^{0.5} \, \text{Pr}_f^{0.38} \left(\frac{\text{Pr}_f}{\text{Pr}_w} \right)^{0.25} ;$$

at $1 \times 10^3 < \text{Re}_f < 2 \times 10^5$ ...(5.12)

$$\text{Nu}_f = 0.25 \text{Re}_f^{0.6} \, \text{Pr}_f^{0.38} \left(\frac{\text{Pr}_f}{\text{Pr}_w} \right)^{0.25} ;$$

where taken as the reference dimension is the diameter of the cylinder, and the subscripts "f" and "w" indicate that the corresponding physical properties are taken based on the temperature of free-stream flow of liquid t_f and the temperature of the liquid near the wall t_w. In the case considered $1 \times 10^3 < \text{Re}_f < 2 \times 10^5$ and calculations should be done with the second equation.

At $t_f = 10°\text{C}$, $v_f = 1.3 \times 10^{-6}$ m^2/s, $\lambda_f = 0.574$ W/m.°C, and the Prandtl number $\text{Pr}_f = 9.5$.

At a temperature $t_w = 50°\text{C}$, $\text{Pr}_w = 3.55$. Consequently,

$$\mathrm{Nu}_f = 0.25 \, (1.54 \times 10^4)^{0.6} \, (9.5)^{0.38} \left(\frac{9.5}{3.55}\right)^{0.25} = 246$$

and the heat-transfer coefficient

$$\alpha = \mathrm{Nu}_f \frac{\lambda_f}{d} = 246 \frac{0.574}{0.02} = 7050 \text{ W} / \text{m}^2 .°\text{C}.$$

Example 5.14. In a heat exchanger liquid sodium is in cross flow past a staggered bank of tubes, having an outside diameter $d = 20$ mm. The mean velocity of the free-stream flow of liquid sodium $w = 1$ m/s, and mean temperature of liquid sodium $t_f = 250°C$.

Determine the mean coefficient of heat transfer from the tubes to the sodium coolant and the mean rate of heat flow on the surface of the tubes on condition that the mean tube surface temperature $t_w = 256°C$.

Solution. With sodium in cross flow past staggered and in-line banks of tubes the mean heat-transfer coefficient can be calculated with the aid of the following equation :

$$\mathrm{Nu}_f = 2\mathrm{Pe}\,_f^{0.5} \qquad \qquad ...(5.13)$$

where, taken as the reference dimension is the tube diameter. Equation (5.13) is valid at $100 \le \mathrm{Pe}_f \le 1000$.

In the case considered, at $t_f = 250°C$, the physical properties of liquid sodium are respectively:

$$v_f = 45 \times 10^{-8} \text{ m}^2/\text{s}; \; \lambda_f = 76.1 \text{ W/m. °C}; \; \mathrm{Pr} = 0.69 \times 10^{-2}.$$

The Reynolds and Peclet numbers

$$\mathrm{Re}_f = \frac{wd}{v_f} = \frac{1 \times 20 \times 10^{-3}}{45 \times 10^{-8}} = 4.44 \times 10^4;$$

$$\mathrm{Pe}_f = \mathrm{Re}_f \mathrm{Pr}_f = 4.44 \times 10^4 \times 0.69 \times 10^{-2} = 306.$$

The Nusselt number and the heat-transfer coefficient

$$\mathrm{Nu}_f = 2(306)^{0.5} = 35;$$

$$\alpha = \mathrm{Nu}_f \frac{\lambda_f}{d} = 35 \frac{76.1}{20 \times 10^{-3}} = 1.34 \times 10^5 \text{ W} / \text{m}^2 .°\text{C}.$$

$$q = \alpha \, (t_w - t_f) = 1.34 \times 10^5 \, (256 - 250) = 8 \times 10^5 \text{ W/m}^2.$$

Example 5.15. Determine the coefficient of heat transfer from the external surface of the tube of an evaporator to boiling water, if the rate of heat flow through the heat-transfer surface $q = 2 \times 10^5$ W/m^2, the liquid is in nucleate boiling under a pressure $p = 2\ 5\ 10^5$ Pa.

Solution. With liquid in pool nucleate boiling the heat-transfer coefficient can be calculated :

with Re$_*$ $\geqslant 10^{-2}$

$$\text{Nu}_* = 0.125\text{Re}_*^{0.65}\text{Pr}^{1/3}; \qquad\qquad ...(5.14a)$$

with Re$_*$ $\leqslant 10^{-2}$

$$\text{Nu}_* = 0.0625\text{Re}_*^{0.5}\text{Pr}^{1/3}. \qquad\qquad ...(5.14b)$$

where

$$\text{Re}_* = \frac{q l_*}{r \rho'' v};$$

$$\text{Nu}_* = \frac{\alpha l_*}{\lambda};$$

$$\text{Pr} = \frac{v}{a};$$

$$l_* = \frac{c_p \rho' \sigma T_s}{(r\rho'')^2}, \text{ m};$$

v, c_p, r, λ, a and σ = respectively the dynamic viscosity coefficient, heat capacity, evaporation, thermal conductivity, thermal diffusivity, surface tension coefficient at the saturation temperature t_s;

ρ' and ρ'' = respectively densities of liquid and vapour at saturation temperature t_s;

T_s = saturation temperature, K.

Equations (5.14a) and (5.14b) are valid at $0.86 \leqslant \text{Pr} \leqslant 7.6$, $10^{-5} \leqslant \text{Re}_* \leqslant 10^4$, and a pressure from 45×10^2 to 175×10^5 Pa.

For water the values of l_* and $l_*/r\rho'' v$, depending on temperature, are given in Table 5.1.

Values of l_*, $\dfrac{l_*}{r\rho'' v}$ and $\dfrac{\lambda}{r\rho'' v}$ in equation (5.14a) and (5.14b)

Table 5.1. Values of water and temperatures.

$t_s,°C$	$l_* \times 10^6$	$\dfrac{l_*}{r\rho''v}10^6$	$\dfrac{\lambda}{r\rho''v}10^2$	$t_s,°C$	$l_* \times 10^6$	$\dfrac{l_*}{r\rho''v}10^6$	$\dfrac{l_*}{r\rho''v}10^2$
	m	m²/W	1/°C		m	m²/W	1/°C
30	16450	276870	1040	190	0.450	0.216	32.2
40	5950	73345	782	200	0.296	0.123	27.5
50	2305	20894	587	210	0.200	0.0718	23.5
60	960	6543	450	220	0.136	0.0426	20.2
70	423	2201	347	230	0.0938	0.0254	17.3
80	197	798	273	240	0.0646	0.0155	15.1
90	96.0	304	216	250	0.0451	0.00989	13.6
100	48.7	122.4	172	260	0.0318	0.00593	11.4
110	25.9	51.8	138	270	0.0224	0.00373	9.80
120	14.2	22.8	110	280	0.0158	0.00243	8.80
130	8.05	10.7	96.0	290	0.0114	0.00153	7.47
140	4.70	5.13	75.0	300	0.00800	0.000911	6.16
150	2.82	2.58	60.5	310	0.00565	0.000609	5.64
160	1.73	1.33	52.6	320	0.00398	0.000388	4.93
170	1.08	0.710	44.5	330	0.00278	0.000249	4.34
180	0.715	0.396	37.5	340	0.00192	0.000158	3.77
				350	0.00126	0.0000989	3.36

In the case considered, at a pressure $p = 2 \times 10^5$ Pa, saturation temperature $t_s = 120.2°C$, $\lambda = 0.686$ W/m.°C.

Pr = 1.47. From Table 5.1 we find :

$l_* = 14.08 \times 10^{-6}$ m and $l_*/r\rho''v = 22.56 \times 10^{-6}$ m²/W.

The Reynolds number

$$\text{Re}_* = \frac{ql_*}{r\rho''v} = 2 \times 10^5 \times 22.56 \times 10^{-5} = 4.51.$$

Since $\text{Re}_* > 10^{-2}$, calculations are carried out with the aid of equation (5.14a). Substituting the values of Re_* and Pr into that equation, we find that:

$$\text{Nu}_* = 0.125 \ (4.51)^{0.65} \ (1.47)^{1/3} = 0.378.$$

The heat-transfer coefficient

$$\alpha = \text{Nu}_* \frac{\lambda}{l_*} = 0.378 \frac{0.686}{14.08 \times 10^{-6}} = 18400 \text{ W} / \text{m}^2 \cdot {}^\circ\text{C}.$$

Example 5.16. Determine the rate of heat flow through the heating surface of a steam generator (boiler unit) with water in nucleate pool boiling, if the water is under a pressure $p = 6.2 \times 10^5$ Pa and the temperature of the heating surface $t_w = 175^\circ\text{C}$.

Solution. With $p = 6.2 \times 10^5$ Pa $t_s = 160^\circ\text{C}$, Pr = 1.1, $\lambda = 0.683$ W/m. $^\circ\text{C}$; from Table 5.1 we find :

$$\lambda/r\rho''v = 0.526 \ 1/^\circ\text{C}; \ l_* = 1.73 \times 10^{-6} \text{ m}.$$

The temperature difference $\Delta t = t_w - t_s = 175 - 160 = 15^\circ\text{C}$.

Then,

$$\frac{\lambda \Delta t}{r\rho''v} = 0.526 \times 15 = 7.9$$

and

$$\frac{\lambda \Delta t}{r\rho''v} \text{Pr}^{1/3} = 7.9 \ (1.1)^{1/3} = 8.15.$$

Since

$$\frac{\lambda \Delta t}{r\rho''v} \text{Pr}^{1/3} > 1.6,$$

calculation equation (5.15a) is used :

$$\text{Nu}_* = 2.63 \times 10^{-3} \ (7.9)^{1.86} \ (1.1)^{0.952} = 0.134.$$

The heat-transfer coefficient and the rate of heat flow :

$$\alpha = \text{Nu}_* \frac{\lambda}{l_*} = 0.134 \frac{0.683}{1.73 \times 10^{-6}} = 52800 \text{ W} / \text{m}^2 \cdot {}^\circ\text{C}$$

and

$$q = \alpha \Delta t = 52800 \times 15 = 7.9 \times 10^5 \text{ W/m}^2.$$

Example 5.17. Determine the critical heat flux with water in pool boiling at a pressure $p = 1 \times 10^5$ Pa.

Solution. With water in pool boiling the critical heat flux can be calculated with the aid of equation (5.16).

$$\text{Re}_{*cr} = 68 \text{Ar}_*^{4/9} \ \text{Pr}^{-1/3}, \qquad \qquad ...(5.16)$$

where

$$\mathrm{Re}_{*cr} = \frac{q_{cr}l_*}{r\rho''v};$$

$$\mathrm{Ar}_* = g\frac{l_*^3}{v^2}\frac{\rho'-\rho''}{\rho'}.$$

The notations of all quantities are the same as in equations (5.14) and (5.15). The equation is applicable at $0.86 \leqslant \mathrm{Pr} \leqslant 13.1$ and pressures falling within $1 \times 10^5 \leqslant p \leqslant 185 \times 10^5$ Pa.

In the case considered

with $p = 1 \times 10^5$ Pa, $t_s = 99.6°C$, $v = 0.296 \times 10^{-6}$ m^2/s, Pr = 1.76, ρ' = 960 kg/m^3, $\rho'' = 0.59$ kg/m^3.

From Table 5.1 we find :

$$l_* = 50.6 \times 10^{-6} \text{ m}; \quad \frac{l_*}{r\rho''v} = 130 \times 10^{-6} \text{ m}^2/\text{W}.$$

The Archimedes number

$$\mathrm{Ar}_* = g\frac{l_*^3}{v^2}\frac{\rho'-\rho''}{\rho'} = 9.81\frac{(5.06\times10^{-5})^3}{(2.96\times10^{-3})^2}\frac{960-0.59}{960} = 14.4.$$

With the aid of equation (5.16) we find :

$$\mathrm{Re}_{*cr} = 68\mathrm{Ar}_*^{4/9}\,\mathrm{Pr}^{-1/3} = 68\,(14.4)^{4/9}\,(1.76)^{-1/3} = 184$$

and

$$q_{cr} = \mathrm{Re}_{*cr}\frac{r\rho''v}{l_*} = 184\frac{1}{130\times10^{-6}} = 1.41\times10^6 \text{ W}/\text{m}^2.$$

PROBLEMS

Problem 5.1. Water flows with a velocity $w = 0.2$ m/s in a tube of diameter $d = 4$ mm and length $l = 200$ mm. The temperature of the tube wall $t_w = 70°C$. What will the temperature at the tube outlet be, if water inlet temperature $t_{f\,1} = 10°C$?

Answer. $t_{f\,2} = 27$ °C.

Problem 5.2. Oil, of the MK grade, flows in a tube with a diameter $d = 10$ mm. Oil temperature at the tube inlet $t_{f\,1} = 80$ °C; the rate of oil flow $G = 120$ kg/h. What should the length of the tube be, if it is desired that at the tube wall temperature t_w = 30°C the temperature of oil at the tube outlet t_{f2} be equal to 76°C?

Answer. l = 1.66 m.

Problem 5.3. With liquid in turbulent flow in a tube how will change the heat-transfer coefficient, if the velocity of the liquid increases 2 and 4 times, while the diameter of the tube and the mean temperatures of the liquid and wall remain unchanged?

Answer. The heat-transfer coefficient will increase respectively 1.74 and 3.04 times.

Problem 5.4. Under the conditions of example 5.4, in order to reduce the heat losses, the shell of the heat exchanger is covered with a layer of a heat insulating material.

Find the amount of heat, q, W/m², lost from the surface of the heat exchanger, if after a 50 mm thick layer of heat insulation was applied, the temperature of the outside surface of the heat exchanger got equal to t_w = 50°C, and the ambient air temperature remained as above, i.e. t_f = 30°C.

Answer. q = 65 W/m².

Problem 5.5. In a boiler house there are two horizontal steam mains with diameters d_1 = 50 mm and d_2 = 150 mm. The two steam mains have the same surface temperature t_w = 450°C. The ambient air temperature t_f = 50°C. The steam mains are laid at a distance from each other, precluding any mutual heat effect.

Find the ratios of the heat-transfer coefficients α_1/α_2 and of the losses of heat from length of 1 m of steam mains q_{l1}/q_{l2}.

Answer. α_1/α_2 = 1.315; q_{l1}/q_{l2} = 0.438.

Problem 5.6. Determine the coefficient of heat transfer from the outside surface of the tube of the evaporator, considered in example 5.14, under the conditon that the rate of heat flow q = 3 × 10⁵ and 4 × 10⁵ W/m², all other conditions remaining unchanged.

Answer. With q = 3 × 10⁵ W/m² α = 24200 W/m².°C;

with q = 4 × 10⁵ W/m² α = 29000 W/m².°C.

Problem 5.7. Determine the heat-transfer coefficient and the temperature of the inner surface of a tube having a diameter d = 38 mm and carrying boiling water, assuming a rate of heat flow q = 2 × 10⁵ W/m², water pressure p = 2.8 MPa and velocity w = 1 m/s.

Answer. α = 29000 W/m².°C; t_w = 237°C.

CHAPTER 6

Unsteady-state Conduction (Transient Conduction)

INTRODUCTION

Before steady-state conditions can be reached in a process, some time must elapse after the heat-transfer process is initiated to allow the unsteady-state conditions to disappear. For example, we determined the heat flux through a wall at steady-state. We did not consider the period during which when one side of the wall was being heated up and the temperatures were increasing.

Unsteady-state heat transfer is important because of the large number of heating and cooling problems occurring industrially. In metallurgical processes it is necessary to predict cooling and heating rates of various geometries of metals in order to predict the time required to reach certain temperatures. In food processing, such as in the canning industry, perishable canned foods are heated by immersion in steam baths or chilled by immersion in cold water. In the paper industry wood logs are immersed in steam baths before processing. In most of these processes the material is suddenly immersed into a fluid of higher or lower temperature.

This chapter is devoted to the heat transfer by conductions in the absence of inner heat sources with temperature of the system considered varying not only from point to point, but also with time. The processes of conduction where temperature varies in time as well as in space, are called unsteady, non-stationary, or transient. Transient conduction takes place in the heating (or cooling) of various blanks and articles, glass manufacture, brick burning, vulcanisation of rubber, and during starting and stopping of various heat-exchangers, power installations, etc.

Among the practical problems of transient conduction, two groups of processes are of particular importance: (i) when the body tends to thermal equilibrium, and (ii) when the temperature of the body is subject to periodic variation.

The first group includes processes involving the heating or cooling of bodies placed in a medium of given thermal state, for example, the heating of an ingot in a furnace, the cooling of bars and pigs in steelworks, the cooling of hardened parts, etc.

The second group embraces processes developing in periodically functioning heaters, for instance, the heat process of regenerators whose packing are periodically and alternately heated by flue gases and cooled by air.

The nature of the curves obtained in heating a homogeneous solid in a medium of constant temperature, t_f is shown in Fig. 6.1.

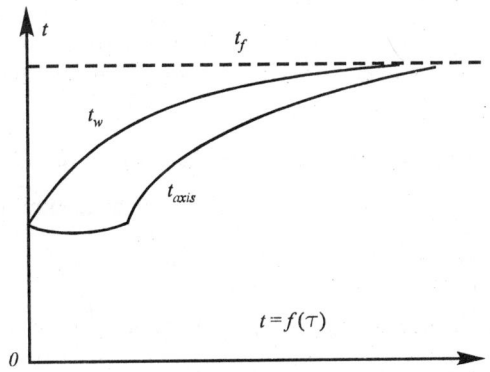

Fig. 6.1. Time variation of temperature within a body.

As the body heats, the temperature at each point asymptotically approaches the temperature of the medium. The temperatures of points near the surface of the body change most rapidly. The differences in the rate of heating of individual points diminish with time and theoretically become zero with a sufficiently large interval of time.

When heat is transferred through a wall and the temperature of one of the heat-carrying agents suddenly changes, all the heat involved will not be transferred through it, and a fraction will be consumed for the change in the internal energy of the wall itself (i.e. the change in its temperature); all the heat will be transferred through the separating wall from one fluid to the other only after steady-state conduction is established.

The examples given indicate that transient thermal processes are always linked with a change in the inner energy, or enthalpy, of the substance concerned.

This chapter will treat only some of the most important problems associated with processes in which the body tends to thermal equilibrium, with the aim of showing the general physical peculiarities of the process of this kind, giving familiarity with the method of solving problems in transient conduction and obtaining mathematical relationships for practical calculations.

DERIVATION OF UNSTEADY-STATE CONDUCTION EQUATION

To derive the equation for unsteady-state condition in one direction in a solid, we refer to Fig. 6.2. Heat is being conducted in the x direction in the cube Δx, Δy, Δz in size. For conduction in the x direction, we write

$$q_x = -kA\frac{\partial T}{\partial x} \qquad \text{... (6.1)}$$

The term $\partial T/\partial x$ means the partial or derivative of T with respect to x with the other variables, y, z, and time t, being held constant. Next, making a heat balance on the cube, we can write

rate of heat input + rate of heat generation = rate of heat output

$$\text{+ rate of heat accumulation} \qquad \text{... (6.2)}$$

The rate of heat input to the cube is

$$\text{rate of heat input} = q_{x|x+\Delta x} = -k(\Delta y \Delta z)\frac{\partial T}{\partial x}\Big|_x \qquad \text{... (6.3)}$$

Also,

$$\text{rate of heat input} = q_{x|x+\Delta x} = -k(\Delta y \Delta z)\frac{\partial T}{\partial x}\Big|_{x+\Delta x} \qquad \text{... (6.4)}$$

The rate of accumulation of heat in the volume $\Delta x, \Delta y, \Delta z$ in time ∂t is

$$\text{rate of heat accumulation} = (\Delta x\, \Delta y\, \Delta z)\rho c_p \frac{\partial T}{\partial t} \qquad \text{... (6.5)}$$

The rate of heat generation in volume $\Delta x, \Delta y, \Delta z$ is

$$\text{rate of heat generation} = (\Delta x\, \Delta y\, \Delta z)\rho c_{\dot q} \qquad \text{... (6.6)}$$

Substituting Eqs. (6.3)–(6.6) into (6.2) and dividing by $\Delta x\, \Delta y\, \Delta z$,

$$\dot q + \frac{-k\left(\frac{\partial T}{\partial x}\Big|_x - \frac{\partial T}{\partial x}\Big|_{x+\Delta x}\right)}{\Delta x} = \rho c_p \frac{\partial T}{\partial t} \qquad \text{... (6.7)}$$

Letting Δx approach zero, we have the second partial of T with respect to x or $\partial^2 T/\partial x^2$ on the left side. Then, rearranging,

$$\frac{\partial T}{\partial t} = \frac{k}{\rho c_p}\frac{\partial^2 T}{\partial x^2} + \frac{\dot{q}}{\rho c_p} = \alpha\frac{\partial^2 T}{\partial x^2} + \frac{\dot{q}}{\rho c_p} \qquad \dots (6.8)$$

where α is $k/\rho c_p$, thermal diffusivity. This derivation assumes constant k, ρ, and c_p. In SI units, α = m^2/s, T = K, t = s, k = W/m.K, ρ = kg/m^3, and c_p = J/kg.K. In English units, α = ft^2/h, T = °F, t = h, k = btu/h.ft. °F, ρ = lbm/ft^3, $\dot{q}$ = btu/h.ft^3, and c_p = btu/lb$_m$.°F.

For conduction in three dimensions, a similar derivation gives

$$\frac{\partial T}{\partial t} = \alpha\left(\frac{\partial^2 T}{\partial x^2} + \frac{\partial^2 T}{\partial y^2} + \frac{\partial^2 T}{\partial z^2}\right) + \frac{\dot{q}}{\rho c_p} \qquad \dots (6.9)$$

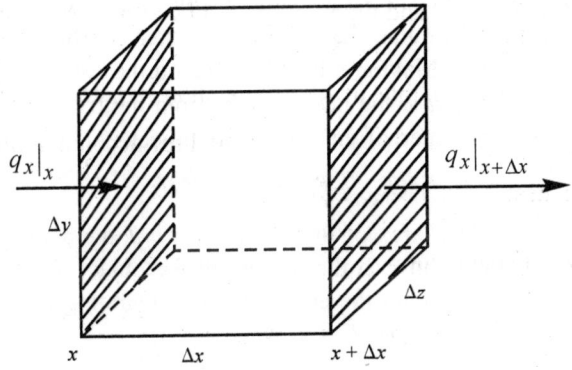

Fig. 6.2. Unsteady-state conduction in one direction.

In many cases, unsteady-state heat conduction is occurring but the rate of heat generation is zero. Then Eqs. (6.8) and (6.9) became

$$\frac{\partial T}{\partial t} = \alpha\frac{\partial^2 T}{\partial x^2} \qquad \dots (6.10)$$

$$\frac{\partial T}{\partial t} = \alpha\left(\frac{\partial^2 T}{\partial x^2} + \frac{\partial^2 T}{\partial y^2} + \frac{\partial^2 T}{\partial z^2}\right) \qquad \dots (6.11)$$

Equations (6.10) and (6.11) relate the temperature T with position x, y, and z and time t. The solutions of Eqs. (6.10) and (6.11) for certain specific cases as well as for the more general cases are considered in much of the remainder of this chapter.

SIMPLIFIED CASE FOR SYSTEMS WITH NEGLIGIBLE INTERNAL RESISTANCE

Basic Equation

We begin our treatment of transient heat conduction by analysing a simplified case. In this situation we consider a solid which has a very high thermal conductivity or very low internal conductive resistance compared to the external surface resistance, where convection occurs from the external fluid to the surface of the solid. Since the internal resistance is very small, the temperature within the solid is essentially uniform at any given time.

An example would be a small, hot cube of steel at T_o K at time $t = 0$, suddenly immersed into a large bath of cold water at T_α which is held constant with time. Assume that the heat-transfer coefficient h in W/m^2. K is constant with time. Making a heat balance on the solid object for a small time interval of time dt s, the heat transfer from the bath to the object must equal the change in internal energy of the object.

$$hA(T_\alpha - T) \, dt = c_p \rho V dT \qquad \qquad ... \text{ (6.12)}$$

where A is the surface area of the object in m^2, T the average temperature of the object at time t in s, ρ the density of the object in kg/m^3, and V the volume in m^3. Rearranging the equation and integrating between the limits of $T = T_0$ when $t = 0$ and $T = T$ when $t = t$,

$$\int_{T=T_0}^{T=T} \frac{dT}{T_\alpha - T} = \frac{hA}{c_p \rho V} \int_{t=0}^{t=t} dt \qquad \qquad ... \text{ (6.13)}$$

$$\frac{T - T_\alpha}{T_0 - T_\alpha} = e^{-(hA/c_p \rho V)t} \qquad \qquad ... \text{ (6.14)}$$

This equation describes the time-temperature history of the solid object. The term $c_p \rho V$ is often called the lumped thermal capacitance of the system. This type of analysis is often called the lumped capacity method or Newtonian heating or cooling method.

Equation for Different Geometries

In using Eq. (6.14) the surface/volume ratio of the object must be known. The basic assumption of negligible internal resistance was made in the derivation. This assumption it reasonably accurate when

$$N_{Bi} = \frac{hx_1}{k} < 0.1 \qquad \qquad ... \text{ (6.15)}$$

where hx_1/k is called the Biot number N_{Bi}, which is dimensionless, and x_1 is a characteristic dimension of the body obtained from $x_1 = V/A$. The Biot number compares the relative values of internal conduction resistance and surface resistance to heat transfer.

For a sphere,

$$x_1 = \frac{V}{A} = \frac{4\pi r^3 / 3}{4\pi r^2} = \frac{r}{3} \qquad \ldots (6.16)$$

For a long cylinder,

$$x_1 = \frac{V}{A} = \frac{\pi D^2 L / 4}{\pi DL} = \frac{D}{4} = \frac{r}{2} \qquad \ldots (6.17)$$

For a long square rod,

$$x_1 = \frac{V}{A} = \frac{(2x)^2 L}{4(2x)L} = \frac{x}{2} \quad (x = 1/2 \ \text{thickness}) \qquad \ldots (6.18)$$

Solved example

Example 6.1. Cooling of a Steel Ball.

A steel ball having a radius of 1.0 in. (25.4 mm) is at a uniform temperature of 800°F (699.9 K). It is suddenly plunged into a medium whose temperature is held constant at 250°F (394.3 K). Assuming a convective coefficient of $h = 2.0$ btu/h.ft^2.°F (11.36 W/m^2.K), calculate the temperature of the ball after 1 h (3600 s). The average physical properties are $k = 25$ btu/h.ft.°F (43.3 W/m. K), $\rho = 490$ lb$_m$/ft^3 (7849 kg/m^3), and $c_p = 0.11$ btu/lb$_m$ °F (0.4606 kJ/kg/ K). Use SI and English units.

Solution. For a sphere from Eq. (6.16),

$$x_1 = \frac{V}{A} = \frac{r}{3} = \frac{\frac{1}{12}}{3} = \frac{1}{36} \text{ft}$$

$$= \frac{25.4}{1000 \times 3} = 8.47 \times 10^{-3} \text{m}$$

From Eq. (6.15) for the Biot number,

$$N_{Bi} = \frac{hx_1}{k} = \frac{2\left(\frac{1}{36}\right)}{25} = 0.00222$$

$$N_{Bi} = \frac{11.36(8.47 \times 10^{-3})}{43.3} = 0.00222$$

This value is < 0.1; hence the lumped capacity method can be used. Then,

$$\frac{hA}{c_p \rho V} = \frac{2}{0.11(490)\left(\frac{1}{36}\right)} = 1.335 \text{ h}^{-1}$$

$$\frac{hA}{c_p \rho V} = \frac{11.36}{(0.4606 \times 1000)(7849)(8.47 \times 10^{-3})} = 3.71 \times 10^{-4} \text{ s}^{-1} (1.335 \text{h}^{-1})$$

Substituting into Eq. (6.14) for $t = 1.0$ h and solving for T.

$$\frac{T - T_x}{T_0 - T_x} = \frac{T - 250° F}{800 - 250} = e^{-(hA/c_p\rho V)t} = e^{-(1.335)(1.0)} \qquad T = 395° F$$

$$\frac{T - 394.3K}{699.9 - 394.3} = e^{-(3.71 \times 10^{-4})(3600)} \qquad T = 474.9 \ K$$

The temperature of the solid at any time t can be calculated from Eq. (6.14). At any time t, the instantaneous rate of heat transfer $q(t)$ in W from the solid of negligible internal resistance can be calculated from

$$q(t) = hA \ (T - T_\infty) \qquad \text{... (6.19)}$$

Substituting the instantaneous temperature T from Eq. (6.14) into Eq. (6.19)

$$q(t) = hA \ (T_0 - T_\infty)e^{-(hA/c_p\rho V)t} \qquad \text{... (6.20)}$$

To determine the total amount of heat Q in W.s or J transferred from the solid from time $t = 0$ to $t = t$, we can integrate Eq. (6.20).

$$Q = \int_{t=0}^{t=t} q(t)dt = \int_{t=0}^{t=t} hA(t_0 - T_\infty)e^{-(hA/c_p\rho V)t} \qquad \text{... (6.21)}$$

$$Q = c_p \rho V(T_0 - T_\infty) \ [1 - e^{-(hA/c_p\rho V)t}] \qquad \text{... (6.21)}$$

UNSTEADY-STATE HEAT CONDUCTION IN VARIOUS GEOMETRIES

Introduction and Analytical Methods

We have already considered a simplified case of negligible internal resistance where the object has a very high thermal conductivity. Now we will consider the more general situation where the internal resistance is not small, and hence the temperature is not constant in the solid. The first case that we shall consider is one where the surface convective resistance is negligible compared to the internal resistance. This could occur because of a very large heat-transfer coefficient at the surface or because of a relatively large conductive resistance in the object.

To illustrate an analytical method of solving this first case, we will derive the equation for unsteady-state conduction in the x direction only in a flat plate of thickness $2H$ as shown in Fig. 6.3. The initial profile of the temperature in the plate at $t=0$ is uniform at $T = To$. At time $t=0$, the ambient temperature is suddenly changed to T_1 and held there. Since there is no convection resistance, the temperature of the surface is also held constant at T_1. Since this is conduction in the x direction, Eq. (6.21) holds.

$$\frac{\partial T}{\partial t} = \alpha \frac{\partial^2 T}{\partial x^2} \qquad \dots (6.23)$$

The initial and boundary conditions are :

$$T = T_0, \qquad t = 0, \qquad x = x$$
$$T = T_1, \qquad t = t, \qquad x = 0$$
$$T = T_1, \qquad t = t, \qquad x = 2H \qquad \dots (6.24)$$

Generally, it is convenient to define a dimensionless temperature Y so that it varies between 0 and 1. Hence,

$$Y = \frac{T_1 - T}{T_1 - T_o} \qquad \dots (6.25)$$

Substituting Eq. (6.25) into (6.23),

$$\frac{\partial Y}{\partial t} = \alpha \frac{\partial^2 Y}{\partial x^2} \qquad \dots (6.26)$$

Redefining the boundary and initial conditions,

$$Y = \frac{T_1 - T_o}{T_1 - T_o} = 1, \qquad t = 0, \qquad x = x$$

$$Y = \frac{T_1 - T_o}{T_1 - T_o} = 0, \quad t = t, \quad x = 0 \qquad \qquad \dots \text{(6.27)}$$

$$Y = \frac{T_1 - T_o}{T_1 - T_o} = 0, \quad t = t, \quad x = 2H$$

A convenient procedure to use to solve Eq. (6.26) is the method of separation of variables, which leads to a product solution

$$Y = e^{-a^2 \alpha t}(A \cos ax + B \sin ax) \qquad \dots \text{(6.28)}$$

where A and B are constants and a is a parameter. Applying the boundary and initial conditions of Eq. (6.27) to solve for these constants in Eq. (6.28), the final solution is an infinite Fourier series.

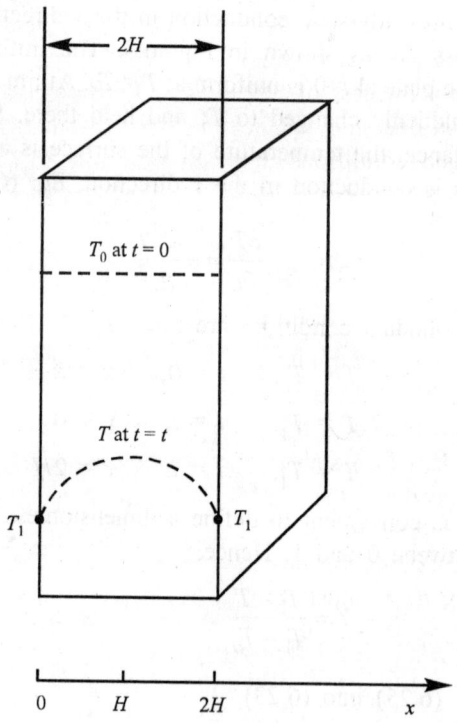

Fig.6.3. Unsteady-state conduction in a flat plate with negligible surface resistance.

$$\frac{T_1 - T}{T_1 - T_o} = \frac{4}{\pi}\left(\frac{1}{1}\exp\frac{-1^2\pi^2\alpha t}{4H^2}\sin\frac{1\pi x}{2H} + \frac{1}{3}\exp\frac{-3^2\pi^2\alpha t}{4H^2}\sin\frac{3\pi x}{2H}\right.$$

$$\left. + \frac{1}{5}\exp\frac{-5^2\pi^2\alpha t}{4H^2}\sin\frac{5\pi x}{2H} + ...\right) \qquad ... (6.29)$$

Hence from Eq. (6.29), the temperature T at any position x and time t can be determined. However, these types of equations are very time consuming to use, and convenient charts have been prepared.

INFINITE FLAT PLATE OR SLAB

The determination of the temperature variation through a relatively thin sheet or slab, through which the heat flow is substantially one-dimensional, is a problem frequently encountered. Solutions have been obtained for several different initial and boundary conditions. Possibly the most useful and illustrative of these is the case of a plate at a uniform temperature, suddenly heated or cooled on one side and insulated on the other.

These prescribed conditions are shown in Fig. 6.4a. The plate is considered to be effectively infinite in the y and z directions; its thermal conductivity k and diffusivity α will be taken as constant and uniform. The initial uniform temperature throughout the plate is designated by $t = t_i$. The face at $x = l$ is suddenly ($\tau = 0$) subjected to heating or cooling by a fluid at t_f flowing over it; an average and uniform value of the heat transfer coefficient h between the plate and the fluid is considered to apply. Insulating the face $x = 0$ is equivalent to specifying that $\partial t/\partial x = 0$ at that location. Thus it may be noted that the problem under study is the same as considering either half of a slab of the same material $2l$ in thickness and simultaneously subjected to the same heating or cooling conditions on both sides (Fig. 6.4b).

Under these conditions the heat flow will be one-dimensional and the temperature history $t(x,\tau)$ must satisfy

$$\frac{\partial^2 t}{\partial x^2} = \frac{1}{\alpha}\frac{\partial t}{\partial \tau} \qquad ... (6.30)$$

In addition the following initial and boundary conditions must be met:

(a) $t - t_f = t_i - t_f$ at $\tau = 0$

(b) $\dfrac{\partial(t - t_f)}{\partial x} = 0$ at $x = 0$ ...(6.31)

(c) $-k\dfrac{\partial(t - t_f)}{\partial x} = h(t - t_f)$ at $x = l$

From the single boundary condition (c) it is seen that, in the case of heating, as shown in Fig. 6.4 ($t_f > t_i$), $\partial t/\partial x$ at $x = l$ is positive; the negative sign then means that heat flow is to the left. In the case of cooling ($t_f < t_i$), $\partial t/\partial x$ at $x = l$ is, of course, negative and heat flow is positive to the right.

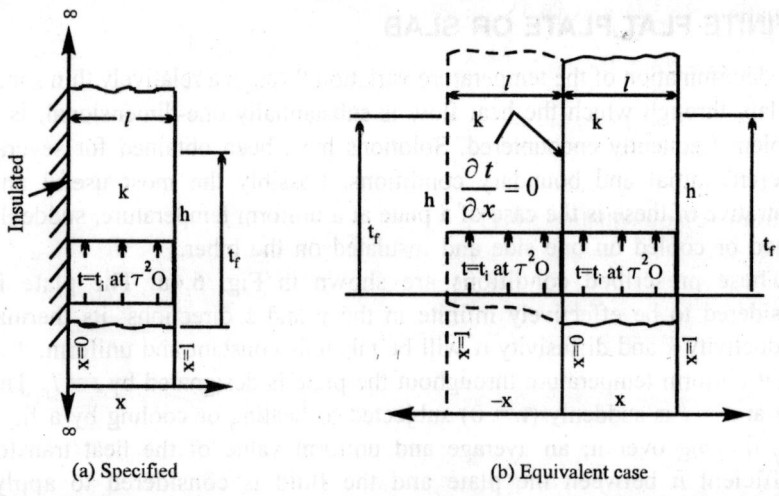

(a) Specified   (b) Equivalent case

Fig. 6.4. Initial and boundary conditions for transient heating of an infinite plate.

Using the separation of variables technique, we assume $t - t_f = XT$, where X is a function of x only, and T is a function of τ only. Substitution in equation (6.30) yields the two ordinary differential equations

$$\frac{d^2 X}{X dx^2} = \frac{1}{\alpha T}\frac{dT}{d\tau} = -\lambda^2$$

or

$$\frac{d^2 X}{dx^2} + \lambda^2 X = 0 \qquad \dots (6.32)$$

and

$$\frac{dT}{d\tau} + \alpha\lambda^2 T = 0 \qquad \dots (6.33)$$

The product of the solutions of equations (6.32) and (6.33) can, for this case, be most conveniently expressed as

$$t - t_f = e^{-\lambda^2 \alpha \tau} (C_1 \cos\lambda x + C_2 \sin\lambda x) \qquad \ldots (6.34)$$

Differentiating with respect to x gives

$$\frac{\partial(t - t_f)}{\partial x} = e^{-\lambda^2 \alpha \tau}(C_2 \lambda \cos \lambda x - C_1 \lambda \sin \lambda x) \qquad \ldots (6.35)$$

We see that in order to satisfy (b) of equation (6.31), $\partial(t - t_f)/\partial x = 0$ at $x = 0$, C_2 must be zero.

Equations (6.34) and (6.35) therefore become

$$t - t_f = e^{-\lambda^2 \alpha \tau} C_1 \cos\lambda x \qquad \ldots (6.36)$$

$$\frac{\partial(t - t_f)}{\partial x} = -e^{-\lambda^2 \alpha \tau} C_1 \lambda \sin \lambda x \qquad \ldots (6.37)$$

Now from (c) of equation (6.31)

$$\left(\frac{\partial(t - t_f)}{\partial x}\right)_{x=l} = \frac{h}{k}(t_f - t_{x=l})$$

or
$$- e^{-\lambda^2 \alpha \tau} C_1 \lambda \sin \lambda l = -\frac{h}{k} e^{-\lambda^2 \alpha \tau} C_1 \cos \lambda l \qquad \ldots (6.38)$$

$$\cot \lambda l = \frac{k}{h}\lambda \qquad \ldots (6.39)$$

Equation (6.39) is a transcendental equation from which λ can be determined. It may be noted that the number of possible solutions is infinite.

Although these cannot be obtained by ordinary algebraic methods, a simple graphical determination is possible. This is indicated in Fig. 6.5. Using λ as the independent variable, first curves of $\cot\lambda l$ and then the curve of $k\lambda/h$ are plotted; there is an infinite set of intersections of these curves corresponding to the roots λ_1, λ_2, λ_3, ... Each value for k/h will give a different set of roots.

The individual roots of equation (6.39) are called eigenvalues (from the German *eigenwerte*). They may be thought of as special values which are applicable only to the solution of a specific problem. In our case they are needed to satisfy (a) of equation (6.31). In order to meet this requirement, consider $(t - t_f)$ equal to an infinite series of terms

$$t - t_f = C_1 e^{-\lambda_1^2 \alpha r} \cos\lambda_1 x + C_2 e^{-\lambda_2^2 \alpha r} \cos\lambda_2 x + C_3 e^{-\lambda_3^2 \alpha r} \cos\lambda_3 x$$

$$+ \ldots + C_n e^{-\lambda_n^2 \alpha r} \cos\lambda_n x$$

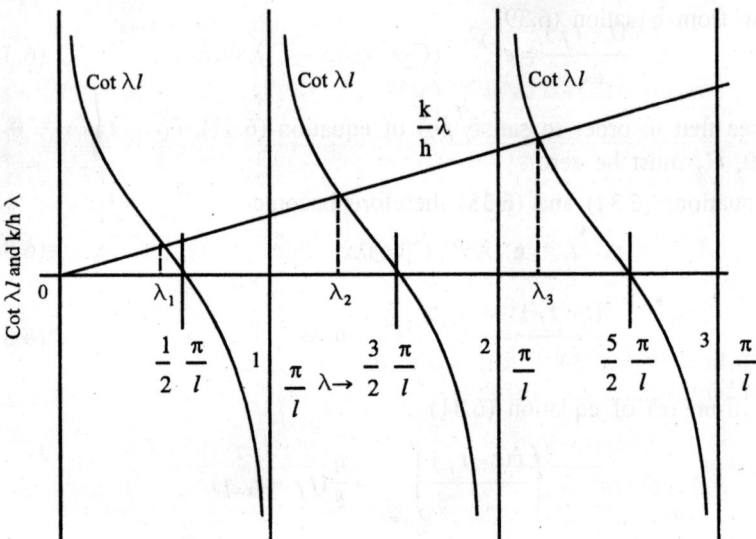

Fig. 6.5. Graphical solution for the roots of equation (6.39).

or

$$t - t_f = \sum_{n=1}^{\infty} C_n e^{-\lambda_n^2 \alpha \tau} \cos\lambda_n x \qquad \ldots (6.40)$$

in which $\lambda_1, \lambda_2, \lambda_3, \ldots, \lambda_n$ have the values determined from equation (6.39) and $C_1, C_2, C_3, \ldots, C_n$ are selected to satisfy the condition that $t - t_f = t_i - t_f$ at $\tau = 0$. The constants C_n can be determined in this case by multiplying both sides of equation (6.40) by $\cos\lambda_m x dx$, where $m \neq n$, and integrating over the interval in which $t_i - t_f$ exists, i.e.,

$$(t_i - t_f) \int_0^l \cos\lambda_m x dx = \sum_{n=1}^{\infty} C_n \int_0^l \cos\lambda_m x \cos\lambda_n x dx \qquad \ldots (6.41)$$

The integral on the left gives

$$\int_0^l \cos\lambda_m x dx = \frac{1}{\lambda_m} \sin\lambda_m l \qquad \ldots (6.42)$$

The integral on the right side leads to

$$\int_0^l \cos\lambda_m x \cos\lambda_n x\, dx = \frac{\lambda_m \sin \lambda_m l \cos \lambda_n l - \lambda_n \cos \lambda_m l \sin \lambda_n l}{(\lambda_m^2 - \lambda_n^2)} \quad ... \quad (6.43)$$

Now from equation (6.39)

$$\lambda_m \sin \lambda_m l = \frac{h}{k} \cos \lambda_m l$$

and

$$\lambda_n \sin \lambda_n l = \frac{h}{k} \cos \lambda_n l$$

so that

$$\lambda_m \sin\lambda_m l \, \cos\lambda_n l = \lambda_n \, \cos\lambda_m l \, \sin\lambda_n l \qquad ... \quad (6.44)$$

From this we see that if $m \neq n$, the integral of the right side vanishes. For each term, therefore, when $m = n$, equation (6.41) yields

$$\frac{(t_i - t_f)}{\lambda_n} \sin \lambda_n l = C_n \left(\frac{l}{2} + \frac{\sin \lambda_{nl} \cos \lambda_n l}{2\lambda_n} \right)$$

and

$$C_n = \frac{2(t_i - t_f)\sin \lambda_n l}{(\lambda_n l + \sin \lambda_n l \cos \lambda_n l)} \qquad ... \quad (6.45)$$

Our final solution is then

$$\frac{t - t_f}{t_i - t_f} = 2\sum_{n=1}^{\infty} \left(\frac{\sin \lambda_n l}{\lambda_n l + \sin \lambda_n l \cos \lambda_n l} \right) e^{-\lambda_n^2 \alpha \tau} \cos \lambda_n x \qquad ... \quad (6.46)$$

It is often necessary to know the rate of heat flow into or out of the plate or the total amount of heat which has been transferred to or from the plate since the beginning of the transient period. The heat rate per unit area at the surface is given by

$$q = -k \left(\frac{\partial(t - t_f)}{\partial x} \right)_{x=l}$$

$$= 2k(t_i - t_f)\sum_{n=1}^{\infty} \left(\frac{\sin \lambda_n l}{\lambda_n l + \sin \lambda_n l \cos \lambda_n l} \right) e^{-\lambda_n^2 \alpha \tau} \lambda_n \sin \lambda_n l \qquad ... \quad (6.47)$$

and the heat transfer per unit area by

$$Q = \int_0^\tau q\,d\tau = 2k(t_i - t_f)\int_0^\tau \sum_{n=1}^{\infty}\left(\frac{\lambda_n \sin^2 \lambda_n l}{\lambda_n l + \sin \lambda_n l \cos \lambda_n l}\right)e^{-\lambda_n^2 \alpha \tau}$$

$$Q = \frac{t_i - t_f}{\alpha} = (2k)\sum_{n=1}^{\infty}\left(\frac{\sin^2 \lambda_n l}{\lambda_n^2 l + \lambda_n \sin \lambda_n l \cos \lambda_n l}\right)(1 - e^{-\lambda_n^2 \alpha \tau}) \quad ...(6.48)$$

In the case of a plate of thickness $2l$ heated or cooled on both sides, equation (6.48) must be multiplied by 2.

CALCULATIONS OF THE TEMPERATURE AND HEAT FLOW IN AN INFINITE PLATE

The temperature variation with time and location in an infinite plate, satisfying the conditions (a), (b), and (c) of equation (6.31), can be determined by evaluating an appropriate number of terms of equation (6.46). Similarly, heat rates and heat flow can be obtained from equations (6.47) and (6.48). Fortunately the series converge rather rapidly and satisfactory accuracy can be obtained by considering only five terms (except for very small values of the time). The resulting solutions, however, would apply only for the particular plate under consideration and the particular boundary heat transfer coefficient used. A different value of l, k, or h would determine a different set of eigenvalues, which would require re-evaluation of the constants in equations (6.46), (6.47) and (6.48). Obviously a more general approach is desirable. This is possible by noting that these three equations may be regarded as functions of three dimensionless groups or variables. If solutions over wide ranges of these dimensionless groups are calculated and presented in chart form, the answers to a specific problem can then be determined simply and quickly from an appropriate chart.

Consider, for example, equation (6.46); by letting $\lambda_n l = N_n$,

$$\frac{t - t_f}{t_i - t_f} = 2\sum_{n=1}^{\infty}\left(\frac{\sin N_n}{N_n + \sin N_n \cos N_n}\right)e^{-N_n^2(\alpha\tau/l^2)}\cos N_n \frac{x}{l} \quad ... (6.49)$$

The temperature ratio on the left is seen to be a function of N_n, $\alpha\tau/l^2$, and the dimensionless position or location ratio, $S = x/l$. N_n and $\alpha\tau/l^2$ are also dimensionless. The group, $\alpha\tau/l^2 = N_{Fo}$, composed of the product of the thermal diffusivity and the time divided by the plate thickness squared, is called the Fourier number or modulus. Its magnitude is a measure of the degree heating or cooling effects have penetrated through the body; if

α/l^2 is small, a large value of τ is required before significant temperature changes through the body occur; if α/l^2 is large, the reverse is, of course, true. N_n is a dimensionless expression for λ_n and is determined from equation (6.39). If we write

$$\cot \lambda_n l = \cot N_n = \frac{k}{h}\lambda_n = \frac{\lambda_n l}{hl/k} = \frac{N_n}{hl/k}$$

or

$$\frac{\cot N_n}{N_n} = \frac{1}{hl/k} = \frac{1}{N_{Bi}} \qquad \qquad \dots (6.50)$$

it is apparent that values of N_n are determined by the reciprocal of $N_{Bi} = hl/k$. This group is also dimensionless and is usually referred to as the Biot number or modulus. It is indicative of the resistance to heat transfer at the surface of a body compared to its internal resistance.

In view of the dependence of N_n on N_{Bi}, $(t - t_f)/(t_i - t_f)$ of equation (6.49) may be considered as a function of the Fourier number, the Biot number, and the position ratio. i.e.,

$$\frac{t - t_f}{t_i - t_f} = P\left(\frac{\alpha\tau}{l^2}, \frac{1}{hl/k}, \frac{x}{l}\right) \qquad \qquad \dots (6.51)$$

For the heat flow at the surface $x = l$, equation (6.48) would be

$$\frac{Q}{t_i - t_f} = \frac{k}{\alpha} P_Q\left(\frac{\alpha\tau}{l^2}, \frac{1}{hl/k}, 1\right) \qquad \qquad \dots (6.52)$$

CYLINDER AND SPHERE

The solutions for the transient temperature variation in an infinitely long cylinder and a sphere, both initially at a uniform temperature, t_i, and suddenly exposed at $\tau = 0$ to convective heating from an ambient fluid at t_f, and be obtained by following the same procedure used for the infinite plate. For the cylinder the temperature is symmetrical about the axis, being a function only of τ and the radial distance from the axis, r. The final solution is

$$\frac{t - t_f}{t_i - t_f} = 2 \sum_{n=1}^{\infty} \frac{1}{N_n} \frac{J_1(N_n)}{J_0^2(N_n) + J_1^2(N_n)} e^{-N_n^2(\alpha\tau/r_1^2)} J_0\left(N_n \frac{r}{r_1}\right) \qquad \dots (6.53)$$

where $N_n = \lambda_n r_1$

r_1 = cylinder radius

λ_n = roots of the eigenfunction equation

$$\lambda_n r_1 J_1(\lambda_n r_1)/J_0(\lambda_n r_1) = hr_1/k$$

J_o = Bessel function of the first kind and zero order

J_1 = Bessel function of the first kind and first order

The solution for the sphere is

$$\frac{t-t_f}{t_i-t_f} = 2\sum_{n=1}^{\infty} \frac{\sin N_n - N_n \cos N_n}{N_n - \sin N_n \cos N_n} e^{-N_n^2(\alpha\tau/r_1^2)} \frac{\sin(N_n r/r_1)}{N_n r/r_1} \qquad ...(6.54)$$

where r_1 = sphere radius

$N_n = \lambda_n r_1$

λ_n = roots of the eigenfunction equation

$\lambda_n r_1/(1 - hr_1/k) = \tan\lambda_n r_1$

SEMI-INFINITE SOLID

We shall mention one more type of solid which has considerable practical importance—the semi-infinite solid. The earth usually serves as the best example; often, however, many much less extensive bodies, such as thick plates and walls, can also be considered from this point of view.

The ideal system is indicated in Fig. 6.6 as a solid bounded by the plane $x = 0$ and extending to infinity in the positive x direction. Consider the initial temperature uniform and equal to t_i; at time $\tau = 0$ the surface at $x = 0$ is suddenly exposed to cooling by an ambient fluid at t_f. An average and uniform heat transfer coefficient, h, is assumed to hold for the entire surface. Also, as before, average thermal properties of the solid are used. Since the heat flow is one dimensional, the differential equation to be satisfied is

$$\frac{\partial^2 t}{\partial x^2} = \frac{1}{\alpha}\frac{\partial t}{\partial \tau}$$

The initial and boundary conditions are the same as for the plate, equation (6.31), except that $t = t_i$ at $x = \infty$ replaces $\partial t/\partial x = 0$ at $x = l$.

The reader is referred to more advanced works for the derivation of the solution which is

$$\frac{t-t_f}{t_i-t_f} = \text{erf}\frac{x}{2\sqrt{\alpha\tau}} + e^{(hx/k)+(h^2/k^2)\alpha\tau}\text{erfc}\left\{\frac{x}{2\sqrt{\alpha\tau}} + \frac{h}{k}\sqrt{\alpha\tau}\right\} \qquad ... (6.54)$$

in which erf ξ designates the error function or probability integral, and

$$\text{erfc } \xi = 1 - \text{erf } \xi = 1 - \frac{2}{\sqrt{\pi}} \int_0^{\xi} e^{-\beta^2} d\beta$$

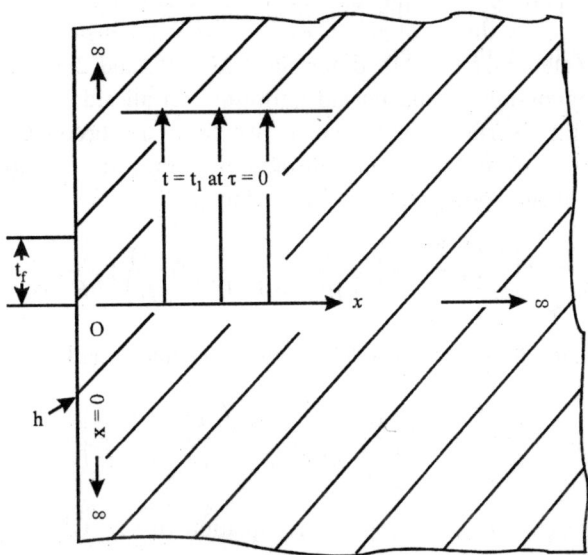

Fig. 6.6. The semi-infinite solid.

Both erf ξ and erfc ξ are available in tables so that temperature calculations using equation (6.55) are easily performed.

Here $Q^* = 1 - (t - t_f)/(t_i - t_f)$ has been plotted against $\log_{10} (h\sqrt{\alpha\tau} / k)$ for the values of $x / 2\sqrt{\alpha\tau}$ from 0 to 1.5 in 0.1 intervals. It is useful to note when $h \rightarrow \infty$ equation (6.55) reduces to

$$\frac{t - t_f}{t_i - t_f} = \text{erf } \frac{x}{2\sqrt{\alpha\tau}} = \frac{2}{\sqrt{\pi}} \int_0^{x/2\sqrt{\alpha\tau}} e^{-\beta^2} d\beta \qquad \dots (6.56)$$

This will be recognised as the case of zero surface resistance or when the surface temperature can be considered as changing instantaneously to t_f at $t = 0$. We shall illustrate the use of equations (6.55) and (6.56) in determining the depth to which the temperature drops to freezing in dry earth following a sudden cold wave.

HEATING AND COOLING OF FINITE BODIES

The solutions presented for the infinite plate and cylinder, the sphere, and the semi-infinite slab can also be used to determine the temperatures in finite solids in which the heat flow is two or three dimensional. Important cases are the finite cylinder and the brick. This results from the fact that the solution of the heat conduction equation for two or three dimensions with constant temperature heating can be considered in most cases as the product of two or three solutions, each of one dimension. Taking as an example a rectangular brick of dimensions $2L$, $2W$, and $2H$, originally at a uniform temperature t_i, and heated uniformly on all sides by a fluid at t_f with an average surface coefficient, h, we form the product of equation (6.51) applied to the three directions to obtain the solution of the temperature at any point, x, y, z. Thus (Fig. 6.7).

$$\frac{t-t_f}{t_i-t_f} = P\left(\frac{\alpha\tau}{L^2},\frac{k}{hL},\frac{x}{L}\right)P\left(\frac{\alpha\tau}{W^2},\frac{k}{hW},\frac{y}{W}\right)P\left(\frac{\alpha\tau}{H^2},\frac{k}{hH},\frac{z}{H}\right) \quad \ldots (6.57)$$

Similarly for the short cylinder of radius r_1 and length L

$$\frac{t-t_f}{t_i-t_f} = C\left(\frac{\alpha\tau}{r_1^2},\frac{k}{hr_1},\frac{r}{r_1}\right)P\left(\frac{\alpha\tau}{L^2},\frac{k}{hL},\frac{x}{L}\right) \quad \ldots (6.58)$$

where $C\,(\alpha\tau/r_1^2,\ k/hr_1, r/r_1)$ represents equation (6.53). Olsen and Schultz outline the product solutions for a large variety of other simple cases. As an illustration we shall calculate the temperature at the centre of a brick which is cooling after being removed from a kiln.

FINITE-DIFFERENCE METHOD OF SCHMIDT

One of the most flexible and revealing methods of solving transient heating problems is a graphical finite-difference method credited to E. Schmidt. Schmidt and others have shown it to be applicable to cylindrical and spherical flow. We have thus far discussed the heating and cooling of bodies of finite internal resistance and initial uniform temperature when subjected to heating by a constant temperature environment. The solution of problems where the surrounding fluid temperature or the body surface temperature varies with time is generally more difficult. The Schmidt method is applicable in both cases and we shall describe it by solving a one-dimensional heat flow problem where the surface temperature varies with time.

Assume that the temperature at the surface of a 1 in. thick SAE 4130 steel plate ($\alpha = 0.25$ ft^2/hr) changes as shown in Fig. 6.8; the plate is initially at a uniform temperature of 80°F. Heat flow will be normal to the surface only and it is desired to determine the temperature throughout the plate as a function of time. The procedure is to divide the plate into small imaginary slabs of thickness Δx parallel to the surface as shown in Fig. 6.9. Temperature changes are determined at the centre of each of these slabs. Since the temperature variation at the surface is known, it is convenient to regard it as the centre of a slab; we may therefore regard the 0 slab as $\Delta x/2$ in thickness or add a fictitious half-slab of the plate material as indicated. A temperature distribution assumed to have been determined for the time $\tau = n\Delta\tau$ after the surface temperature began to change is also drawn in Fig. 6.9 to illustrate the derivation following. Δy is a finite increment of time and n is the number of increments that have elapsed since the beginning of heating. The temperature curve is drawn as a series of straight lines because of convenience and to suggest the finite difference nature of the method. Actually a curve faired through the points determined would be more correct.

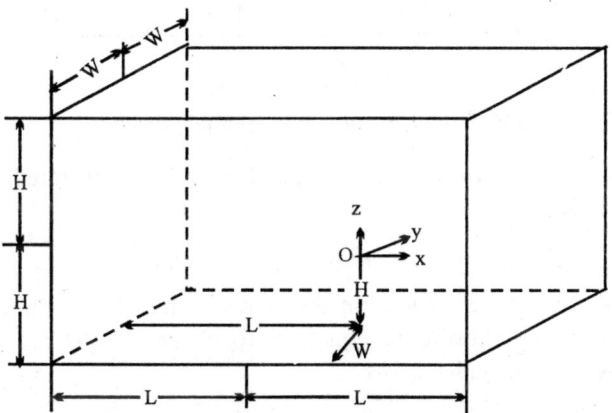

Fig. 6.7. Co-ordinate system and dimensions of brick for equation (6.59).

The equation for calculating the temperature at the centre of any slab, m, at the end of the next time interval (from $n\Delta\tau$ to $(n + 1)\Delta\tau$) will be derived by setting up a heat balance on the slab. Referring to Fig. 6.23 the heat flow per unit area in time $\Delta\tau$ in the left face of slab m can be represented by

$$q_{mL} = -k\frac{(t_{m,n} - t_{m-1,n})}{\Delta x}\Delta\tau \qquad \text{... (6.59)}$$

and out the right face by

$$q_{mR} = -k\frac{(t_{m+1,n} - t_{m,n})}{\Delta x}\Delta\tau \qquad \text{... (6.60)}$$

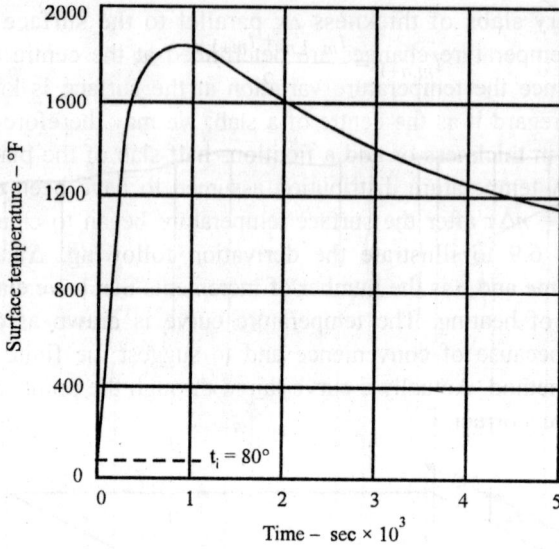

Fig. 6.8. Plate surface temperature history.

The difference between these two quantities is the heat stored in the slab and must equal

$$q_{mL} - q_{mR} = wc\Delta x \, (t_{m,\,n+1} - t_{m,n}) \qquad \text{... (6.61)}$$

where $t_{m,\,n+1}$ is the temperature at the centre of the slab at the end of the time increment $\Delta\tau$. Substituting equations (6.59) and (6.60) into (6.61) and dividing by Δx, $\Delta\tau$, and k gives

$$\left\{ \frac{\dfrac{t_{m+1,n}-t_{m,n}}{\Delta x} - \dfrac{t_{m,n}-t_{m-1,n}}{\Delta x}}{\Delta x} \right\} = \frac{wc}{k}\left(\frac{t_{m,\,n+1}-t_{m,n}}{\Delta\tau} \right) \qquad \text{... (6.61)}$$

It should be recognised that equation (6.62) could also have been obtained by writing the differential equation of heat conduction for one dimension in finite difference form.

Writing α for k/wc and multiplying each side by Δx^2 equation (6.62) becomes

$$t_{m-1,n} + t_{m+1,n} - 2t_{m,n} = \frac{\Delta x^2}{\alpha\Delta\tau}(t_{m,n+1} - t_{m,n}) \qquad \text{... (6.63)}$$

We now observe that by setting

$$\Delta x^2/\alpha\Delta\tau = 2 \qquad \text{... (6.64)}$$

$t_{m,\,n}$ is eliminated and equation (6.63) reduces to

$$t_{m,n+1} = \frac{t_{m-1,n} + t_{m+1,n}}{2} \qquad \dots (6.65)$$

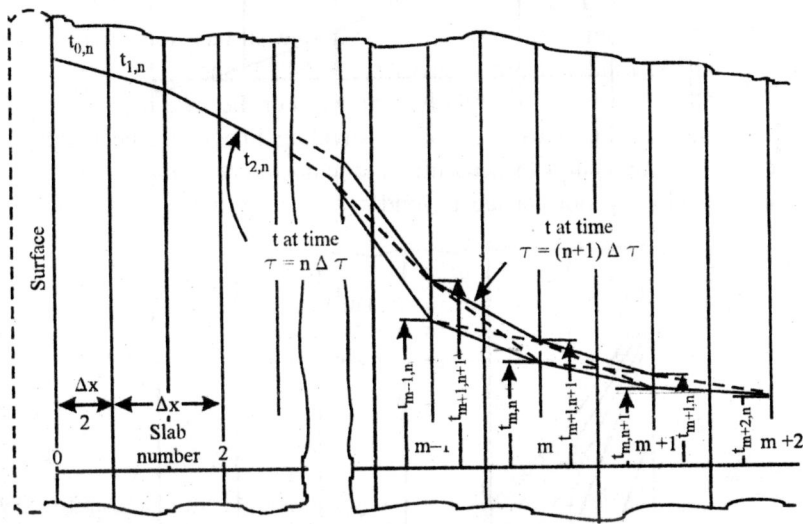

Fig. 6.9. Division of solid into incremental slabs for graphical determination of temperature distribution at instants of time differing by finite increments.

Equation (6.65) tells us that when Δx and Δr are selected so as to satisfy equation (6.64), the temperature at the centre of any slab m at time $(n + 1)\,\Delta \tau$ is the arithmetic mean of the temperatures at the centre of the slab on the left and the slab on the right at time $n\Delta \tau$. This is the key to the Schmidt method.

Considering therefore slab m of Fig. 6.9, we can determine the temperature $t_{m,n+1}$ at the centre at time $(n +1)\,\Delta\tau$ by the simple graphical procedure of drawing a straight line (dashed in the figure) between he temperatures in slabs $m - 1$ and $m + 1$ at time $n\Delta\tau$. Similarly for slab $m - 1$ the temperature at time $(n + 1)\,\Delta\tau$ is determined from the center temperatures of slabs $m - 2$ and m at time $n\Delta\tau$. This process is followed for all slabs except the one at the surface in which the temperature changes in known. In this way the temperature distribution throughout the body at time $(n + 1)\,\Delta\tau$ is determined. The same procedure is then repeated for the time $(n + 2)\Delta\tau$.

In applying this to the 1 in. thick plate, the surface temperature history of which is given in Fig. 6.23, we shall use a time increment of 0.2×10^{-3} sec. For $\Delta x^2/\alpha\Delta\tau = 2$ our slab thickness must equal 0.002 in. The temperature distributions determined after the first 5 time increments are shown in Fig. 6.10. At the end of the first increment $\tau = \Delta\tau$ the surface temperature from Fig. 6.8 has risen to 1200 F. Applying equation (6.66) to slabs 0 and 2 at $\tau = 0$, we find the temperature at the centre of slab 1 still at 80 F $= t_{1,1}$. The same is true for slabs 2,3 etc. At the end of the second time period, $r = 2\Delta\tau$, the surface temperature has risen to 1640 F. Equation (6.65) gives $t_{1,2} = 640$ F and $t_{2,2} = t_{2,3} = \ldots$ 80 F. The other curves were similarly obtained. For the sake of clarity only the centrelines of the slabs and not the dividing lines have been drawn. Dashed construction lines to determine the temperature changes have been omitted and plotting not carried beyond $\tau = 5\Delta\tau$ for the same reason.

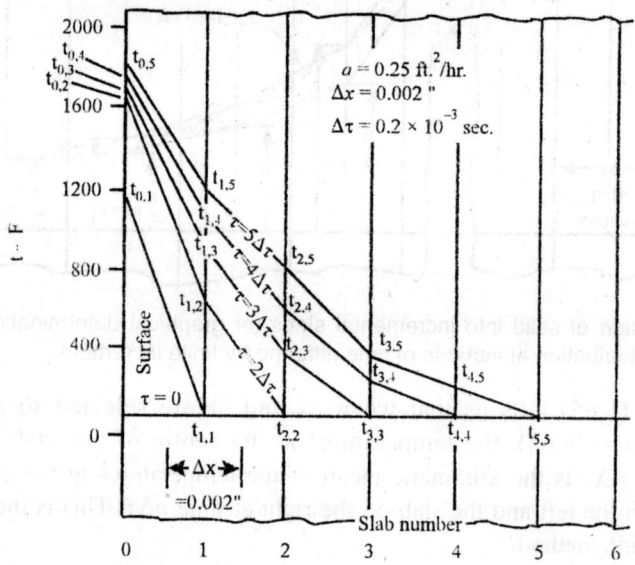

Fig. 6.10. Temperature distribution in a steel plate using Schmidt method.

It may be noted in Fig. 6.10 that, at the end of 1×10^{-3} sec, no temperature change had occurred beyond a distance of 0.01 in. from the heated surface of the plate. In fact, it will be found that only after several hundred time intervals have elapsed will any change occur near the outer surface.

Furthermore, if the surface temperature continues to drop slowly toward its initial value as suggested in Fig. 6.8 (as it will if heating is discontinued), the eventual temperature change near the outer surface will be barely measurable. In such a case the plate can be treated as a semi-infinite solid.

SOLVED EXAMPLES

Example 6.1. A steel plate of thickness $2\delta = 400$ mm is heated in a furnace inside which a constant temperature $t_f = 800°C$ is maintained. When placed into the furnace the plate had a uniform temperature $t_0 = 30°C$. The local coefficient of heat transfer to the surface of the heated plate remains constant and equal to $\alpha = 200$ W/m^2.°C. The two other dimensions of the plate. width and length, are large compared with its thickness and the temperature field of the plate can be considered as unidimensional.

Determine the amount of heat transferred to 1 m^2 of the plate in 2 hours after heating is initiated. The thermal conductivity of steel $\lambda = 37.2$ W/m.°C and the thermal diffusivity $a = 7 \times 10^{-6}$ m^2/s; the density of steel $\rho = 7800$ kg/m^2.

Solution. The calculation of the amount of heat rejected from (or transferred to) the plate in the course of cooling (heating) during a time interval ranging from $\tau = 0$ to τ practically resolves into the finding the mean dimensional temperature at the moment of time τ, i.e. it can be calculated by the following equation :

$$Q = Q_t (1 - \overline{\Theta}),\qquad ...(6.1)$$

where Q_t is the total amount of heat which can be rejected from or transferred to the plate during a time interval ranging from $\tau = 0$ to $\tau = \infty$:

$$Q_t = 2\delta f\rho c \ (t_0 - t_f), \ J,\qquad ...(6.2)$$

where f is the area of one side of the plate.

For a plate the mean dimensionless temperature at the moment of time τ can be calculated from the equation

$$\overline{\Theta} = \sum_{n=1}^{\infty} \frac{2\sin^2 \varepsilon_n}{\varepsilon_n^2 + \varepsilon_n \sin\varepsilon_n \cos\varepsilon_n} \ \exp(-\varepsilon_n^2 Fo).\qquad ...(6.3)$$

Substituting the corresponding values of the quantities given in equating the problem into equation (6.2), we obtain:

$$Q_t = 0.4 \times 7800 \times 682 (800 - 30) = 1630 \times 10^3 \text{ kJ/m}^2,$$

where

$$c = \frac{\lambda}{\rho a} = \frac{37.2}{7800 \times 7 \times 10^{-6}} = 682 \text{ J / kg.°C.}$$

Calculate the dimensionless terms Fo and Bi:

$$Fo = \frac{a\tau}{\delta^2} = \frac{7 \times 10^{-6} \times 7200}{0.2^2} = 1.26;$$

$$Bi = \frac{a\delta}{\lambda} = \frac{200 \times 0.2}{37.2} = 1.075$$

The magnitude of the dimensionless term Fo > 0.3 and for calculations with a sufficient accuracy we can use the first term of the sum (6.3). Using the value of Bi we can find the value of ε_1 by substituting this value of ε_1 into equation (6.3), we find :

$$\overline{\Theta} = 0.098.$$

Substituting the calculated values of Q_t and $\overline{\Theta}$ into equation (6.1), we find that

$$Q = 1630 \times 10^3 \,(1 - 0.098) = 1470 \times 10^3 \text{ kJ/m}^2.$$

Example 6.2. A thin plate of length $l_0 = 2$ m and width $a = 1.5$ m is exposed to a flow of air parallel to its surface (Fig. 6.11). The velocity and temperature of the free-stream flow of air are respectively $w_0 = 3$ m/s and $t_0 = 20°C$. The temperature at the surface of the plate $t_w = 90°C$.

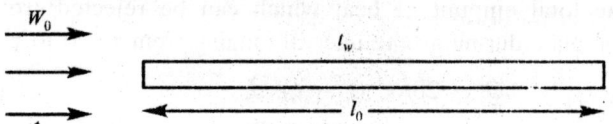

Fig. 6.11. Solution to Example 6.2.

Determine the lengthwise mean local coefficient of heat transfer from the plate to the air and the amount of heat transferred.

Solution. For the air at a temperature $t_0 = 20°C$ $v = 15.06 \times 10^{-6}$ m²/s, $\lambda = 2.59 \times 10^{-2}$ W/m. °C, Pr = 0.703. The Reynolds number

$$Re = \frac{w_0 l_0}{v} = \frac{3 \times 2}{15.06 \times 10^{-6}} = 3.98 \times 10^5 < 5 \times 10^5.$$

Consequently, in the boundary layer the pattern of flow is laminar. Under these conditions the lengthwise mean local coefficient of heat transfer from the plate to the air can be calculated from :

$$Nu = 0.67 \, Re^{1/2} Pr^{1/3}, \qquad\qquad ...(6.4)$$

where

$$Nu = \frac{a l_0}{\lambda}; \quad Re = \frac{w_0 l_0}{v},$$

and the physical properties are selected for the temperature of the free-stream flow, t_0.

In the case considered

$$\text{Nu} = 0.67(3.98 \times 10^5)^{1/2}(0.703)^{1/3} = 375$$

and the local heat-transfer coefficient

$$\alpha = \text{Nu}\frac{\lambda}{l_0} = 375 \, \frac{2.59 \times 10^{-2}}{2} = 4.87 \ \text{W} / \text{m}^2.°\text{C}.$$

The amount of heat transferred from the both sides of the plate

$$Q = \alpha \, (t_w - t_0) \, F = 4.87 \, (90 - 20) \times 2 \times 2 \times 1.5 = 2050 \ \text{W}.$$

Example 6.3. For the conditions of example 6.2 calculate the thickness of the hydrodynamic boundary layer and the values of the local heat-transfer coefficients at various distances from the front edge of the plate: $x = 0.1l_0$, $0.2l_0$, and $1.0l_0$. Plot the graph showing the dependence of the thickness of the hydrodynamic boundary layer δ_1 and local heat-transfer coefficient on the relative distance x/l_0.

Solution. In accordance with the conditions of example 6.2, the transfer of heat takes place in the laminar boundary layer. The thickness of the laminar boundary layer and the local heat-transfer coefficient at a distance x from the front edge of the plate are determined from the following eqn.:

$$\delta_1 = \frac{4.64x}{\sqrt{\text{Re}_x}} \qquad \qquad ...(6.5)$$

and

$$\text{Nu}_x = 0.335\text{Re}_x^{1/2} \, \text{Pr}^{1/3}, \qquad \qquad ...(6.6)$$

where

$$\text{Nu}_x = \frac{\alpha_x x}{\lambda} \quad \text{and} \quad \text{Re}_x = \frac{w_0 x}{v}$$

At a distance from the front edge $x = 0.1 \, l_0$

$$\text{Re}_x = \frac{w_0(0.1l_0)}{v} = \frac{3 \times 0.2}{15.06 \times 10^{-6}} = 3.98 \times 10^4;$$

$$\delta_l = \frac{4.64 \times 0.2}{\sqrt{3.98 \times 10^4}} = 4.66 \times 10^{-3} \ \text{m};$$

$$\text{Nu}_x = 0.335 \, (3.98 \times 10^4)^{1/2} \, (0.703)^{1/3} = 59.5$$

$$\alpha_x = \mathrm{Nu}_x \frac{\lambda}{x} = 59.5 \frac{2.59 \times 10^{-2}}{0.2} = 7.73 \ \mathrm{W}/\mathrm{m}^2.°C.$$

In a similar way, we can calculate the unknown quantities for other values of the relative distance x/l_0. The calculation results are compiled in the Table given below, and are presented graphically in Fig. 6.12.

x/l_0	0.1	0.2	0.5	1.0
δ_p, mm	4.66	6.58	10.4	14.7
α_x, W/m².°C	7.73	5.65	3.45	2.44

Example 6.4. A flat plate of length $l = 1$m is exposed to air flow parallel to its surface. The velocity and temperature of the free-stream air flow w_0 = 80 m/s and $t_0 = 10°C$. A turbulising grid is placed upstream of the plate, resulting in that the fluid is in turbulent flow in the boundary layer over the whole length of the plate.

Calculate the mean local coefficient of heat transfer from the surface of the plate and the value of the local heat-transfer coefficient on the back edge. Also calculate the hydrodynamic boundary layer on the back edge of the plate.

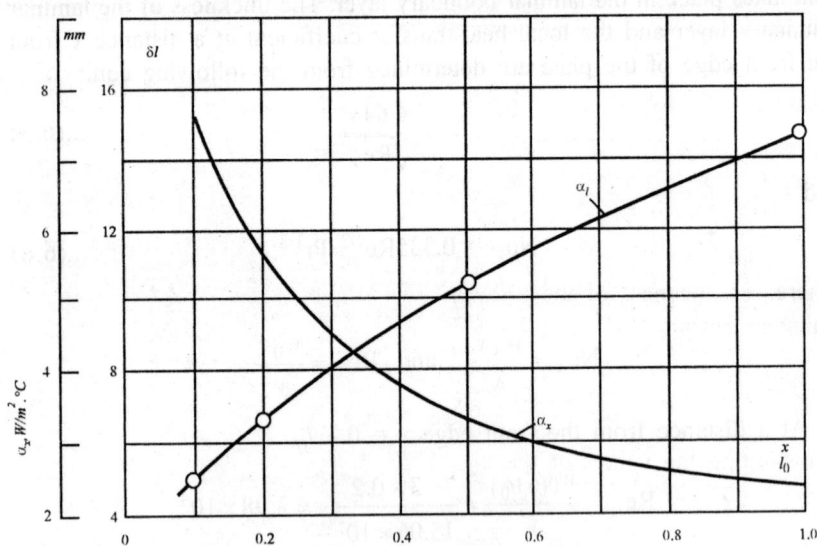

Fig. 6.12. Solution to Example 6.2.

Solution. With a temperature of the free-stream flow $t_0 = 10°C$ the physical properties of air are: $v = 14.16 \times 10^{-6}$ m²/s, $\lambda = 2.51 \times 10^{-2}$ W/m.°C.

The Reynolds number

$$Re = \frac{w_0 l_0}{\nu} = \frac{80 \times 1.0}{14.6 \times 10^{-6}} = 5.65 \times 10^6 > 5 \times 10^5$$

The pattern of flow in the boundary layer on the plate is turbulent.

With the plate exposed to air flow the mean heat-transfer coefficient for the turbulent boundary layer can be calculated from the following equation:

$$Nu = 0.032 Re^{0.8} \qquad \qquad ...(6.7)$$

Substituting the obtained value of the Reynolds number into equation 6.7, we get:

$$Nu = 0.032 \ (5.65 \times 10^6)^{0.8} = 8050 \text{ and}$$

$$\alpha = Nu \frac{\lambda}{l_0} = 8050 \frac{2.51 \times 10^{-2}}{1.0} = 202 \ W/m^2.°C.$$

For a plate in a longitudinal air flow with a turbulent boundary layer the local heat-transfer coefficient can be calculated from the following equation:

$$Nu_x = 0.0255 \ Re_x^{0.8} \qquad \qquad ...(6.8)$$

where $Nu_x = \alpha_x x/\lambda$ and $Re_x = w_0 x/\nu$.

The value of the local coefficient of heat transfer on the back edge of the plate can be found, assuming $x = l_0$, then $Re_x = 5.65 \times 10^6$, $Nu_x = 0.255 \ (5.65 \times 10^6)^{0.8} = 6280$ and

$$\alpha_{x=l_0} = Nu_{x=l_0} \frac{\lambda}{l_0} = 6280 \frac{2.51 \times 10^{-2}}{1.0} = 157.5 \ W/m^2.°C.$$

The local thickness of the turbulent hydrodynamic boundary layer can be calculated from :

$$\delta_t = \frac{0.37x}{\sqrt[5]{Re_x}} \qquad \qquad ...(6.9)$$

Substituting the values of the known quantities, we obtain at $x = l_0$:

$$\delta_t = \frac{0.37 \times 1.0}{\sqrt[5]{5.65 \times 10^6}} = 0.0165 \ m.$$

Example 6.5. A thin plate of length $l = 0.2$ m is exposed to a longitudinal air flow. The velocity and temperature of the free-stream flow respectively $w_0 = 150$ m/s and $t_0 = 20°C$.

Determine the mean heat-transfer coefficient and the rate of heat flow from the surface of the plate, if the surface temperature of the plate t_w = 50°C. Calculate, assuming the pattern of flow to be turbulent in the boundary layer over the entire length of the plate.

Solution. At the temperature of the free-stream flow t_0 = 20°C the physical properties of air are:

$$v = 15.06 \times 10^{-6} \text{ m}^2/\text{s}; \quad \lambda = 2.59 \times 10^{-2} \text{ W/m.°C}; \quad c_p = 1.0 \text{ kJ/kg.°C}.$$

The Reynolds number

$$\text{Re} = \frac{w_0 l}{v} = \frac{150 \times 0.2}{15.06 \times 10^{-6}} = 1.99 \times 10^6.$$

The Mach number

$$M = \frac{w}{a} = \frac{150}{344} = 0.436,$$

where the velocity of the sound in air

$$a = 20.1\sqrt{T_0} = 20.1\sqrt{293} = 344 \text{ m/s}.$$

The equation (6.7) is valid for calculating heat transfer from a plate exposed to air flow at a high sub-sonic velocity at $10^5 <$ Re $< 2 \times 10^6$ and $0.25 <$ M < 0.8, provided the heat-transfer coefficient is related to the difference between the temperature of the wall and adiabatic wall temperature $t_{a.w}$:

$$t_{a.w} = t_0 + r\frac{w_0^2}{2c_p}$$

where the recovery factor for a plate in a longitudinal flow with a boundary turbulent layer can be assumed equal to $r = \sqrt[3]{\text{Pr}}$.

For air at t_0 = 20°C $r = \sqrt[3]{0.703}$ = 0.89.

In the case considered

$$\text{Nu} = 0.032\text{Re}^{0.8} = 0.032 \ (1.99 \times 10^6)^{0.8} = 3500$$

and

$$\alpha = \text{Nu}\frac{\lambda}{l} = 3500\frac{2.59 \times 10^{-2}}{0.2} = 454 \text{ W/m}^2.°\text{C}.$$

The adiabatic wall temperature

$$t_{a.w} = 20 + 0.89 \frac{150^2}{2 \times 1 \times 10^3} = 30^\circ C.$$

and the rate of heat flow

$$q = \alpha \, (t_w - t_{a.w}) = 454 \, (50 - 30) = 9080 \ \text{W/m}^2.$$

PROBLEMS

Problem. 6.1. A cylindrical steel ingot with a diameter $d = 80$ mm and length $l = 160$ mm (Fig. 6.13) was first heated uniformly to a temperature $t_0 = 800$ °C. The billet is then cooled in the open air which is at a temperature $\tau_f = 30$°C.

Determine the temperature at the centre of the ingot $t_{x=0; \, r=0}$, and at the centre of its side $t_{r=0; x=l/2}$ in a time interval $\tau = 30$ min after cooling is initiated.

The thermal conductivity and thermal diffusivity of steel are respectively $\lambda = 23.3$ W/m.°C and $a = 6.11 \times 10^{-6}$ m²/s.

The local coefficient of heat transfer from the surface of the billet to the cooling air $\alpha = 118$ W/m². °C.

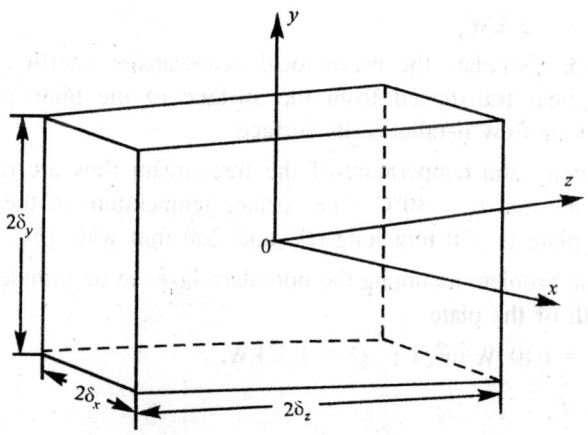

Fig. 6.13. Solution to Problem 6.1.

Answer. $t_{r=0; \, x=0} = 55$°C; $t_{r=0; \, x=l/2} = 50$°C.

Problem 6.2. Under the conditions for cooling a steel billet assumed in problem (6.1), determine the temperature at the centre of the billet and at the middle of its side surface, if the dimensions of the billet are doubled, i.e., $d = 160$ mm and $l = 320$ mm, all other conditions remaining as in problem 6.1.

Answer. $t_{r=0;\ x=0} = 211\ °C$; $t_{r=0;\ x=1/2} = 153\ °C$.

Problem 6.3. A rectangular billet, or slab, measuring 480 × 360 × 280 mm in size, is heated in a furnace at a constant temperature $t_f = 800°C$. Before the beginning of heating all points of the billet were at the same temperature $t_0 = 20°C$. The local coefficient of heat transfer from surroundings to the sides of the billet remains constant throughout the process of heating and equal to 200 W/m².°C.

The thermal conductivity, thermal diffusivity and density of steel are respectively : λ = 37.2 W/m.°C; $a = 7 × 10^{-6}$ m²/s; ρ = 7800 kg/m³.

Determine the amount of heat transferred to the billet in 2.5 hours after heating is initiated.

Answer. $Q = 189 × 10^3$ kJ.

Problem 6.4. A flat plate is exposed to air flow parallel to its surface. The velocity and temperature of the free-stream air flow are respectively $w_0 = 6$ m/s and $t_0 = 20°C$.

Calculate the quantity of heat transferred from the plate to the air, if the surface temperature of the plate $t_w = 80°C$, the length of the plate in the direction of flow $l =$ m, and the dimension of the plate across flow $b = 0.9$ m.

Answer. Q = 1.3 kW.

Problem 6.5. Calculate the mean local heat-transfer coefficient and the amount of heat transferred from the surface of the plate placed in a longitudinal air flow parallel to its surface.

The velocity and temperature of the free-stream flow are respectively: $w_0 = 200$ m/s and $t_0 = 30°C$. The surface temperature of the plate $t_w = 90°C$. The plate is 120 mm long (l) and 200 mm wide (b).

Solve the problem assuming the boundary layer to be turbulent over the entire length of the plate.

Answer. α = 640 W/m².°C; Q = 1.3 kW.

CHAPTER 7

Radiation

INTRODUCTION

In conduction heat is transferred from one part of a body to another, and the intervening material is heated. In convection the heat is transferred by the actual mixing of materials and by conduction. In radiant heat transfer the medium through which the heat is transferred usually is not heated. Radiation heat transfer is the transfer of heat by electromagnetic radiation.

Thermal radiation is a form of electromagnetic radiation similar to X-rays, light waves, gamma rays, and so on, differing only in wavelength. It obeys the same laws as light: travels in straight lines, can be transmitted through spaces and vacuum, and so on. It is an important mode of heat transfer and is especially important where large temperature differences occur, as for example, in a furnace with boiler tubes, in radiant dryers, and in an oven baking food. Radiation often occurs in combination with conduction and convection. In an elementary sense the mechanism of radiant heat transfer is composed of three distinct steps or phases :

1. The thermal energy of a hot source, such as the wall of a furnace at T_1, is converted into the energy of electromagnetic radiation waves.

2. These waves travel through the intervening space in straight lines and strike a cold object at T_2 such as a furnace tube containing water to be heated.

3. The electromagnetic waves that strike the body are absorbed by the body and converted back to thermal energy or heat.

MONOCHROMATIC AND TOTAL EMISSIVE POWER

Although energy is transmitted by radiation at all wavelengths throughout the electromagnetic spectrum manifestation in the form of heating effects occurs only over a very limited range. This range extends from wavelengths around 2000 angstroms (0.2μ, where $\mu = 10^{-4}$ cm) to approximately 50,000 angstroms (5μ). Notice that the human eye is only sensitive to radiation in the lower part of this range, i.e., from 0.38μ to 0.76μ. The rate at which a given body emits radiation depends upon the nature of its surface. The total radiant energy emitted per unit time per unit area is defined as the total emissive power and is denoted by E. It is the quantity directly proportional to the fourth power of the absolute temperature.

The amount of energy emitted by a body generally varies with wavelength or frequency in the manner shown in Fig. 7.1. The ordinates of such a curve will vary with the temperature and surface condition of the body.

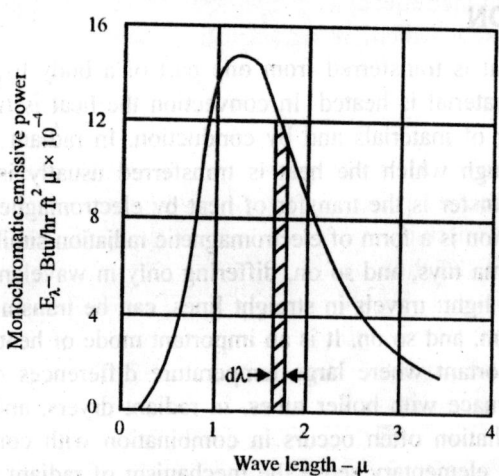

Fig. 7.1. Distribution of the energy in the spectrum of a tungsten lamp at 4000°F.

To describe this variation it is convenient to define a monochromatic emissive power, E_λ, such that the amount of energy emitted per unit time and area in the spectral range λ to $\lambda + d\lambda$ is given by $E_\lambda d\lambda$. Thus

$$E = \int_0^\infty E_\lambda d\lambda \qquad \dots (7.1)$$

ABSORPTIVITY, REFLECTIVITY AND TRANSMISSIVITY

Consider now radiation falling or incident upon a surface; some will be absorbed, some reflected, and unless the body is very thick or perfectly opaque, some will be transmitted. The sum of these must, of course, equal the total incident radiation. It is possible to define a monochromatic absorptive power, A_λ, in a manner similar to E_λ. A more useful quantity, however, is the monochromatic absorptivity α_λ, defined as the fraction of the radiant energy, incident on the surface in the spectral range λ to $\lambda +$ $d\lambda$, which is absorbed. α_λ is, therefore, a dimensionless ratio which is less than one; for a given body it usually varies markedly with wavelength and, to some extent, with temperature. The absorptivity α then follows as the fraction of the total radiant energy, incident on the surface, which is absorbed. If G_λ denotes the spectral distribution of the incident radiation and G the total incident energy, the absorptivity would be determined by

$$\alpha = \frac{1}{G} \int_0^\infty \alpha_\lambda G_\lambda d\lambda \qquad \text{... (7.2)}$$

Similarly, defining the reflectivity ρ and transmissivity τ as the fractions of the incident radiation which are reflected and transmitted, respectively, it follows that

$$\alpha + \rho + \tau = 1 \qquad \text{... (7.3)}$$

Equations similar to (7.2) would hold for the reflectivity ρ and the monochromatic reflectivity ρ_λ, as well as for the transmissivity τ and the monochromatic transmissivity, τ_λ.

The absorption of radiation by liquids and solids occurs in a very thin region near the surface. For good conductors of electricity this is of the order of 1μ. In electrical insulators this thickness may be as much as $10^3\mu$. However, this is still such a short distance (about 0.05 in.) that, in practically all cases, the following equation applies :

$$\alpha + \rho = 1 \qquad \text{... (7.4)}$$

KIRCHHOFF'S LAW

A very simple and useful relation exists between the emissive and absorptive powers of a surface. Suppose body 1 is an unbounded plane which emits only rays of radiation of the wavelength Λ and absorbs only rays of the same wavelength. Parallel to this body place a plane 2 which emits and absorbs rays of all wavelengths. The outer surfaces of each plane are assumed to be backed with perfect reflectors. If both bodies are at the same temperature, each must absorb as much energy as it emits in order to maintain this state. Considering first all the rays which body 2 emits of wavelength λ different from Λ, we see that they will be perfectly reflected from 1 and must all be re-absorbed by 2. Since this holds for all values of λ different from Λ, the invariability of the temperature of body 2 requires that it absorb as much of the energy of wavelength Λ as it emits. Thus at wavelength Λ_2 $E_{\Lambda 2}$ must equal the amount of the energy emitted by body

T, $E_{\Lambda 1}$, which is absorbed by 2 plus the amount of $E_{\Lambda 2}$ reflected by 1 and re-absorbed by 2. Of the quantity emitted by 1, $E_{\Lambda 1}$, body 2 absorbs $\alpha_{\Lambda 2} E_{\Lambda 1}$ and reflects $(1 - \alpha_{\Lambda 2}) E_{\Lambda 1}$; of this body 1 absorbs $\alpha_{\Lambda 1} (1 - \alpha_{\Lambda 2}) E_{\Lambda 1}$ and reflects $(1 - \alpha_{\Lambda 1})(1 - \alpha_{\Lambda 2}) E_{\Lambda 1}$ to 2, which then absorbs $\alpha_{\Lambda 2} (1 - \alpha_{\Lambda 1})(1 - \alpha_{\Lambda 2}) E_{\Lambda 1}$. Letting $k = (1 - \alpha_{\Lambda 1})(1 - \alpha_{\Lambda 2})$ and carrying out this process until all of $E_{\Lambda 1}$ is absorbed, the amount absorbed by body 2 will be

$$(q/A)_{1-2} = \alpha_{\Lambda 2} E_{\Lambda 1} (1 + k + k^2 + k^3 + \ldots)$$

$$= \frac{\alpha_{\Lambda 2} E_{\Lambda 1}}{1 - k} \qquad \ldots (7.5)$$

since $k < 1$ and the sum of the geometric series in the parentheses is then $1/(1 - k)$. A similar analysis will show that the amount of $E_{\Lambda 2}$ re-absorbed by 2 will be

$$(q/A)_{2-2} = \frac{\alpha_{\Lambda 2}(1 - \alpha_{\Lambda 1}) E_{\Lambda 2}}{1 - k} \qquad \ldots (7.6)$$

For equilibrium $E_{\Lambda 2}$ must equal $(q/A)_{1-2} + (q/A)_{2-2}$, which gives

$$E_{\Lambda 2} = \frac{\alpha_{\Lambda 2} E_{\Lambda 1}}{1 - k} + \frac{\alpha_{\Lambda 2}(1 - \alpha_{\Lambda 1})}{1 - k} E_{\Lambda 2}$$

This can be reduced to

$$\frac{E_{\Lambda 1}}{\alpha_{\Lambda 1}} = \frac{E_{\Lambda 2}}{\alpha_{\Lambda 2}} \qquad \ldots (7.7)$$

Now since the temperature and the wavelength Λ in the above derivation are arbitrary, it follows that for radiation at the same wavelength and temperature the ratio of the emissive power to the absorptivity is the same for all bodies. This relationship was first demonstrated by Gustav Kirchhoff in 1859 and is referred to as Kirchhoff's law.

REGULAR AND DIFFUSE SURFACES; RADIATION INTENSITY

As with light rays the fraction of incident radiation which is reflected from a completely smooth surface leaves at an angle with the normal to the surface equal to the angle made by the incident energy. In this case the reflection in specular or regular. No surface is perfectly smooth and most are actually quite rough, causing the reflected portion of incident radiation to be distributed in all directions. This kind of reflection is called diffuse.

More specifically, a diffuse surface is one from which the radiant flux density to all directions in space is constant. To illustrate what is meant by this, let dA_1 in Fig. 7.2 be an element of a diffusely radiating surface.

Describe about dA_1 a hemisphere of radius R and let dA_2 located at P be a small element of the surface of this hemisphere; the radius OP to this element makes an angle θ with the radius ON which is normal to dA_1. The rate q at which radiant energy is incident upon dA_2 is proportional to dA_2, dA_1, and to $1/R^2$ because, looking from P toward dA_1, a surface dA_1 $\cos\theta$ is seen which, for equal space distribution of radiation, would appear as bright as the area dA_1 seen from N. We can write, therefore,

$$q = i dA_1 \cos\theta \frac{dA_2}{R^2} \qquad \ldots (7.8)$$

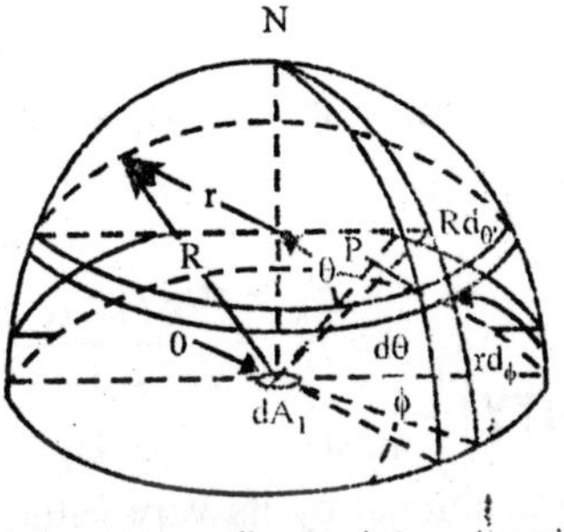

Fig. 7.2. Radiation from a small area into a hemispherical space.

where i denotes the intensity of radiation from dA_1 in space. It is uniform in all directions for a diffuse surface. Alternately a surface radiating in this fashion is said to obey Lambert's cosine law. Since dA_2/R^2 is the element of solid angle $d\omega$ which the area dA_2 subtends, i may be defined as

$$i = \frac{q}{dA_1 d\omega \cos\theta} \qquad \ldots (7.9)$$

From this we see that i represents the rate at which dA_1 radiates energy in a given direction per unit solid angle and per unit of its own area as projected on a plane perpendicular to the given direction. This definition applies to the total radiation from a surface. We may also—as in the case of emissive power—designate a monochromatic intensity of radiation, denoted by i_λ, as the intensity of radiation in the wavelength range $d\lambda$ at wavelength λ.

When i is constant in all directions (approximately true for many surfaces and exact for a black surface), a very simple relation exists between it and the total emissive power E. Referring to Fig 7.2, we see that the area of dA_2 is given by $Rd\theta r d\phi = R^2 \sin\theta d\theta d\phi$. The area of the ring including dA_2 is $2\pi R^2 \sin\theta d\theta$ and the rate at which energy is radiated through this ring from dA_1 is

$$dq_{ring} = idA_1 \frac{\cos\theta 2\pi R^2 \sin\theta d\theta}{R^2} = 2\pi idA_1 \cos\theta \sin\theta d\theta \qquad ...(7.10)$$

(Note that dq is equal to i multiplied by (1) the projected area of dA_1 in the direction θ and (2) the solid angle subtended by the area of the ring.) Integration of this equation from $\theta = 0$ to $\theta = \pi/2$ will give the total radiation through the hemispherical surface from dA_1. Thus

$$EdA_1 = 2\pi idA_1 \int_0^{\pi/2} \cos\theta \sin\theta d\theta = dA_1 \pi i \sin^2\theta \Big|_0^{\pi/2} = \pi idA_1$$

and

$$E = \pi i \qquad ... (7.11)$$

Similarly, we would find

$$E_\lambda = \pi i_\lambda \qquad ... (7.12)$$

RADIATION DENSITY

Although radiation is transitional by its very nature, it is useful to consider the energy per unit volume in a stream of radiation. This is referred to as the density of radiation and is denoted by the symbol ψ. Its magnitude is most easily determined by considering the radiation which passes through a given volume in unit time. If in Fig. 7.2 the radius of the hemisphere is made very large, we may regard all the rays from dA_1 to dA_2 as being parallel. The radiant energy emitted—say in 1 second from dA_1 in the direction OP—will be contained, therefore, in a parallelepiped whose length is the velocity of light c. The cross-sectional area of this parallelepiped would be $dA_1 \cos\theta$ and its volume

$$dv = cdA_1 \cos\theta$$

If the radiation leaving the surface is black, the quantity of energy leaving dA_1 in this direction in unit time would be

$$dE_b = idA_1 \cos\theta \frac{dA_2}{R^2}$$

The energy density due to this radiation anywhere inside this volume is then

$$d\psi = \frac{dE_b}{dv} = \frac{i}{c} \frac{dA_2}{R^2} \qquad ... (7.13)$$

Equation (7.13) gives the energy density in the space immediately above, due only to the radiation in one direction. Since radiation is leaving dA_1 in all directions, the total density near the surface would be

$$\psi = \int d\psi = \frac{i}{cR^2} \int dA_2 = \frac{i}{cR^2} 2\pi R^2 = \frac{2\pi i}{c} \qquad \ldots (7.14)$$

Expressed in terms of the emissive power this is

$$\psi = \frac{2E_b}{c} \qquad \ldots (7.15)$$

ISOTHERMAL ENCLOSURE

Although no totally black surfaces exist in nature, it is possible to describe a system in which the energy density is black body radiation. This consists of a cavity where the walls and contents are at a common temperature, which we shall call an isothermal enclosure (the term *hohlraum* is also frequently used). On the basis of thermodynamic reasoning it can be established that the stream of radiation in any given direction in an isothermal enclosure is the same as in any other direction. If this were not so, it would be possible to introduce similar absorbers and one would get hotter than the other by absorbing energy from the larger stream. This implies that a source and sink could be developed and used to drive an engine to obtain work, a result contradicting the second law. Similarly it can be shown that the stream of radiation in any given direction in all isothermal enclosure at the same temperature must be the same. Since one of these enclosures might have black surfaces, we see that the radiation in any isothermal enclosure travelling in any one direction must be the same as that emitted by a black surface or body in any direction.

The energy density in the enclosure, however, will be twice that produced just in front of a black surface from radiation emitted by that surface alone. This, of course, is because energy streams are travelling in all directions in an enclosure, whereas above a flat surface emitted energy travels in only a hemisphere of directions. Hence for the total energy density ψ in the enclosure, we would obtain from equation (7.15)

$$\psi = \frac{4E_b}{c} = \frac{4\pi i}{c} \qquad \ldots (7.16)$$

An isothermal enclosure provides a means for studying black body radiation. Such an enclosure can be made with a tube heated electrically by a coil of wire wound around it. A small hole is made through the wall and the radiation streaming from it approximates very closely that which would be emitted by a black surface at the temperature of the inner region of the tube.

RADIATION PRESSURE

The concept of an expanding isothermal enclosure was employed by Boltzmann to derive the law for the variation of black body radiation with temperatures. The procedure consisted in accounting for the work done on the piston of an ideal engine by radiation pressure. Before describing the analysis, however, we must briefly consider the pressure exerted by radiation. It has been possible to explain the charges of momentum of material bodies caused by electromagnetic field itself to have momentum. Known effects can be accounted for if each unit of moving energy is assumed to have a mass of $1/u^2$, where u is the velocity of energy travel. Thus for black body radiation the number of energy units per unit volume ψ are moving with a velocity c so that the momentum per unit volume would be

$$m = \frac{\psi}{c^2}c = \frac{\psi}{c} \qquad \dots (7.17)$$

If a stream of radiation of energy density ψ falls normally on a black surface, $(\psi/c)c = \psi$ units of momentum will reach unit area of the surface per unit of time. Since all the radiation will be absorbed and force equals rate of change of momentum, the pressure will equal the momentum delivered by the radiation absorbed by unit area in unit time, i.e., $p = \psi$. Taking the solar constant as 400 Btu/hr ft^2 (this is the amount of the sun's radiation received per unit time per unit area perpendicular to the sun's rays and at a distance from the sun equal to the mean radius of the earth's orbit) and $c = 3 \times 10^{10}$ cm/sec $= 9.835 \times 10^8$ ft/sec, the radiation density in a column of solar radiation 1 ft^2 in cross-section and 9.835×10^8 ft long would be

$$\psi_{solar} = (400/3600)/9.835 \times 10^8 = 1.13 \times 10^{-10} \text{ Btu/ft}^3$$

Expressing this in ft lbs, we obtain

$$\psi_{solar} = 1.13 \times 10^{-10} (778) = 8.8 \times 10^{-8} \text{ ft lb/ft}^3$$

and

$$p_{solar} = 8.8 \times 10^{-8} \text{ lb/ft}^2$$

The existence of radiation pressure and agreement with the theoretical magnitude have been experimentally verified. In an isothermal enclosure where radiation is streaming toward the surface and also away from it with the same intensity in all directions, it can be shown that the average pressure due to radiation is

$$p = \tfrac{1}{3}\psi \qquad \qquad ...(7.18)$$

ABSORPTIVITY AND BLACK BODIES

When thermal radiation (like light waves) falls upon a body, part is absorbed by the body in the form of heat, part is reflected back into space, and part may be actually transmitted through the body. For most cases in process engineering, bodies are opaque to transmission, so this will be neglected. Hence, for opaque bodies,

$$\alpha + \rho = 1.0 \qquad \qquad ... (7.19)$$

where α is absorptivity or fraction absorbed and ρ is reflectivity or fraction reflected.

A black body is defined as one that absorbs all radiant energy and reflects none. Hence, ρ is 0 and $\alpha = 1.0$ for a black body. Actually, in practice there are no perfect black bodies, but a close approximation to this is a small hole in a hollow body, as shown in Fig. 7.3. The inside surface of the hollow body is blackened by charcoal. The radiation enters the hole and impinges on the rear wall; part is absorbed there and part is reflected in all directions. The reflected rays impinge again, part is absorbed, and the process continues. Hence, essentially all of the energy entering is absorbed, and the area of the hole acts as a perfect black body. The surface of the inside walls are "rough" and rays are scattered in all directions, unlike a mirror, where they are reflected at a definite angle.

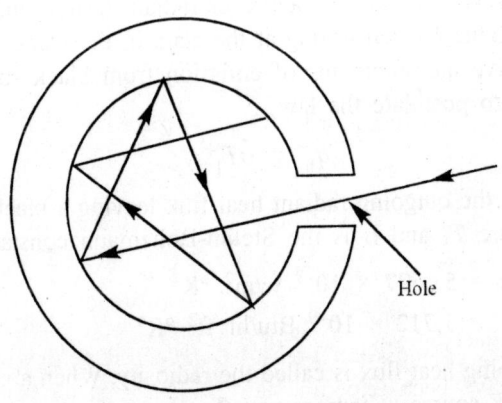

Fig. 7.3. Concept of a perfect black body.

As stated previously, a black body absorbs all radiant energy falling on it and reflects none. Such a black body also emits radiation, depending on its temperature, and does not reflect any. The ratio of the emissive power of a surface to that of a black body is called emissivity ε and it is 1.0 for a black body. Kirchhoff's law states that at the same temperature T_1, α_1 and ε_1 of a given surface are the same or

$$\alpha_1 = \varepsilon_1 \qquad \qquad \dots (7.20)$$

Equation (7.20) holds for any black or non-black solid surface.

TOTAL EMISSION FROM BLACK SURFACES

Black Surfaces

Many surfaces encountered in technology can be idealised as being black to photons and/or molecules. The word black merely denotes that no photons or molecules are reflected from the surface; all photons are absorbed and all molecules are either adsorbed or condense. Whether or not a surface is black or nearly so can be determined, for example, by directing a stream of molecules from a source in a vacuum chamber onto a surface and determining whether or not they are promptly reflected with same energies they possessed before striking the surface. Much the same can be done with photons from a radiant source. An obvious consequence of our definition of a black surface is that all photons or molecules, leaving the surface are emitted by the surface.

Radiant Fluxes

All have experienced feeling the warmth of a sun on the skin, feeling heat radiated across a room by a fire or radiant heater, and feeling the cooling of the forehead when gazing at the stars in the black sky on a clear night. Quantitative measurements of emission from black surfaces led J. Stefan, in 1879, to postulate the law

$$q_b^+ = \sigma T_1^4 \qquad \qquad \dots (7.21)$$

where q_b^+ the outgoing radiant heat flux leaving a black surface at absolute temperature T_1 and σ is the Stefan-Boltzmann constant,

$$\sigma = 5.6697 \times 10^{-8} \text{ w/m}^2 \text{ }^\circ K^4$$
$$= 1.712 \times 10^{-9} \text{ Btu/hr ft}^2 \text{ }^\circ R^4$$

The outgoing heat flux is called the radiosity. When a surface is surrounded by a black source at temperature T_2, the surface experiences an incoming flux q_b^-, given by

$$q_b^- = \sigma T_2^4 \qquad \qquad ... (7.22)$$

The incoming flux is called the irradiation.

The net outward radiant flux from a surface is simply the radiosity minus the irradiation,

$$q = q^+ - q^- \qquad \qquad ... (7.23)$$

Thus if a surface is black at temperature T_1 and is completely surrounded by a black source at temperature T_2, the net flux from surface 1 is

$$q_1 = \sigma T_1^4 - \sigma T_2^4 \qquad \qquad ... (7.24)$$

Notice that, when the surface and source are at the same temperature T_1, Eq. (7.24) gives $q_1 = 0$, and the system can attain a state of thermodynamic equilibrium.

For the present, Stefan's law, Eq. (7.21), is best regarded as based on experimental observation. Later, we will regard this law from a microscopic viewpoint in which each photon leaving the surface is an energy carrier, and, by summing over all such photons, we will find the required expression for the outgoing radiant heat flux. Notice also that, in writing down Eqs. (7.23) and (7.24), we have assumed that an isothermal black surface radiates at the rate $q_b^+ = \sigma T_1^4$ regardless of the irradiation falling upon the surface; we suppose that q_b^+ is σT_1^4 even when irradiation is zero. No experiment yet performed has shown this assumption to be in error.

The outgoing flux, when it comes from a black body so that no reflected photons make up part of it, is due to emission only. Thus the radiosity q_b^+ of a black body is also called the black body emissive power. Table 7.1 shows the black body emissive power for a few temperatures to emphasise that radiation particularly important at elevated temperatures. Even at room temperature the flux is of a magnitude comparable to natural convection heat transfer fluxes or fluxes due to conduction in a gas.

Table 7.1. Black body emission.

Source temperature	Emission	
	W/m^2	$Btu/hr\ ft^2$
300°K (room temperature)	459	146
1000°K (cherry-red hot)	56,700	18,000
3000°K (lamp filament)	4,590,000	1,460.000
5700°K (sun temperature)	60,000,000	19,000.000

Example 7.1. A catalytic afterburner is part of an automobile exhaust system. The incoming stream of 300 lb/hr exhaust gas contains a mass fraction m_1 of CO equal to 0.04, and preheated air is added to this stream to oxidise the CO to CO_2 inside the device. A system design requirement is that 60% of the heat liberated by the combustion must be radiated away by the afterburner case. How much radiator area is required if the case is allowed to attain a temperature of 1400°F? Assume the case surface to be black, the surrounds to be black and at 140°F, and the heat of combustion of CO to CO_2 to be 4350 Btu/lb CO.

Solution. The total heat release is

$$\dot{Q} = \dot{m} m_1 \Delta \hat{h}_{comb} = (300)(0.04)(4350)$$

$$\dot{Q} = 52,200 \text{ Btu / hr}$$

The radiosity of the black radiator would be

$$q^+ = \sigma T_1^4 = (1.712 \times 10^{-9})(1860)^4 = 2.05 \times 10^4 \text{ Btu/hr ft}^2$$

The irradiation is almost negligible by comparison;

$$q^- = \sigma T_2^4 = (1.712 \times 10^{-9}) (600)^4 = 222 \text{ Btu/hr ft}^2$$

The net flux is therefore

$$q = q^+ - q^- = 20,500 - 222 = 20,300 \text{ Btu/hr ft}^2$$

A radiating area is then found from

$$\dot{Q}_r = qA, \quad A = \dot{Q}_r / q = (0.60)(52,200) / 20,300$$

$$A = 1.54 \text{ ft}^2$$

Molecular Fluxes

The molecular flux bombarding a surface is denoted as J^- (molecules/m^2 sec). When a gas is in thermodynamic equilibrium at temperature T,

$$J^- = \frac{1}{4} N \bar{v} \qquad \qquad \text{... (7.25a)}$$

where N is the number of molecules per unit volume given by the ideal gas law and $\bar{v}$ is the mean molecular speed.

$$N = \frac{P}{kT} \qquad \bar{v} = \left(\frac{8kT}{\pi m} \right)^{1/2}$$

The quantity k, Boltzmann's constant, is the universal gas constant divided by Avogadro's number, that is, the gas constant per molecule. Table 7.2 gives numerical values for k and Avogadro's number. The molecular mass m is obtained by multiplying the value for the atomic mass unit by the molecular weight.

Table 7.2. Selected physical constants and conversion factors.

Quantity	Symbol	Values
Atomic mass unit	amu	1.66043×10^{-27} kg
Velocity of light	c	$2.997925 \times 10^{+8}$ metre/sec
Planck constant	h	6.6256×10^{-34} joule sec
Boltzmann constant	k	1.38054×10^{-23} joule/°K
Avogadro number	N_{Av}	6.02252×10^{26} molecules/kg mole
Length		10^6 μ/metre
		10^7 A°/metre
Energy		10^7 erg/joule
		6.2418×10^{18} ev/joule
Pressure		1.01325×10^5 newtons/m² atm
		760 torr/atm

Now picture the gas (say water vapour) contained in an enclosure, the walls of which are coated with the condensed phase (water or ice). At thermodynamic equilibrium the system is isothermal, and molecules must leave the walls at the same rate at which they arrive. The flux of molecules emitted by the condensed phase on the surface (assumed to be black) is

$$J_b^+ = \frac{1}{4} N \bar{v} \qquad \qquad \dots (7.25b)$$

Observe that Eq. (7.25b) has thus been derived in a thermodynamic manner; no physical model for the emission process has been postulated. Notice also that Eq. (7.25b) is the molecular analog to Eq. (7.21) which applied to photon emission, and it too may be regarded for now as a law based on experiment. As was the case for photons, experiments on the evaporation of various substances into a vacuum have shown that the law is valid even when the incident flux is zero.

The flux of energy striking a black surface being bombarded by molecules is obtained by multiplying the flux of molecules by the energy that they bring to the surface. As shown the result for a monoatomic gas is

$$q^- = \tfrac{1}{4} N \bar{v}(2kT)$$

A polyatomic gas carries additional energy so we can write more generally

$$q^- = \tfrac{1}{4} N \bar{v}(m\hat{u} + \tfrac{1}{2}kT) \qquad \text{... (7.26)}$$

where $\hat{u}$ is the internal energy per unit mass. For a monoatomic gas $m\hat{u} = 3kT/2$.

Example 7.2. Determine the rate of evaporation from an aluminium surface into a vacuum if its surface temperature is 889°C. The Handbook of Chemistry and Physics gives the saturation vapour pressure of aluminium at 889°C as 1.00×10^{-3} mm Hg (1 mm Hg = 1 torr).

Solution. The saturation vapour pressure is precisely the pressure of pure vapour which would be in equilibrium with the condensed phase surface at the specified temperature. Therefore we may use Eq. (7.25b) directly

$$J_b^+ = \tfrac{1}{4} N \bar{v}$$

Assuming a monoatomic gas ($M = 27$) we obtain, in SI units,

$$\bar{v} = \left(\frac{8kT}{\pi m} \right)^{1/2}$$

$$\bar{v} = \left[\frac{(8)(1.38 \times 10^{-23})(1162)}{(3.14)(1.66 \times 10^{-27})(27)} \right]^{1/2} = 0.95 \times 10^3 \, \text{m/sec}$$

The molar concentration is

$$c = \frac{N}{N_{Av}} = \frac{P}{RT} = \frac{(10^{-3}/760)(1.013 \times 10^5)}{(8.3 \times 10^3)(1162)} = 1.38 \times 10^{-8} \, \text{kg mole/m}^3$$

Upon substitution there results

$$\frac{J_b^+}{N_{Av}} = \tfrac{1}{4}(0.95 \times 10^3)(1.38 \times 10^{-8}) = 3.28 \times 10^{-6} \, \text{kg mole/m}^2\text{sec}$$

TRANSPORT BETWEEN INFINITE PARALLEL BLACK WALLS

Radiant Energy Transport

We have already noted in Eq. (7.24) that the net flux from one black surface entirely surrounded by another is simply $\sigma T_1^4 - \sigma T_2^4$. Thus the net flux of radiation between infinite parallel black walls is also given by this simple relation.

Free Molecule Conduction

Exactly the same can be written for two parallel plates exchanging mass. The net flux is simply

$$\frac{1}{4} N_1 \bar{v}_1 - \frac{1}{4} N_2 \bar{v}_2 = J^+ - J^- \qquad \ldots (7.27)$$

where N_1 is computed using the saturation vapour pressure at surface temperature T_1 of the one plate, and N_2 is similarly computed for the second plate. But Eq. (7.27) is not particularly useful. For example, it describes transfer of mass across the gap between two plates when both plates are of the same substance; the one at the higher temperature undergoes net evaporation and the other net condensation. Such a situation is not of much practical importance.

The more usual engineering situation is that of two parallel walls containing a superheated gas with no net transfer of mass. Suppose that the pressure is low enough for the distance between the plates to be small compared to the mean free path so that "free molecule" conduction will take place. One wall, the left one, is heated to temperature T_1, while the other wall, the right one is colder at temperature T_2. We assume that the surfaces are black to incident molecules (though not necessarily to photons). Then the molecules are adsorbed on the walls long enough for those which are re-emitted to have acquired the statistical nature of a gas in thermodynamic equilibrium with the wall.

The situation described above occurs quite commonly in the vacuum jackets around pipes and containers used to hold cryogenic liquids such as liquid helium, hydrogen, and nitrogen. An evacuated space around the container is used to reduce the boil-off losses of the cryogens. The space is evacuated to reduce the number density so that there are not too many energy carriers, and layers of shielding with spacing less than the mean free path are used to prevent the energy carriers from crossing directly to the cold inner wall in contact with the cryogen from the relatively hot outer wall heated by the surrounds at room temperature.

Once one wall is heated and the other cooled, the number of molecules per unit volume in each space divides itself into two populations, one consisting of molecules going from left to right N_1^+ and one going from right to left N_2^-. The sum of the two parts equals the whole population

$$N = N_1^+ + N_2^- \qquad \ldots (7.28)$$

The number of molecules crossing per unit area per unit time from left to right J^+ equals the number crossing from right to left J^-, because we are supposing that steadystate prevails with no net mass transfer. We assume that the wall material have a very low vapour pressure so that molecules of the wall material are not subliming from one wall and condensing on the other. (High vacuum systems are constructed from stainless steel rather than a metal such as zinc). It can therefore be written as

$$J^- = J^+ \qquad \qquad \text{... (7.29)}$$

In Eqs. (7.25a,b) the quantity N was the total population. However, in the case when T_1 was equal to T_2, the quantity N was twice the population N_1^+ of the molecules going one way. Thus Eqs. (7.25a,b) may be rewritten generally as

$$J^- = \frac{1}{2} N_2^- \bar{v}_2 \qquad \qquad \text{... (7.30a)}$$

$$J^+ = \frac{1}{2} N_2^+ \bar{v}_1 \qquad \qquad \text{... (7.30b)}$$

We now assume that, because the surfaces are black, the emitted molecules from have a velocity distribution identical to that characterising ec um ission. Therefor

$$\bar{v}_1 = \left(\frac{8kT_1}{\pi m} \right)^{1/2}, \quad \bar{v}_2 = \left(\frac{8kT_2}{m} \right)^{1/2}$$

Equations (7.28) to (7.30b) are a set of four simultaneous algebraic equations in four unknowns which must be solved. Substituting Eqs. (7.30a) and (7.30b) into (7.29) and using Eq. (7.28) to eliminate N_2^- gives

$$N_1^+ = N \frac{\bar{v}_2}{\bar{v}_1 + \bar{v}_2} = N \frac{T_2^{1/2}}{T_1^{1/2} + T_2^{1/2}} \qquad \qquad \text{... (7.31)}$$

Equation (7.30b) then gives

$$J^+ = \frac{1}{4} N \left(\frac{8kT_M}{\pi m} \right)^{1/2}, \quad T_M^{1/2} = \frac{2 T_1^{1/2} T_2^{1/2}}{T_1^{1/2} + T_2^{1/2}} \qquad \qquad \text{... (7.32)}$$

The net flux transferred is then the flux crossing from left to right q_1^+ minus that crossing from right to left. From Eq. (7.26), assuming c_v constant, we obtain

$$q_{net} = J^+ (mc_v T_1 + \frac{1}{2} kT_1) - J^- (mc_v T_2 + \frac{1}{2} kT_2)$$

$$q_{net} = \frac{1}{4} N \left(\frac{8kT_M}{\pi m} \right)^{1/2} (mc_v + \frac{1}{2}k)(T_1 - T_2)$$

Indroducting the ratio of specific heats and the relation $c_p = c_v + k/m$, it can also be written as

$$q_{net} = \frac{1}{4} N \left(\frac{8kT_M}{\pi m} \right)^{1/2} \left(\frac{\gamma + 1}{2\gamma} mc_p \right)(T_1 - T_2) \qquad \ldots (7.33)$$

This result gives the stedy-state power transfer between two parallel plates when there is no net mass transfer. Note that k/m is R/M and Nm is density ρ.

Example 7.3. A vacuum system is evacuated to 10^{-5} mm Hg pressure. Calculate the free molecule conduction between two layers of superinsulation 0.1 in. apart when one layer is at 70°F and the other is at 69°F.

Solution. Again our first step is to compute $\bar{v}$, this time for $T_M = 69.5°F = 294°K$.

$$\bar{v} = \left(\frac{8kT_M}{\pi m} \right)^{1/2} = \left(\frac{(8)(1.38 \times 10^{-23})(294)}{(3.14)(1.66 \times 10^{-27})(29)} \right)^{1/2} = 4.63 \times 10^2 \, m/sec$$

or

$$\bar{v} = 1520 \text{ ft/sec} = 5.47 \times 10^6 \text{ ft/hr}$$

As we have said, in Eq. (7.33) the grouping Nm is just the density ρ,

$$\rho = \frac{PM}{RT} = \frac{(10^{-5})(14.7)(144)(29)}{(760)(1545)(529)} = 0.987 \times 10^{-9} \text{ lb/ft}^3$$

For air $c_p = 0.24$ and $\gamma = 1.4$. Substituting these numbers in Eq. (7.33) we obtain

$$q = \frac{1}{4}(0.987 \times 10^{-9})(5.47 \times 10^6)\left(\frac{2.4}{2.8} \right)(0.24)(70 - 69)$$

$$= 2.78 \times 10^{-4} \text{ Btu/hr ft}^2$$

We see that evacuation to 10^{-5} mm Hg effectively eliminates conduction at room temperature, for if the mean free path were short compared to the gap we would have

$$q = \frac{k\Delta T}{L} = \frac{(0.015)(70-69)}{(0.1)/(12)} = 1.8 \text{ Btu}/\text{hr ft}^2$$

But, in fact, the mean free path is about 5 m for this situation, and q has the much lower value of 2.78×10^{-4} Btu/hr ft^2.

TRANSFER BETWEEN FINITE BLACK WALLS

Shape Factor

In the development which follows, the shape factor concept is presented in the context of radiation (photon) transport. However, the concept is equally applicable to molecular transport in, for example, a vacuum chamber.

Consider two finite walls such as those shown in Fig 7.4. To find the one-way rate of energy (power) transfer by thermal radiation from surface 1 to surface 2 we introduce the shape factor concept. The shape factor F_{1-2} is the fraction of power leaving surface 1 which is intercepted by surface 2. We can therefore write simply

$$Q^+_{1-2} = A_1 \sigma T_1^4 F_{1-2} = A_1 F_{1-2} \sigma T_1^4 \qquad \dots (7.34)$$

Thus, $A_1 F_{1-2}$ is given by the double area-integral

$$A_1 F_{1-2} = \int_{A_1} \int_{A_2} \frac{\cos\theta_1 \cos\theta_2 \, dA_2 \, dA_1}{\pi r_{1-2}^2} \qquad \dots (7.35)$$

where dA_1 and dA_2 are elements of area as shown in Fig. 7.4. Fig. 7.5 and Fig. 7.6 present values for adjacent and opposite rectangles. The power transfer from 2 to 1 is

$$\dot{Q}^+_{2-1} = A_2 F_{2-1} \sigma T_2^4$$

The net transfer is consequently

$$\dot{Q}^+_{1-2} = A_1 F_{1-2} \sigma T_1^4 - A_2 F_{2-1} \sigma T_2^4 \qquad \dots (7.36)$$

Note the symmetry in the double integral of Eq. (7.35). We see that the 1 and 2 subscripts are interchangeable; therefore

$$A_1 F_{1-2} = A_2 F_{2-1} \qquad \dots (7.37)$$

and Eq. (7.36) may be written

$$\dot{Q}_{1-2} = (\sigma T_1^4 - \sigma T_2^4) A_1 F_{1-2} \qquad \dots (7.38)$$

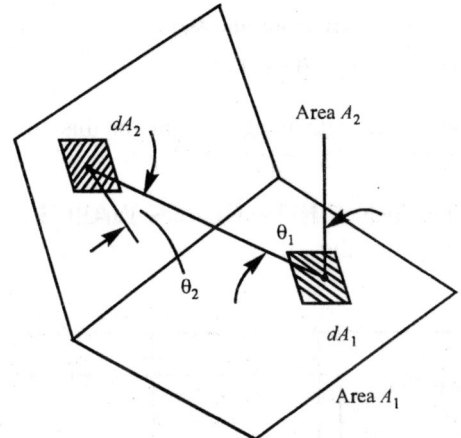

Fig. 7.4. Two finite areas.

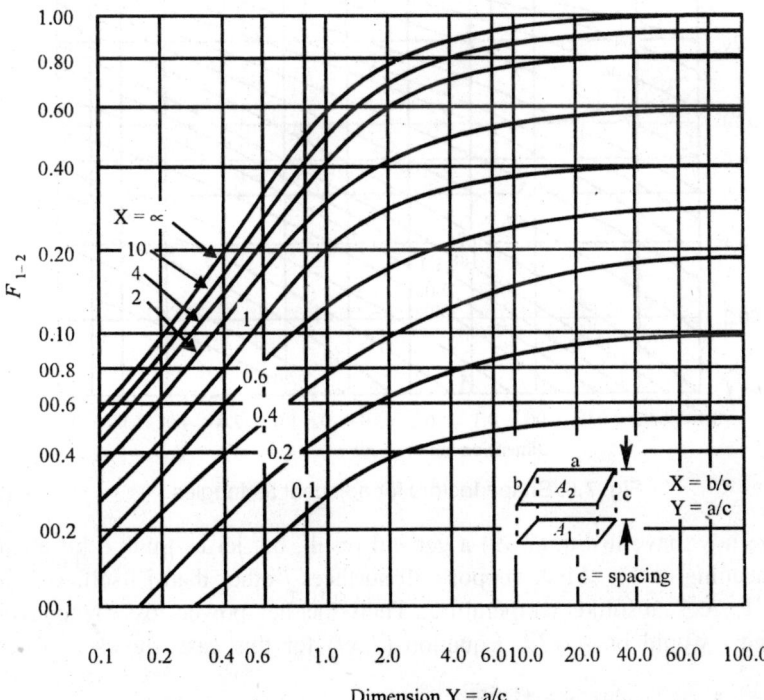

Dimension Y = a/c

Fig. 7.5. Shape factors for opposite rectangles.

Imagine a complete enclosure such as an empty room. Suppose that the surfaces of the enclosure are m in number, each isothermal and black.

Consider the finite surface, for example, the left-hand wall. The net power transfer by radiation between that surface and all the others is

$$\dot{Q}_i = \sum_{j=1}^{m} \dot{Q}_{i-j} = \sum_{j=1}^{m} (\sigma T_i^4 - \sigma T_j^4) A_i F_{i-j} \qquad \ldots (7.39)$$

The sum may be imagined to include surface i itself, for F_{i-i} is not zero for a concave surface, but the difference in σT^4 is zero for this term in any event.

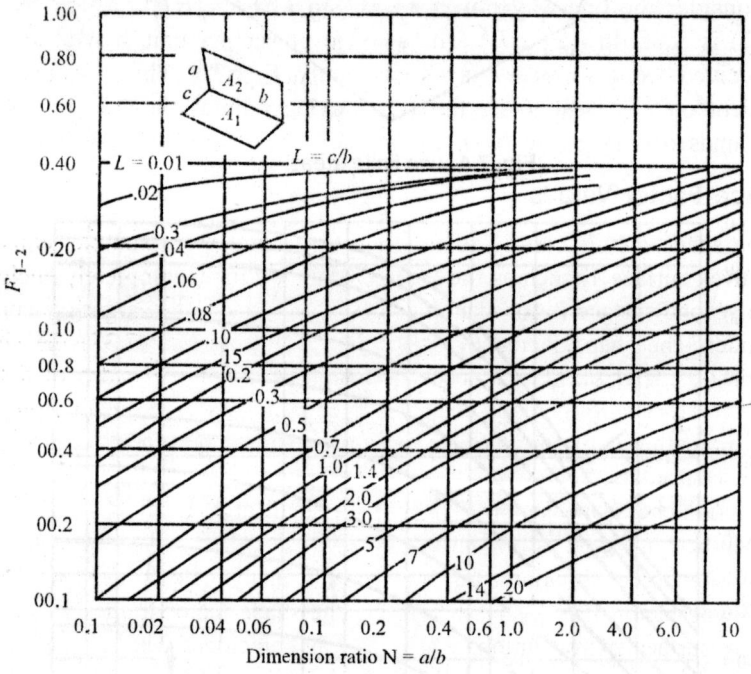

Fig. 7.6. Shape factors for adjacent rectangles.

We now have in Eq. (7.39) a general result, but let us think a bit about the meaning of F_{i-j}. First, suppose all surfaces j other than i itself were at zero degrees absolute temperature. Then the net power loss by convex surface i would be $A_i \sigma T_i^4$. Equation (7.39) for this case shows

$$\sum_{j=1}^{m} F_{i-j} = 1 \qquad \ldots (7.40)$$

We see that F_{i-j}, a positive number, cannot be greater than unity. Now consider each term separately in Eq. (7.38) or (7.39). The term

$\sigma T_i^4 A_i F_{i-j}$ is the one-way power from surface i to j. The total power leaving the surface is $A_i \sigma T_i^4$. The shape factor F_{i-j} is thus the fraction of the power radiated by surface i which is intercepted by surface j. Seen in this light it is obvious why the shape factors in a complete enclosure must sum to unity.

As we have said, the shape factor F_{i-j} is the fraction of power radiated by surface i which is intercepted by surface j. But another way, and perhaps a better way, of saying the same thing is to say that the shape factor F_{i-j} is the fraction of the surrounds of surface i taken up by surface j. Consider the one-way power from j to i, $A_j F_{j-i} \sigma T_j^4$, which by Eq. (7.37) is equal to $A_i F_{i-j} \sigma T_j^4$. The average power per unit area of surface i received from surface j is consequently $F_{i-j} \sigma T_j^4$. This average flux incident on i is seen to be the flux leaving j times the fraction of the surrounds of i taken up by j, F_{i-j}.

Shape Factor Values

Some shape factors can be found from inspection. For example, consider a convex surface 1, such as the outside of a long tube, completely enclosed by a second surface 2, for example, the inner surface of a larger concentric cylinder. The shape factor F_{1-2} is unity, and from Eq. (7.37) the shape factor F_{2-1} is A_1/A_2. For a second example, consider the sides of a long equilateral triangle (perhaps an A-frame cabin). The shape factor from one side to another is by virtue of Eq. (7.40), one-half. By inspection, one-third is the shape factor between two sides of a regular tetrahedron. Also by inspection the shape factor F_{1-2} from a small area 1 adjacent to a large perpendicular area 2 is one-half.

In some other situations, Eq. (7.35) can be analytically integrated to yield the shape factor. It is easily shown that the shape factor from a small area to any surface of revolution, whose axis is coincident with the normal of the small area, is simply $\sin^2 \theta$, where θ is the angle from the normal of the area to the limiting ray tangent to or at the edge of the surface. For example, the shape factor from a small detector to a telescope mirror or lens can be found from this relation as can also the shape factor from a downward facing area on a satellite to the planet about which it orbits. The shape factor from a small area to an infinitely long object parallel to the area is equal to $(\sin\theta_a - \sin\theta_b)$, where θ_a and θ_b are the angles between the normal and the two limiting rays. The shape factor from an area A_1 on the inside of a sphere wall of radius R to another area A_2 also on the inside of the sphere wall is simply $A_2/4\pi R^2$.

In many other cases of practical interest, analytical or numerical results may be obtained from the definition Eq. (7.35) or its equivalent

$$A_1 F_{1-2} = \int_{A_1} \int_{2\pi} \cos\theta_1 \frac{d\omega_{1-2}}{\pi} dA_1$$

where ω denotes solid angle. The values in Figs. 7.5 and 7.6 were obtained in this manner.

Shape Factor Algebra

The information stored in graphs, such as Figs. 7.5 and 7.6, is more useful than might be thought at first. The applicability of the data can be extended by the use of shape factor algebra. Fig. 7.7 shows a simple example.

By clever manipulation of the limits of integration, a much more powerful theorem of shape factor algebra may be proven. For example, referring to Fig. 7.8, for two adjacent or opposite rectangles, sub-divided by a plane perpendicular to them, the therorem states

$$A_1 F_{1-4} = A_2 F_{2-3}$$

To illustrate the use of the above relation consider how we would determine F_{1-4} for the geometry of Fig. 7.8. We start by writing

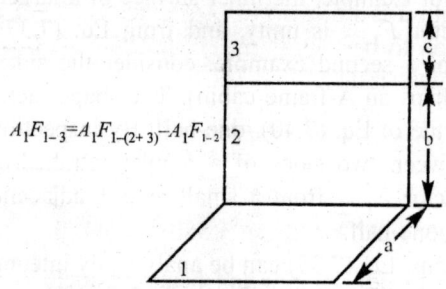

$$A_1 F_{1-3} = A_1 F_{1-(2+3)} - A_1 F_{1-2}$$

Fig. 7.7. A simple case of shape factor algebra.

$$A_{(1+2)} F_{(1+2)-(3+4)} = A_1 F_{1-3} + A_1 F_{1-4} + A_2 F_{2-3} + A_2 F_{2-4}$$

Then we introduce $A_1 F_{1-4} = A_2 F_{2-3}$

$$A_{(1+2)} F_{(1+2)-(3+4)} = A_1 F_{1-3} + 2A_1 F_{1-4} + A_1 F_{2-4}$$

so that we may solve for the desired quantity,

$$A_1 F_{1-4} = \tfrac{1}{2} \left[A_{(1+2)} F_{(1+2)-(3+4)} - A_1 F_{1-3} - A_2 F_{2-4} \right]$$

in terms of values obtainable from Fig. 7.6.

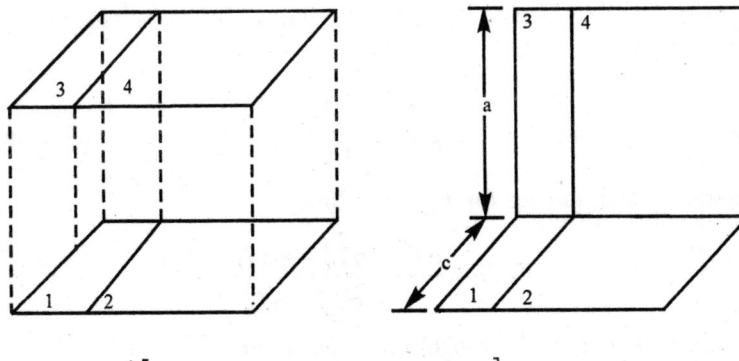

$$A_1F_{1-4} = \frac{1}{2}\left[A_{(1+2)-(3+4)} - A_1F_{1-3} - A_2F_{2-4}\right]$$

Fig. 7.8. A more complex example of shape factor algebra.

Example 7.4. Calculate the irradiation a 1 metre square plane area on a space vehicle immediately adjacent and perpendicular to a solar cell array 1.5 metres long. The solar cells are at 120°F and may be considered black. *Solution.* In this case we can use Fig. 7.6 directly. The length b is 1 metre as is the length c. Length a is 1.5 metres. Consequently, $L = 1$ and $N = 1.5$. The shape factor is seen to be approximately 0.225. The flux incident on surface 1, the 1 × 1 area, is then

$$q_1^- = \frac{\dot{Q}_{2-1}^+}{A_1} = \frac{A_2F_{2-1}\sigma T_2^4}{A_1} = F_{1-2}\sigma T_2^4$$

$$q_1^- = (0.225)\,(1.712 \times 10^{-9})(580)^4 = 44 \text{ Btu/hr ft}^2$$

Example 7.5. We wish to design a radiant heating panel on the ceiling of a hospital room so that the face of a bedridden patient would not be uncomfortably cold in 60°F air. The ceiling is 6 ft above the bed and the panel operates at 180°F. Even though painted a pleasant off-white matt colour the infrared emissivity can be 0.85 or 0.9 by virtue of the low temperature (Tables 7.3 and 7.6) so we can reasonably assume the panel to be black for heat radiation. As a rough design criterion we choose to size the panel so that the patient's face, modelled as a dry adiabatic black surface, attains a 90°F equilibrium temperature. *Solution.* We first make a heat balance on the patient's face; for a dry adiabatic surface

$$\dot{Q}_{in} = \dot{Q}_{out}$$

$$A_{\text{Panel}} \, F_{\text{P--F}} \sigma T_p^4 + A_{\text{Room}} \, F_{\text{R--F}} \sigma T_R^4 = A_{\text{Face}} \, \sigma T_F^4 + h_c A_{\text{Face}}(T_F - T_R)$$

We note

$$A_{\text{Panel}} F_{\text{P--F}} = A_{\text{Face}} F_{F-P}, \quad A_{\text{Room}} F_{\text{R--F}} = A_{\text{Face}} F_{F-R}$$

$$F_{F-P} + F_{F-R} = 1$$

Substituting and solving for F_{F-P} we find

$$F_{F-P} = \frac{\sigma T_F^4 - \sigma T_R^4 + h_c(T_F - T_R)}{\sigma T_P^4 - \sigma T_R^4}$$

The value of h_c is to be roughly 0.7 Btu/ft^2 hr°F.

$$F_{F-P} = \frac{(157 - 126) + 0.7(30)}{(287 - 126)} = \frac{52}{161} = 0.323$$

This shape factor requires say a circular disk subtending a half angle of $\sin^2\theta = 0.323$, $\theta = 34.7°$. A circular area would therefore have to have radius

$$R = 6 \tan (34.7°) = 4.15 \text{ ft}$$

and the resulting panel area is about 54 ft^2.

TRANSFER BETWEEN FINITE GRAY SURFACES

Non-black Opaque Surfaces

Although many surfaces are nearly black, others are not. Bare metals, in particular, reflect 90–98% of the irradiation. Ceramics, such as aluminium, beryllium, and magnesium oxides, reflect appreciably when irradiated with high temperature radiation. The fraction of the irradiation which is reflected is called the hemispherical reflectance ρ, and if the surface is opaque, the remainder α is absorbed. These two fractions sum to unity.

$$\alpha + \rho = 1 \qquad\qquad \text{... (7.41)}$$

The power absorbed per unit area when a non-black surface is irradiated by a surrounding black surface at T_2 is then

$$q_{\text{absorbed}} = \alpha q^- = \alpha \sigma T_2^4$$

Experimental observations indicate that a non-black surface emits radiation at a rate less than a black surface would at the same temperature. This fraction of the black body emissive power emitted by a non-black surface is called the hemispherical emittance ε. The radiation leaving surface 1 is thus composed of an emitted flux $\varepsilon_1 \sigma T_1^4$ and a reflected flux $\rho_1 q_1^- = \rho_1 \sigma T_2^4$,

$$q_1^+ = \varepsilon_1 \sigma T_1^4 + \rho_1 \sigma T_2^4$$

The net radiative flux leaving surface 1 is $q_1 = q_1^+ - q_1^-$,

$$q_1 = (\varepsilon_1 \sigma T_1^4 + \rho_1 \sigma T_1^4) - \sigma T_2^4 \qquad \ldots (7.42a)$$

or, rearranging,

$$q_1 = \varepsilon_1 \sigma T_1^4 - (1 - \rho_1)\, \sigma T_2^4$$

But from Eq. (7.41) $\alpha = 1 - \rho$; therefore

$$q_1 = \varepsilon_1 \sigma T_1^4 - \alpha_1 \sigma T_2^4 \qquad \ldots (7.42b)$$

We see that the net radiative flux leaving a surface can be regarded as either the difference between the radiosity q^+ and the irradiation q^-, Eq. (7.42a), or as the difference between the emitted flux $\varepsilon_1 \sigma T_1^4$ and the absorbed flux $\alpha_1 \sigma T_2^4$, Eq. (7.42b).

The second law of thermodynamics states that no energy can flow from a colder body to a warmer one. Thus ε can be no larger than unity, for if ε were to exceed unity, $\varepsilon \sigma T_1^4 - \alpha \sigma T_2^4$ could be greater than zero for $T_1 < T_2$. Recall that the absorptivity α cannot exceed unity.

An important special case, which has been exploited widely in engineering practice, is that of a gray surface. A surface is said to be gray when its absorptivity equals its emittance, that is, $\alpha = \varepsilon$. In general, surfaces have properties which are a function of the energy of the photons being emitted or absorbed. Gray surfaces are special ones whose properties are independent of photon energy. For the present we will restrict our attention to systems having gray surfaces and develop some useful engineering relations.

Diffuse Reflection

A further restriction on the development that follows is that the reflection is perfectly diffuse. Reflection from rough ceramics or sintered metals may often be regarded as perfectly diffuse to a good approximation. However, even mirrorlike surfaces, called specular surfaces, may be regarded as perfectly diffuse when they are diffusely irradiated.

A perfectly diffuse surface has no preferred direction of reflection; its brightness does not change with direction of view. The important practical consequence of diffuseness is that the same shape factor which applies to radiation transfer between black surfaces also applies to radiation transfer between non-black perfectly diffuse surfaces.

Transfer in an Enclosure

Consider an opaque surface on the inside of a space vehicle, a room, an industrial furnace, or other such enclosure. Let the surface be sub-divided and the sub-divisions assigned numbers, $i = 1, 2, ..., m$. The total incoming radiation for the i surface from all the other surfaces, including i itself, if it is convace, is

$$A_i q_i^- = \sum_{j=1}^{m} q_j^+ A_j F_{j-i}$$

Introducing the shape factor reciprocal relation, Eq. (7.37), gives

$$A_i q_i^- = \sum_{j=1}^{m} A_i F_{i-j} q_j^+$$

or

$$q_i^- = \sum_{j=1}^{m} F_{i-j} q_j^+ \qquad ...(7.43)$$

As before, the radiosity q_j^+ can be written as a sum of emitted and reflected fluxes.

$$q_j^+ = \varepsilon_j \sigma \, T_j^4 + \rho_j q_j^- \qquad ... (7.44)$$

Substituting Eq. (7.44) into Eq. (7.43) and rearranging yields

$$q_i^- - \sum_{j=i}^{m} F_{i-j} \rho_j q_j^- = \sum_{j=1}^{m} F_{i-j} \varepsilon_j \sigma T_j^4 \qquad ... (7.45a)$$

Using the delta function δ_{ij}, which is zero when $i \neq j$ and unity when $i = j$, permits us to write

$$\sum_{j=1}^{m} (\delta_{ij} - F_{i-j} \rho_j) q_j^- = \sum_{j=1}^{m} F_{i-j} \varepsilon_j \sigma T_j^4 \qquad ... (7.45b)$$

This set of m simultaneous linear algebraic equations in the m unknowns $q_j^-, j = 1, 2, . . ., m$ can be solved by any of the methods of linear algebra, for example, Cramer's rule, successive elimination, or matrix inversion. Alternatively iteration can be used in a numerical solution procedure.

Any set of linear algebraic equations can be interpreted in terms of a linear electrical network. The Oppenheim radiation network interprets heat flux $\dot{Q}$ as current and radiosity q^+ as voltage. The net radiative power leaving surface i is

$$Q_i = A_i q_i = A_i q_i^+ - A_i q_i^-$$

Substituting for q_i^- from Eq. (7.43), and using the property of shape factors summing to unity, yields

$$\dot{Q}_i = A_i q_i^+ - A_i \sum_{j=1}^{m} F_{i-j} q_j^+$$

$$\dot{Q}_i = \sum_{j=1}^{m} A_i F_{i-j} (q_i^+ - q_j^+) \qquad \dots (7.46)$$

To relate the heat flow $\dot{Q}_i$ to the surface temperature T_i, we obtain q_i^- from Eq. (7.44), replacing the subscript j by subscript i, and substitute it into

$$Q_i = A_i q_i^+ - A_i q_i^-$$

to obtain

$$\dot{Q}_i = A_i q_i^+ - A_1 \frac{q_i^+ - \varepsilon_i \sigma T_i^4}{\rho_i}$$

$$= \frac{(1-\rho_i)A_i}{\rho_i} \left[\frac{\varepsilon_i}{1-\rho_i} \sigma T_i^4 - q_i^+ \right]$$

But $1 - \rho_i = \alpha_i$, and $\alpha_i = \varepsilon_i$ for a gray surface; thus

$$\dot{Q}_i = \frac{(1-\rho_i)A_i}{\rho_i} \left[\sigma T_i^4 - q_i^+ \right] \qquad \dots (7.47)$$

Equations (7.46) and (7.47) then permit a network to be constructed as shown in Fig. 7.9.

The source voltage V_i is

$$V_i = \sigma T_i^4 \qquad \dots (7.48)$$

The total radiant power $\dot{Q}_i$ is represented by the current flowing from source voltage V_i through the surface resistance R_i to the surface voltage q_i^+, where

$$R_i = \frac{\rho_i}{(1-\rho_i)A_i} = \frac{1-\varepsilon_i}{\varepsilon_i A_i} \qquad \dots (7.49)$$

Likewise, the network interpretation of Eq. (7.46) is that of a shape factor resistance R_{ij} connecting two surface nodes with voltages q_i^+ and q_j^+, respectively

$$R_{ij} = \frac{1}{A_i F_{ij}} \qquad \ldots (7.50)$$

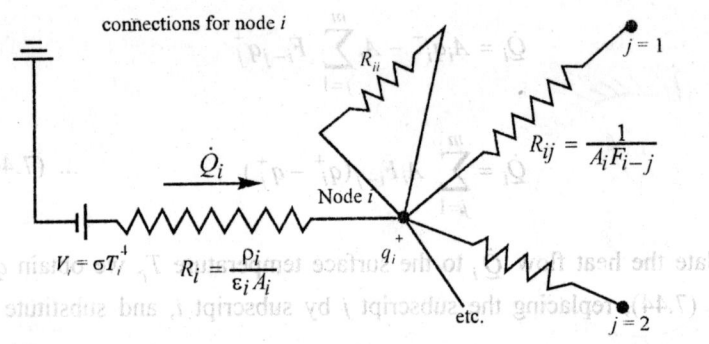

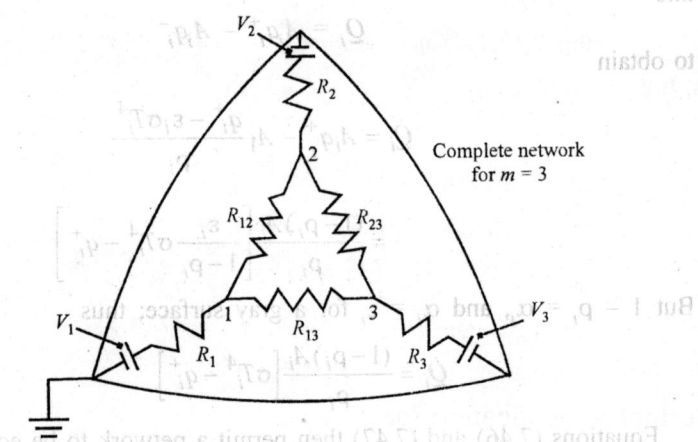

Fig. 7.9. Radiation network.

As an example, consider the radiant power transfer between two long concentric cylinders, the inner one at temperature T_1 with area $A_1 = \pi D_1 L$. By inspection $F_{1-2} = 1$. The network is shown in Fig. 7.10. The net power transfer is the current flowing in the network

$$\dot{Q} = \frac{1}{R_1 + R_{12} + R_2}(\sigma T_1^4 - \sigma T_2^4)$$

$$\dot{Q} = \frac{1}{\frac{1-\varepsilon_1}{\varepsilon_1 A_1} + \frac{1}{A_1} + \frac{1-\varepsilon_2}{\varepsilon_2 A_2}} (\sigma T_1^4 - \sigma T_2^4)$$

$$\dot{Q} = \frac{\varepsilon_1 A_1}{1 + \frac{\varepsilon_1 A_1}{\varepsilon_2 A_2}(1-\varepsilon_2)} (\sigma T_1^4 - \sigma T_2^4) \qquad \dots (7.51)$$

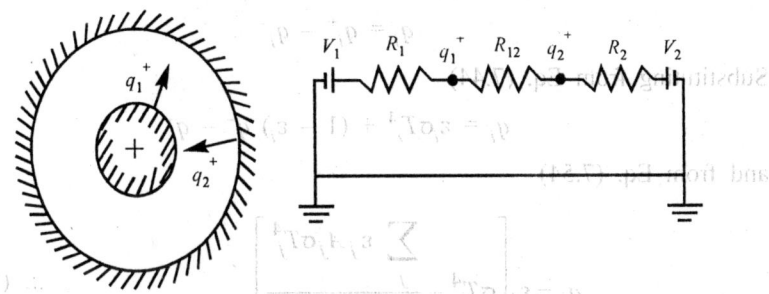

Fig. 7.10. Radiation network for two concentric cylinders.

This result reduces to a particularly simple one when

$$(\varepsilon_1 A_1/\varepsilon_2 A_2)(1-\varepsilon_2)$$

is very small, perhaps because $\varepsilon_1 A_1/\varepsilon_2 A_2$ is small or because $1-\varepsilon_2$ is small. Then we obtain

$$\dot{Q} = \varepsilon_1 A_1(\sigma T_1^4 - \sigma T_2^4) \qquad \dots (7.52)$$

This equation is of great practical importance, for it applies to a common situation.

Further Simplification for an Enclosure

Radiant transfer between surfaces which lie upon the inside wall of a sphere is particularly simple to calculate because the shape factor F_{i-j} degenerates to a function of surface j only.

$$F_{i-j} = \frac{A_j}{A_{\text{total}}}, \qquad A_{\text{total}} = 4\pi R^2 \qquad \dots (7.53)$$

In this case Eq. (7.43) yields a constant q^- independent of i. Equation (7.45a) can thus be solved for q^-

$$q^- = \frac{\sum\limits_{j=1}^{m} \varepsilon_j A_j \sigma T_j^4}{A_{\text{total}} - \sum\limits_{j=1}^{m} A_j \rho_j} = \frac{\sum\limits_{j=1}^{m} \varepsilon_j A_j \sigma T_j^4}{\sum\limits_{j=1}^{m} \varepsilon_j A_j} \qquad \text{... (7.54)}$$

The net radiative flux leaving surface i is

$$q_i = q_i^+ - q_i^-$$

Substituting from Eq. (7.44)

$$q_i = \varepsilon_i \sigma T_i^4 + (1 - \varepsilon_i) q^- - q^-$$

and from Eq. (7.54)

$$q_i = \varepsilon_i \left[\sigma T_i^4 - \frac{\sum\limits_{j} \varepsilon_j A_j \sigma T_j^4}{\sum\limits_{j} \varepsilon_j A_j} \right] \qquad \text{... (7.55)}$$

Multiplying both sides by A_i, changing the index in the numerator from j to k, and rearranging gives

$$\dot{Q}_i = \sum_{k} A_i T_{i-k} (\sigma F_k^4 - \sigma T_k^4) \equiv \sum_{k} \dot{Q}_{i-k}$$

where $F_{i-k} \equiv \varepsilon_i \varepsilon_k A_k / \sum\limits_{j} \varepsilon_j A_j$. The quantity q_{i-k} may be interpreted as the net transfer between i and k. Replacing A_k by $F_{i-k} A_{\text{total}}$ gives

$$\dot{Q}_{i-k} = A_i F_{i-k} \frac{\varepsilon_i \varepsilon_k}{\frac{1}{A_{\text{total}}} \sum\limits_{j} \varepsilon_j A_j} (\sigma T_i^4 - \sigma T_k^4) \qquad \text{... (7.56a)}$$

where the denominator is ε_{avg}, the area weighted average emittance of the enclosure.

$$\dot{Q}_{i-k} = A_i F_{i-k} (\sigma T_1^4 - \sigma T_k^4) \left(\frac{\varepsilon_1 \varepsilon_k}{\varepsilon_{\text{avg}}} \right) \qquad \text{... (7.56b)}$$

Another simple result is obtained when ε_k coincides with ε_{avg} or when ε_i coincides with ε_{avg}. In the former case

$$\dot{Q}_{i\text{-}k} = A_i F_{i\text{-}k} \varepsilon_i \, (\sigma T_i^4 - \sigma T_k^4), \quad \varepsilon_k = \varepsilon_{avg} \qquad \dots (7.57)$$

while in the latter case

$$\dot{Q}_{i\text{-}k} = A_i F_{i\text{-}k} \varepsilon_k \, (\sigma T_i^4 - \sigma T_k^4), \quad \varepsilon_i = \varepsilon_{avg} \qquad \dots (7.58)$$

In summary, we have discussed about a fairly powerful general method for obtaining an engineering answer for radiant power transfer in an enclosure. When the enclosure is not too elongated, Eq. (7.56b) or one of two simpler ones found from it, Eqs. (7.57) and (7.58), may be used. In each case the total power from any surface i is

$$\dot{Q}_i = \sum_{k=1}^{m} \dot{Q}_{i\text{-}k} \qquad \dots (7.59)$$

Example 7.5. Let us estimate the net heat flux on the floor of a long furnace 10 ft by 10 ft with side walls at 2500°F and a roof at 2000°F. Such furnaces are commonly used in stress relieving, annealing, enamelling, melting, and other heating processes. The floor is 500°F and all surfaces are gray and diffuse and have an emittance 0.5.

Solution. We superimpose the network on a sketch of the furnace in Fig. 7.11.

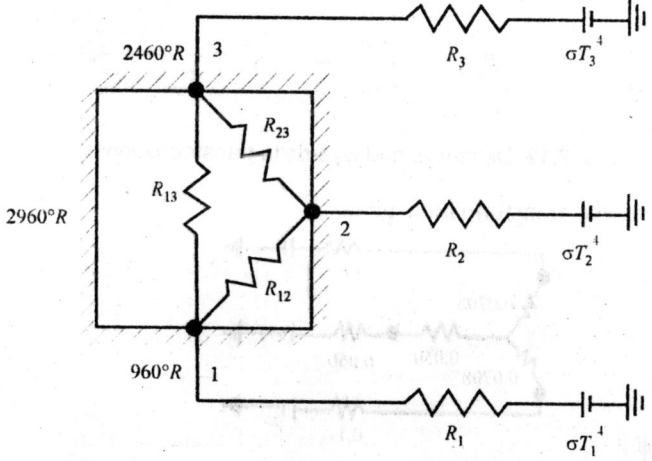

Fig. 7.11. Radiation network superposed on a furnace.

For a rough estimate we take each wall to be a single node. By symmetry the side walls can be treated together. The shape factor $F_{1\text{-}3}$ is 0.414 from Fig.

$$R_3 = R_1 = \frac{1-\varepsilon_1}{\varepsilon_1 A_1} = \frac{0.50}{(0.50)(10)} = 0.1$$

$$R_2 = \frac{1-\varepsilon_2}{\varepsilon_2 A_2} = \frac{0.50}{(0.50)(20)} = 0.05$$

$$R_{13} = \frac{1}{A_1 F_{1-3}} = \frac{1}{(10)(0.414)} = 0.2415$$

$$R_{12} = R_{23} = \frac{1}{A_1 F_{1-2}} = \frac{1}{10(1-0.414)} = 0.1707$$

The network is simplified by means of the delta-wye transformation shown in Fig. 7.12.

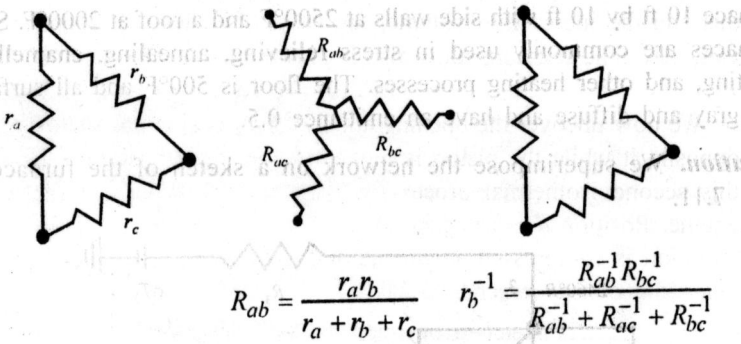

$$R_{ab} = \frac{r_a r_b}{r_a + r_b + r_c} \qquad r_b^{-1} = \frac{R_{ab}^{-1} R_{bc}^{-1}}{R_{ab}^{-1} + R_{ac}^{-1} + R_{bc}^{-1}}$$

Fig. 7.12. Delta-wye and wye-delta transformations.

First a delta-wye transformation gives

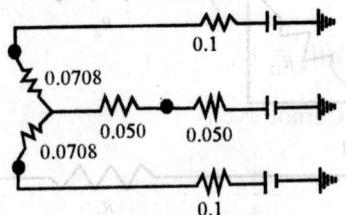

STEFAN-BOLTZMANN LAW

We are now ready to describe Boltzmann's analysis of an ideal engine being radiation as the working substance and operating on a Carnot cycle. The engine consists of a cylinder of non-heat conducting material, a frictionless piston of the same material, and an opening in the head through which radiation may enter or leave (Fig. 7.13a). The cylinder is evaluated and an isothermal enclosure S_1 maintained at temperature T_1 is placed opposite the opening in the head. The cylinder will then fill with radiation from S_1 to a density ψ_1. The initial pressure is then $p_1 = \frac{1}{3}\psi_1$ and the volume v_a. We now cause the piston to move slowly upward to position b, the pressure remaining constant by the addition of energy from S_1. Since the energy in the cylinder increases by $\psi_1 (v_b - v_a)$ and the work done is $p_1(v_b - v_a)$ $= \frac{1}{3}\psi_1 (v_b - v_a)$, the energy coming in must equal

$$Q_1 = \frac{4}{3}\psi_1(v_b - v_a) \qquad \qquad \dots (7.60)$$

When the piston reaches position b, the opening is covered, S_1 removed, and further expansion to position c takes place. The external work done during this adiabatic process is supplied by the radiation. Both because of this and the increase in volume, the energy density will decrease to some value ψ_2. The pressure must also decrease to p_2 corresponding to ψ_2 and T_2.

We now uncover the opening and place an isothermal enclosure S_2 at T_2 opposite it while the piston is moved slowly from position c to d. During this second isothermal process, radiant energy in amount Q_2 leaves the engine. Position d is selected so that when the opening is closed and the radiation is compressed adiabatically, the initial volume v_a will be reached when the density reaches ψ_1.

This cycle is sketched on pv coordinates in Fig. 7.13b. If the pressure change is assumed to be very small, we may write $(T_1 - T_2) = dT$, $(\psi_1 - \psi_2) = d\psi$ and $dp = (\frac{1}{3})d\psi$. Furthermore the net work of the cycle, dW_k, would be simply

$$dW_k = (v_b - v_a)dp = \frac{1}{3}(v_b - v_a)d\psi$$

But since for a Carnot cycle $\eta = dW_k/Q_1 = (T_1 - T_2)/T_1$, we have

$$\frac{dW_k}{Q_1} = \frac{\frac{1}{3}(v_b - v_a)d\psi}{\frac{4}{3}(v_b - v_a)\psi} = \frac{dT}{T}$$

Dropping the subscripts leaves

$$\frac{d\psi}{\psi} = 4\frac{dT}{T}$$

and integration gives

$$\psi = KT^4 \qquad \qquad \dots (7.61)$$

where K is a constant. The relationship between the emissive power of a black body and the energy density within an enclosure at the same temperature is given by equation (7.16); combining this with equation (7.61) gives

$$E_b = \frac{Kc}{4} T^4 = \sigma T^4 \qquad \qquad \dots (7.62)$$

This equation has been identified earlier as the Stefan-Boltzmann law. σ is known as the Stefan-Boltzmann constant for which the value 0.1714×10^{-8} Btu/hr ft$^2 R^4$ is recommended.

SPECTRAL DISTRIBUTION OF RADIATION

The Stefan-Boltzmann law for the total emissive power E_b of a black body has been well confirmed experimentally and one might also expect it to reveal something about monochromatic emission, $E_{\lambda b}$. Let us pursue this possibility by considering what happens to the spectral distribution of a beam of radiation when it is reflected from a moving surface as the piston in the engine of Fig. 7.13a.

When a beam of radiation of wavelength λ strikes a moving surface, the reflected beam experiences a Doppler effect in that its wave length is increased. If the angle of the incident wave is θ and the surface is moving at a velocity u much less than that of light, the increase in wavelength $d\lambda$ can be shown to be

$$d\lambda = 2 \frac{u}{c} \lambda \cos\theta \qquad \qquad \dots (12.63)$$

To determine the time rate of wavelength changes at any given wave length, assume the walls of the cylinder to be perfectly reflecting and that at any time the length is l (Fig. 7.14). A beam of radiation of wavelength λ then preserves its angle θ with the axis and strikes the piston $c \cos\theta/2l$ times a second or $c \cos\theta \Delta \tau/2l$ times in the time increment $\Delta \tau$. From equation (7.63) the increase in wavelength in this time interval would then be

$$\Delta\lambda = \frac{d\lambda c \cos\theta \Delta\tau}{2l} = \frac{u\lambda}{l} \cos^2\theta \Delta\tau$$

and as $\Delta\tau$ approaches an infinitesimal

$$\frac{d\lambda}{d\tau} = \frac{u}{l}\lambda \cos^2 \theta \qquad \ldots (7.64)$$

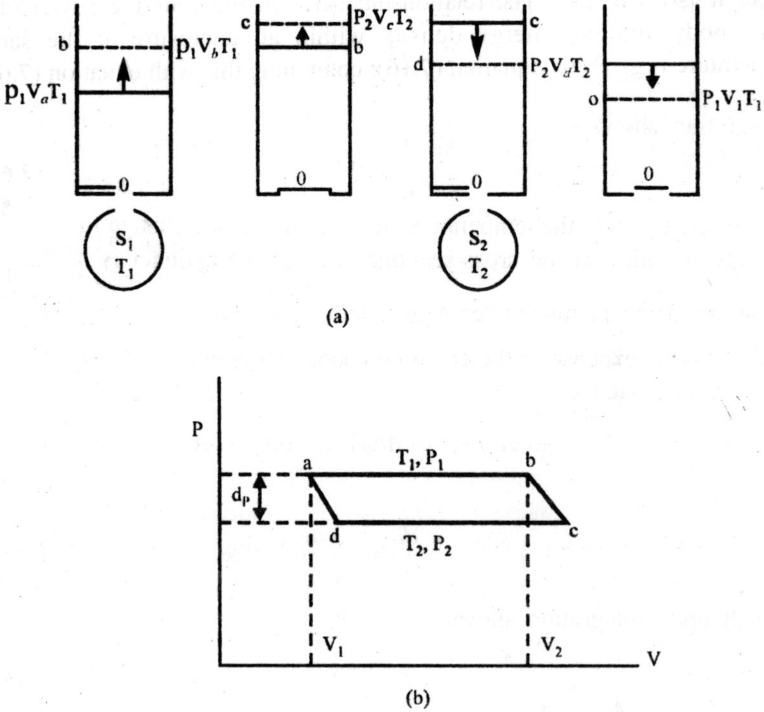

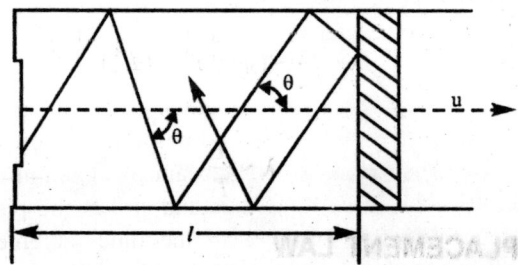

(a)

(b)

Fig. 7.13. Radiation engine operating on a Carnot cycle.

Fig. 7.14. Reflection of radiation of wavelength λ in an engine cylinder.

As in the evaluation of the radiation pressure in order to calculate the average rate of change of λ, the variation with θ must be determined. This can be shown to be the same as for the radiation pressure so that the average rate of increase of waves of length λ is

$$\frac{d\lambda}{d\tau} = \frac{u\lambda}{3l} \qquad \qquad ... (7.65)$$

Substituting $u = dl/d\tau$ leads to

$$\frac{d\lambda}{\lambda} = \frac{1}{3}\frac{dl}{l}$$

Integration shows

$$\lambda \sim l^{\frac{1}{3}} \qquad \qquad ... (7.66)$$

To relate this to the temperature, let us consider the change in the energy density ψ with l. If the cross-sectional area of the cylinder is A, the work done when the piston moves a distance dl is $pAdl = \frac{1}{3}\psi Adl$. This work is done at the expense of the enclosed energy and would, therefore, be equal to $-d(Al\psi)$. Hence

$$\frac{1}{3}\psi Adl = -d(Al\psi) = -Ald\psi - A\psi dl$$

or

$$\frac{dl}{l} = -\frac{3}{4}\frac{d\psi}{\psi}$$

which upon integration shows

$$l \sim \psi^{-\frac{3}{4}} \qquad \qquad ... (7.67)$$

Now since ψ is proportional to T^4 we find from equations (7.66) and (7.67) that

$$\lambda \sim l^{\frac{1}{3}} \sim (\psi^{-\frac{3}{4}})^{\frac{1}{3}} \sim (T^4)^{-\frac{1}{4}}$$

and

$$\lambda \sim \frac{1}{T} \qquad \qquad ... (7.68)$$

WIEN DISPLACEMENT LAW

The above result will enable us to establish an important relationship between the spectral distribution of black body radiation at different temperatures. Consider what happens to the energy per unit volume $\psi_\lambda d\lambda$ associated with waves of length between λ and $\lambda + d\lambda$ during an adiabatic expansion. For any particular range $d\lambda_1$ at temperature T_1, and adiabatic expansion to temperature T_2 will, according to equation (7.68), result in a change in the wavelength λ_1 to λ_2, and $\lambda_1 + d\lambda_1$ to $\lambda_2 + d\lambda_2$ such that

$$\frac{\lambda_2}{\lambda_1} = \frac{T_1}{T_2}, \quad \frac{\lambda_2 + d\lambda_2}{\lambda_1 + d\lambda_1} = \frac{T_1}{T_2}$$

which means that

$$\frac{d\lambda_2}{d\lambda_1} = \frac{T_1}{T_2} \qquad \qquad \dots (7.69)$$

Although the wavelength range is increased, the amount of energy associated with it decreases. This will be in the same ratio as the total energies because the same arguments involved in obtaining equation (7.67) could be applied to radiation in the range $d\lambda_1$ only. Hence the ratio of the energies in the wavelength ranges $d\lambda_1$ and $d\lambda_2$ must equal

$$\frac{\psi_{\lambda_2} d\lambda_2}{\psi_{\lambda_1} d\lambda_1} = \frac{\psi_2}{\psi_1} = \frac{T_2^4}{T_1^4}$$

Substituting for $d\lambda_2/d\lambda_1$ from equation (7.69) leads to

$$\frac{\psi_{\lambda_2}}{\psi_{\lambda_1}} = \frac{T_2^5}{T_1^5} \qquad \qquad \dots (7.70)$$

and since $E_{\lambda b}$ is proportional to ψ_λ at the same temperature

$$\frac{E_{\lambda b_2}}{E_{\lambda b_1}} = \frac{T_2^5}{T_1^5} \qquad \qquad \dots (7.71)$$

Thus for black body radiation at two absolute temperatures T_1 and T_2, we see from equations (7.68) and (7.71) that if the wave of length λ_2 at T_2 is displaced from that of length λ_1 at T_1, such that $\lambda_2 T_2 = \lambda_1 T_1$, the monochromatic emissive powers at these two wavelengths are directly proportional to the fifth powers of the absolute temperatures. This very useful relationship is known as the Wien displacement law.

In looking at this another way, note that when $\lambda_1 T_1 = \lambda_2 T_2 = \lambda T$, equation (7.71) shows

$$\frac{E_{\lambda b_1}}{T_1^5} = \frac{E_{\lambda b_2}}{T_2^5} = \frac{E_{\lambda b}}{T^5}$$

This indicates that for a given value of λT, the ratio of $E_{\lambda b}/T^5$ for all temperatures is the same; furthermore if $E_{\lambda b}/T^5$ is plotted versus λT, all the points should lie on a single curve. The validity of this conclusion has been demonstrated with experimental measurements. Fig. 7.15 shows the energy distribution curves of black body radiation for several temperatures obtained by Lummer and Pringsheim. That these data can be represented by a single curve when plotted as $E_{\lambda b}/T^5$ versus λT is clearly shown in Fig. 7.16. This also means that

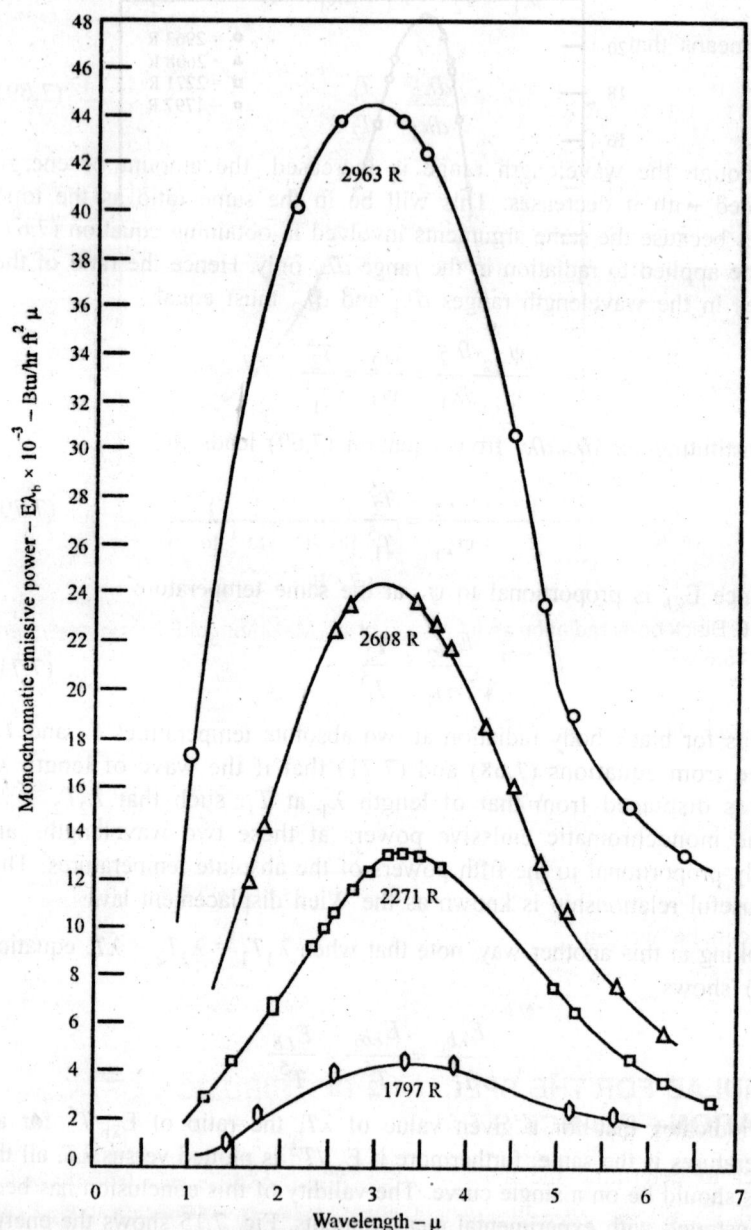

Fig. 7.15. Spectral distribution of energy in black body radiation at several temperatures.

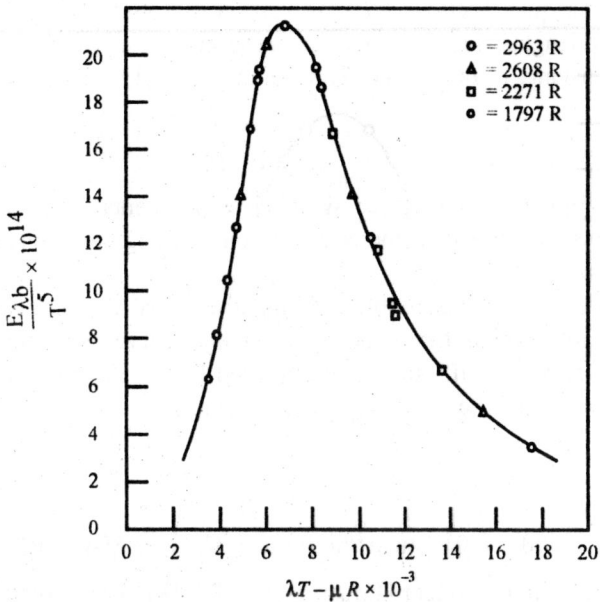

Fig. 7.16. Black body radiation as a function of λT. Maximum of E_λ/T^5 occurs when $\lambda T = 5215.6 \ \mu$ R.

$$\frac{E_{\lambda b}}{T^5} = f(\lambda T)$$

or,

$$E_{\lambda b} = T^5 f(\lambda T) = \frac{(\lambda T)^5 f(\lambda T)}{\lambda^5}$$

which can be written

$$E_{\lambda b} = \frac{1}{\lambda^5} F(\lambda T) \qquad \qquad ... \ (7.72)$$

FORMULAE FOR THE SPECTRAL DISTRIBUTION OF RADIATION—PLANCK'S LAW

In order to determine the unknown function $F(\lambda T)$ is was necessary to consider the process of emission and absorption. A formula proposed by Wien involving two constants

$$E_{\lambda b} = C_1 \lambda^{-5} e^{-c_2/\lambda T} \qquad \qquad ... \ (7.73)$$

proved to be quite accurate up to wavelengths around 2μ, but was low for larger wavelengths. Jeans derived an expression for $E_{\lambda b}$, involving only the known constants c and k (the Boltzmann constant); i,e.,

$$E_{\lambda b} = \frac{2\pi k}{c} \frac{T}{\lambda^4} \qquad \ldots (7.74)$$

This formula, which agreed with the experimental data only at large wavelengths, was based on the equipartition of energy principle of classical mechanics.

In working with this problem Plank discovered a modification of the Wien formula which fitted the experimental curve. Seeking to modify classical theory to justify this expression led him to the introduction of the quantum concept. The similarity of his formula

$$E_{\lambda b} = \frac{C_1 \lambda^{-5}}{e^{c_2/\lambda T} - 1} \qquad \ldots (7.75)$$

$(C_1 = 1.1870 \times 10^8$ Btu μ^4/hr ft^2 and $C_2 = 25,896$ μ R$)$

to that of Wien is evident. Equation (7.75) obviously satisfied equation (7.72). That it is also consistent with the Stefan-Boltzmann law can be proved by showing the integral $E_b = \int_0^\infty E_{\lambda b} d\lambda$ is equal to σT^4.

SPECTRAL DISTRIBUTIONS

Distribution Concept

The free molecule transfer only from a total macroscopic point of view, since most engineering is based upon such a view. In the remainder of the chapter a second pass is made through the subject matter from a more microscopic point of view so that non-gray walls can be treated.

Not all the molecules emitted by a wall have the same velocity nor do all the photons emitted have the same energy. The differences in behaviour of molecules or photons due to differences in energy are used to good advantage by engineers. For example, a photon with an energy of less than 1 ev (the work done on a charge equal to the electron's charge in forcing it through a potential difference of 1 volt, 1 joule = 6.2418 × 10^{18} ev) produces no useful energy in a silicon solar cell but merely raises its temperature and thus decreases its efficiency. The engineer employs a filter which reflects the photons of undesirable energies. Ordinary window glass transmits photons very well; approximately 90% pass through, and only 2% are absorbed, when the photons have energies between approximately 0.45 and 4 ev; but the glass absorbs photons, again approximately 90% of them when the photons have somewhat lower energies, and essentially none are transmitted. In a hothouse the glass transmits during the day solar radiation needed by the plants for photosynthesis and both during the day and night prevents cooling of the plants by radiation to the cold sky, because the glass is opaque to thermal photons and emits nearly the same energy back to the plants as is emitted by the plants to the glass. These phenomena which depend upon different behaviours for photons of different energies are said to be non-gray or spectrally selective. We will show how such phenomena can be treated using the concept of a distribution function.

Velocity, Energy, and Momentum Relations

We are concerned primarily with two types of carriers in this introductory treatment, photons and monoatomic ideal gas molecules. In special circumstances we will treat polyatomic molecules. For other carriers such as electrons, polyatomic molecules in the general case, and ions, the reader should consult more advanced texts. The monoatomic molecule in a large container may have any velocity v (we exclude velocities approaching the speed of light) and at that velocity has momentum mv and energy $1/2\ mv^2$ (neglecting electronic excitation which occurs at high temperatures), where m is the mass of the molecule (1 amu equals 1.6603×10^{-27} kg, $m = M$.amu).

A photon in vacuum has velocity $c = 2.997925 \times 10^8$ m/sec, momentum hv_f/c, and energy hv_f, where h is Planck's constant, $h = 6.6256 \times 10^{-34}$ joule sec, and v_f is the frequency in cycles per second, which may have any value. The relation between energy E and frequency is known as Einstein's photoelectric law. Albert Einstein was able to show that the action of light on a surface is to cause electrons to be emitted only when the photons have more energy than E, the surface work function.

The frequency of a 1 ev photon is enormous. For this reason, wavenumber v is often used instead of v_f; wavenumber is simply frequency divided by the velocity of light. It consequently has units of reciprocal length; a commonly used unit is cm^{-1}. The reciprocal of wavenumber has units of length and it is consequently called wavelength λ. Commonly used units are angstroms, 1 Å $= 10^{-10}$ metre, and microns, 1 $\mu = 10^{-6}$ metre. Table 7.3 summarises these relations (recall Table 7.2).

The concept of the average internal energy of a molecule is perhaps the most complicated one in Table 7.3. We have separated the average total energy into two parts, the average translational kinetic energy,

$$\bar{E}_t = 1/2m\bar{v}^2 = 3/2kT \qquad \qquad ...(7.76)$$

Table 7.3. Momentum and energy relations.

	Monoatomic molecule	Photon
Velocity	v	c
Momentum	$p_m = mv$	$p_m = hv_f/c = hv = h/\lambda$
Energy	$E_t = 1/2\ mv^2$	$E = hv_f = hcv = hc/\lambda$
	$\overline{E}_t = 1/2m\overline{v}^2 = 3/2kT$	
	$\overline{E}_{vr} = m\hat{u} - 3/2kT$	

and the average energy polyatomic molecules carry due to vibrations and rotations,

$$\overline{E}_{vr} = mu - 3/2kT = mc_v T - 3/2kT \qquad ...(7.77)$$

However, a simplifying factor is that many polyatomic gases, particularly diatomic ones, are vibrationally unexcited at ordinary temperatures. If they are vibrationally unexcited but rotationally excited,

$$mc_v = (3 + N_r)\ (1/2k) \qquad ...(7.78)$$

where N_r is the number of rotational degrees of freedom (zero for a monoatomic gas, two for a linear molecule, and three for a non-linear one). Whether or not this is so can be inferred from the ratio of specific heats $\gamma = c_p/c_v$. For an ideal gas

$$mc_p = mc_v + k \qquad (c_p = c_v + R, R = \Re/M) \qquad ...(7.79)$$

$$\gamma = 1 + \frac{k}{mc_v} = 1 + \frac{2}{3 + N_r} \qquad ...(7.80)$$

If γ drops below this value at high temperatures, it is a sign that the vibrational modes are becoming excited.

For example, consider molecular nitrogen N_2, which is imagined to be a linear molecule and can thus rotate about two mutually perpendicular axes:

$$\gamma \doteq 1 + \frac{2}{5} = 1.4$$

The value holds up to approximately 1000°C. Or consider water vapour, a non-linear molecule with three degrees of rotation :

$$\gamma \doteq 1 + \frac{2}{6} = 1.33$$

This value holds true for moderate pressures and temperatures up to roughly 200°C.

Distributions at Thermodynamic Equilibrium

Picture an enclosure at thermodynamic equilibrium filled with a monoatomic gas or filled with photons (or both, since interactions will not occur when the molecules are not electronically excited). The molecules are distributed on the average uniformly over the volume with a total number per unit volume $N = P/kT$. (Recall $k = 1.38054 \times 10^{-23}$ joule/°K). They are colliding randomly and, as a result on the average moving uniformly in all directions (that is, if a small volume of them were suddenly plucked from the enclosure and placed at the centre of a large sphere, and the molecules continued in motion without further collisions the number eventually hitting the sphere per unit area would be uniform over the sphere). The number of molecules dN whose energies lie between E and $E + dE$ is given by Maxwell-Boltzmann statistics as

$$dN = N \frac{2}{\sqrt{\Pi}} \left(\frac{E}{kT} \right)^{1/2} \exp\left(-\frac{E}{kT} \right) \frac{dE}{kT} \qquad ...(7.81)$$

Bose-Einstein statistics gives the number of photons per unit volume dN whose energies lie between E and $E + dE$

$$dN = \frac{8\pi}{(hc/kT)^3} \frac{(E/kT)^2}{\exp(E/kT) - 1} \frac{dE}{kT} \qquad ...(7.82)$$

We take Eqs. (7.81) and (7.82) as the foundation for what follows. The student is instructed to take them on faith. There is no attempt here to derive them. A good course or text in statistical thermodynamics or statistical mechanics is recommended for those who would like to appreciate their foundation. Let us note that Eq. (7.81) will be used in what follows to derive the Planck and Stefan radiation laws which are experimentally observed to be valid.

It should be apparent that the number of molecules between two energies E_1 (say 0.01 ev) and E_2 (say 0.1 ev) can be found by integrating from E_1 to E_2 with respect to E. Changing variables of integration from E to $E^* = E/kT$ gives

$$\Delta N = N \frac{2}{\sqrt{\pi}} \int_{E_1/kT}^{E_2/kT} E^{*1/2} \exp(-E^*) \, dE^* \qquad ...(7.83)$$

The student may readily carry out such an integration, for example, using a Simpson routine on a digital computer. However, one is more often interested in how many carriers crossing a surface per unit time have energies between E_1 and E_2 and how much power they transfer.

Surface Fluxes

Consider a surface of area dA located inside our enclosure at thermodynamic equilibrium. Particles coming up through the surface in a particular direction have a velocity normal to the surface of $v \cos \theta$ where θ is the angle between the normal and direction of interest. Fig. 7.17 shows this projection.

But how many particles come in this direction? To answer such a question we imagine a sphere with its centre at the centroid of our small surface, and we mark off an area dA_s on the sphere. Then we can more properly ask, how many particles come from dA in directions which cross the small area dA_s on the sphere?

The number of particles per unit volume in the energy range between E and $E + dE$ is dN. The number per unit volume times the normal velocity component times the area is then the number crossing the surface per unit time. The fraction of these particles which would cross dA_s is dA_s/A_s where A_s is the total area of our imagined sphere $4\pi r^2$; because, as we have said, the particle trajectories are uniformly distributed in direction. Denoting the flow across the plane containing dA and $d\dot{N}$, we can write

$$dN = dNv \cos \theta \, dA \left(\frac{dA_s}{A_s} \right) \qquad ...(7.84)$$

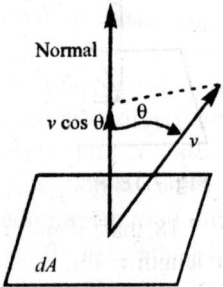

Fig. 7.17. Normal velocity component.

Solid Angles

In thinking about particles travelling in directions crossing a small area dA_s on a sphere we find it convenient to use the concept of a solid angle. Recall that a plane angle in radians (unitless) is the arc length on a circle divided by the radius of the circle and that there are consequently 2π radians in a complete circle. In the same manner let a solid angle $d\omega$ in steradians (unitless) be the area on a sphere divided by the radius squared.

$$d\omega = \frac{dA_s}{r^2} \qquad ..(7.85)$$

There are then 4π steradians in a sphere. We refer to the particles crossing an area dA_s on a sphere as having directions lying within the solid angle $d\omega$ given by Eq. (7.85). The term dA_s/A_s in Eq. (7.84) can then be written

$$\frac{dA_s}{A_s} = \frac{r^2 d\omega}{4\pi r^2} = \frac{d\omega}{4\pi} \qquad ...(7.86)$$

Equation (7.48) is therefore written

$$d\dot{N} = dN \, v \cos\theta \, dA \frac{d\omega}{4\pi} \qquad ...(7.87)$$

It is well to have a coordinate system with which to specify direction. Polar coordinates are convenient. Fig. 7.18 shows polar coordinates adopted here to be a polar angle θ measured from the z-axis, the surface normal, and an azimuthal angle ϕ measured from the x-axis to the projection of the particle trajectory in the x-y plane.

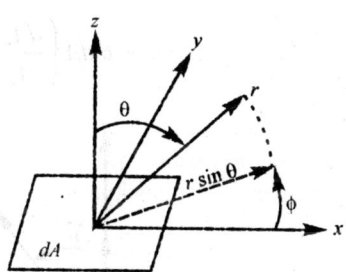

Fig. 7.18. Polar coordinates.

It is apparent in Fig. 7.18 that, if one were to vary ϕ by $d\phi$, the lever arm in the x-y plane of length $r \sin\theta$ would strike an arc of $(r \sin\theta) \, d\phi$, perpendicular to the plane containing the surface normal and r. Varying θ by $d\theta$ would strike an arc of length $r \, d\theta$ lying in the plane. The area of the sphere of radius r struck out would be $r \, d\theta \, r \sin\theta \, d\phi$. Consequently, from Eq. (7.85) the solid angle swept out by varying θ by $d\theta$ and ϕ by $d\phi$ is

$$d\omega = \frac{r d\theta \, r \sin\theta \, d\phi}{r^2} = \sin\theta \, d\theta \, d\phi \qquad ...(7.88)$$

To sweep out a sphere one would vary θ from 0 to π and ϕ from 0 to 2π. We can readily verify that

$$\int_0^{4\pi} d\omega = \int_0^{2\pi} \int_0^{\pi} \sin\theta \, d\theta \, d\phi = 4\pi \qquad ...(7.89)$$

Similarly, to sweep out a hemisphere θ would be varied from 0 to $\pi/2$ and ϕ from 0 to 2π.

Radiant Intensity

To obtain the power-crossing our surface dA we simply multiply $d\dot{N}$ given by Eq. (7.87) by the energy carried by each particle. There results for photons (velocity $v = c$).

$$d\dot{Q} = E d\dot{N}$$

$$d\dot{Q} = (hcv)(dN\, c\cos\theta\, dA\, d\omega\, /\, 4\pi) \qquad ...(7.90)$$

where dN is given by Eq. (7.82). Rearrangement of Eq. (7.90) yields

$$I_p = \frac{d\dot{Q}}{(dA\cos\theta)\,d\omega\,dv} = \frac{2hc^2v^3}{\exp\,(hcv\,/\,kT)-1}$$

a quantity independent of direction. It is termed the Planckian spectral radiant intensity in engineering literature (Planckian radiance in physics literature). The left-hand equation of the two relations above, when rearranged, plays the same role in radiative transport as Fourier's law does in diffusive transport,

$$d\dot{Q} \equiv IdA\cos\theta\, d\omega\, dv \qquad ...(7.91)$$

The relation is written without the subscript P because the source need not be Planckian in nature for a spectral radiant intensity to be defined. The right-hand expression gives the Planckian or black body spectral radiant intensity

$$I_P(v,T) = \frac{2hc^2v^3}{\exp\,(hcv\,/\,kT)-1} \qquad ...(7.92)$$

The term black body is used because thermodynamic equilibrium radiation is approached in the laboratory by having a small opening in a suitably designed large cavity with isothermal walls. When the cavity is cold, it appears black to the eye, because nearly every photon entering the cavity from the room is absorbed. (In an ideal cavity every photon entering would be absorbed).

Radiant Flux

To obtain the total flow of power crossing dA in all directions to the hemisphere above it and for photons of all possible wave numbers from zero to infinity we simply integrate Eq. (7.91) over $d\omega$ and dv. Consider first the integration with respect to solid angle over the hemisphere.

Substituting Eq. ((7.88) into Eq. (7.91) and integrating over a hemisphere gives the heat flow per unit area per wavenumber as

$$q^+(v) = \frac{dq^+}{dv} = \int_0^{2\pi} \int_0^{\pi/2} I^+ \cos\theta \sin\theta \, d\theta \, d\phi \qquad ...(7.93)$$

where the $+$ denotes that the flux is leaving the surface and q is, as before, $\dot{Q}/A$. The quantity $q^+(v) \, dv$ is the flux, power per unit area, crossing dA whose photon wavenumbers lie between v and $v + dv$. In engineering terminology $q^+(v)$ is termed the spectral radiosity; spectral because of the narrow band of the spectrum dv in question, and radiosity because the photons leaving in all possible directions are included.

We know a Planckian radiator has an intensity I^+ independent of direction. Such a source is said to be diffuse. For a diffuse source I^+ may be factored out from under the integral in Eq. (7.93). The integration then yields simply π, as shown below,

$$\int_0^{2\pi} \cos\theta \, d\omega = \int_0^{2\pi} \int_0^{\pi/2} \cos\theta \sin\theta \, d\theta d\phi = 2\pi \int_0^{\pi/2} \cos\theta \sin\theta \, d\theta$$

Let $u = \sin^2\theta$, $du = 2 \sin\theta \cos\theta \, d\theta$.

$$\int_0^{2\pi} \cos\theta \, d\omega = \pi \int_0^1 du = \pi \qquad ...(7.94)$$

For a perfectly diffuse source Eq. (7.93) thus becomes

$$q^+(v) = \pi I^+ \qquad ...(7.95)$$

To obtain the total flux, we integrate $q^+(v)$ over all possible photon wavenumbers from zero to infinity. Note that the probability of zero wavenumber photons is zero, from Eq. (7.92), like $2kTcv^2$ in the limit, and that for infinite wavenumber is also zero, like $2hc^2v^3 \exp(-hcv/kT)$. From a practical point of view we do not need to start at zero or integrate all the way to infinity. To be rigorous, however, we include all possibilities by integrating over the complete range.

$$q_p{}^+ = \int_0^\infty \pi I_p(v,T) \, dv$$

$$q_p{}^+ = \int_0^\infty \frac{2\pi hc^2v^3}{\exp(hcv/kT)-1} \, dv$$

Changing variables of integration from v to $\xi = hcv/kT$ yields

$$q_p^+ = \frac{2\pi k^4 T^4}{h^3 c^2} \int_{\zeta=0}^{\infty} \frac{\zeta^3 d\zeta}{\exp{(\zeta)} - 1}$$

The definite integral has the value $\pi^4/15$; therefore

$$q_p^+ = \frac{2\pi^5 k^4}{15h^3 c^2} T^4 \qquad \qquad ...(7.96)$$

The coefficient multiplying T^4 is the Stefan-Boltzmann constant σ

$$\sigma = \frac{2\pi^5 k^4}{15h^3 c^2} = 5.6697 \times 10^{-8} \text{ w}/\text{m}^2 {}^\circ\text{K}^4 \qquad \qquad ...(7.97)$$

$$= 1.712 \times 10^{-9} \text{ Btu}/\text{hr ft}^2 {}^\circ\text{R}^4$$

External Fraction

We have found that the total flux, power per unit area, leaving the opening in a black body cavity is σT^4. To obtain the total flux of radiation of photons with wavenumbers greater than a certain value v_1 we would integrate not from zero to infinity but from v_1 to infinity. For example, photons more energetic than 1.1 ev operate a silicon solar cell, so that we might wish to know what fraction of the radiant power is carried by photons more energetic than this value. The fraction f would be

$$f = \frac{\int_{v_1}^{\infty} \pi I_p dv}{\sigma T^4} = \frac{\int_{v_1}^{\infty} \dfrac{2\pi h c^2 v^3}{\exp(hcv/kT) - 1} dv}{(2\pi^5 k^4/15h^3 c^2) T^4} \qquad \qquad ...(7.98)$$

Again, changing variables from v to $\zeta = hcv/kT$ gives

$$f\left(\frac{hcv_1}{kT}\right) = \frac{15}{\pi^4} \int_{\zeta=hcv_1/kT}^{\infty} \frac{\zeta^3 d\zeta}{\exp{(\zeta)} - 1} \qquad \qquad ...(7.99)$$

Table 7.4 shows this function. It is, in statistics terminology, the ogive to the Planck distribution. To obtain the fraction of the power between two wavenumbers v_1 and v_2 ($v_2 > v_1$) we simply subtract the fraction above the higher wavenumber from that above the lower one

$$\Delta f = f\left(\frac{hcv_1}{kT}\right) - f\left(\frac{hcv_2}{kT}\right) \qquad \qquad ...(7.100)$$

For example, 90% of black body radiant power is carried by photons with wavenumbers greater than $v_1 = 1.53 \ kT/hc$, while 10% of the power is carried by photons with wavenumbers less than $v_2 = 6.55\ kT/hc$. Thus 80% of Planckian radiant power is carried by photons with wavenumbers between $v_1 = 1.53\ kT/hc$ and $v = 6.55\ kT/hc$.

Table 7.4. External fraction for the Planck distribution.

$\dfrac{hcv}{kT}$	f	$\dfrac{hcv}{kT}$	f
0.0	1.00	3.50	0.50
0.628	0.99	3.75	0.45
0.811	0.98	4.02	0.40
1.06	0.96	4.30	0.35
1.24	0.94	4.61	0.30
1.53	0.90	4.97	0.25
1.83	0.85	5.37	0.20
2.10	0.80	5.88	0.15
2.34	0.75	6.55	0.10
2.58	0.70	7.35	0.06
2.80	0.65	7.97	0.04
3.03	0.60	8.96	0.02
3.26	0.55	9.90	0.01

We can use the above numbers to characterise the wavenumber range of greatest interest for a given temperature. For example, consider a room temperature black body at $T = 300°K$, a red-hot black body at $1000°K$, a black body at a temperature typical for a lamp filament of $3000°K$, and a black body at a temperature of the outer layers of our sun, $5700°K$. Table 7.5 shows the wavenumbers and wavelengths encompassing 80% of the radiation. The values are found from the preceding values $hcv/kT = 1.53, 6.55$, and the value of $hc/k \doteq 1.44$ cm $°K$. Note that the lower wavenumber corresponds to the longer wavelength.

Table 7.5. Wavenumbers and wavelengths encompassing 80% of Planckian radiation.

Source temperature, °K	Wavenumber range, cm^{-1}	Wavelength range, microns
300	319–1365	7.3–31.4
1000	1062–4550	2.2–9.4
3000	3190–13,650	0.73–3.1
5700	6050–25,900	0.39–1.7

Molecular Flux

Equation (7.87) gives the number of molecules per unit time crossing an area dA inside solid angle $d\omega$ and with energies between E and $E + dE$. If we wish to know the total number of molecules per unit area crossing in all directions and with all possible energies, we integrate over all directions in the hemisphere and over all possible energies from zero to infinity. We have already found that

$$\int_{2\pi} \cos\theta \, d\omega = \pi$$

Hence we obtain the result after the first integration simply by replacing $\cos\theta \, d\omega$ in Eq. (7.87) with π. Denoting the flux as J^+, we have

$$dJ^+ = \frac{1}{4} v \frac{dN}{dE} dE \qquad \qquad ...(7.101)$$

Substituting Eq. (7.81), using $E = 1/2\, mv^2$ to express v as $(2E/m)^{1/2}$, and integrating with respect to E gives

$$J^+ = \int_0^\infty \frac{1}{4}\left(\frac{2E}{m}\right)^{1/2} N \frac{2}{\pi^{1/2}} \left(\frac{E}{kT}\right)^{1/2} \exp\left(-\frac{E}{kT}\right) \frac{dE}{kT}$$

$$= \frac{1}{4}\left(\frac{8kT}{\pi m}\right)^{1/2} N \int_0^\infty x e^{-x} dx, \qquad \text{where } x = \frac{E}{kT}$$

$$J^+ = \frac{1}{4} N \left(\frac{8kT}{\pi m}\right)^{1/2}$$

Comparison of Eq. (7.101) with this result shows that the quantity $(8kT/\pi m)^{1/2}$ is simply the mean speed. We therefore write the molecular flux in a particularly simple form

$$J^+ = \frac{1}{4} N\bar{v}, \qquad \bar{v} = \left(\frac{8kT}{\pi m}\right)^{1/2} \qquad \qquad ...(7.102)$$

To obtain the power transferred by a monoatomic molecular flux we multiply the flux of carriers by the energy carried. From Eqs. (7.87) and (7.81)

$$d\dot{Q}^+ = E d\dot{N} = E dN\, v \cos\theta \, dA \frac{d\omega}{4\pi}$$

$$dq^+ = \frac{dQ^+}{dA} = EN\frac{2}{\pi^{1/2}}\left(\frac{E}{kT}\right)^{1/2}\exp\left(-\frac{E}{kT}\right)\frac{dE}{kT}v\cos\theta\frac{d\omega}{4\pi}$$

Again we substitute for v the relation $(2E/m)^{1/2}$ and integrate over a hemisphere with respect to $d\omega$ and over all possible energies from zero to infinity.

$$q^+ = \frac{1}{4\pi}\int_{2\pi}\cos\theta\,d\omega\frac{2N}{\pi^{1/2}}\int_0^\infty E\left(\frac{E}{kT}\right)^{1/2}\left(\frac{2E}{m}\right)^{1/2}\exp\left(-\frac{E}{kT}\right)\frac{dE}{kT}$$

$$= \frac{1}{4}\left(\frac{8kT}{\pi m}\right)^{1/2}kTN\int_0^\infty x^2 e^{-x}dx$$

$$q^+ = \frac{1}{4}\bar{v}N(2kT) = kJ^+(2kT) \qquad \qquad ...(7.103)$$

The average energy per carrier in the volume is not $2kT$; it may be found by multiplying Eq. (7.81) by E and integrating with respect to E from zero to infinity. For a monoatomic gas it is $3/2\ kT$. However, the more energetic molecules in the volume are faster and cross an area dA more frequently. It is for this reason that the energy flux is obtained by multiplying the molecular flux by $2kT$ rather than a lesser value.

As discussed earlier, polyatomic molecules carry energy not only in their translational kinetic energy but also in their rotations and vibrations. These latter energies are quantised; that is, only certain discrete values are allowed, although in the case of rotation the energy levels may be so closely spaced that they may often be regarded as continuous. The average internal energy of such a molecule is mu, where $\hat{u}$ is the internal energy per unit mass. Of this total, the amount contained in the rotational and vibrational energies is on the average $mu - 3/2\ kT$, since the translational energy on the average is $3/2\ kT$. If the vibrational-rotational energy is independent of the translational energy (a good approximation), then the power crossing (in one direction) a plane in a polyatomic gas is

$$q^+ = J^+\left(m\hat{u} + \frac{1}{2}kT\right) \qquad \qquad ...(7.104)$$

A complicating factor is the fact that during a molecule-surface collision, the vibrational energies are usually transferred much less efficiently than are the translational and rotational energies. As might be imagined, there is a probabilistic nature to surface-molecule interactions. Some molecules which impinge upon a surface stick for a rather long time, during which period their energies are accommodated to the distribution characteristic of the surface temperature. Others stick not at all or only very briefly so that little energy is transferred. In the main, a high fraction, say 90%, of the incident molecules stick long enough to many types of dirty engineering surfaces to give up their translational and rotational energies and acquire those characteristic of the surface. We have already noted that many molecules are vibrationally unexcited at ordinary temperatures; thus this complication does not affect transport in those cases.

SPECTRAL VARIATIONS IN WALL CHARACTERISTICS

Surface System Concept

Thus far in our second pass we have obtained the fluxes of photons or molecules across an imaginary surface inside a large enclosure at thermodynamic equilibrium. We wish to use rates under the ideal conditions of thermodynamic equilibrium as a measure of the transport rates at a wall. We restrict ourselves in this introductory treatment to a type of wall whose temperature is sufficiently uniform so that we can, if need be, measure it unambiguously; that is, we consider here only solid walls and rule out perhaps a poorly conducting porous wall such as a fibreglass filter pad. For a solid opaque wall we construct two imaginary surfaces, the s-surface in the medium just outside the wall, and the m-surface sufficiently deep within the wall so that no photons or molecules penetrate from the surface and so that fourier's law of conduction will hold there. We apply the first law of thermodynamics (the energy accounting principle) to the system between the s- and m-surfaces. The volume between these boundaries constitutes a surface system. Such a system is often called (loosely) in engineering, a *surface*. Emphasis is placed on thermal radiation in what follows, but the same type of analysis can be applied for molecules.

Photon energy fluxes crossing the s-surface are the outgoing flux (spectral radiosity) $q_r^+(v)$ and the incoming flux, the spectral irradiation, $q_r^-(v)$. The net spectral radiant flux going away from the wall system is then

$$q_{r,s}(v) = q_r^+ (v) - q_r^- (v) \qquad ...(7.105)$$

There could also be some conduction at the s-surface $q_{c,s}$. At the m-surface there is only conduction in the absence of convection or mass transfer. The heat flux in this instance is

$$q_m = -\left(k \frac{\partial T}{\partial y} \right)_m \qquad ...(7.106)$$

where y is measured in the direction of the outward normal to the surface. The mass between the s- and m-surfaces is ordinarily quite small, so that even during a fast transient it is not likely that the rate of energy stored in this mass is worth considering. We can therefore write, in the absence of convection,

$$-\left(k\frac{\partial T}{\partial y}\right)_m = q_{r,s} + q_{c,s} \qquad ...(7.107)$$

This energy balance states simply that the energy reaching a surface by conduction is radiated and conducted away across the s-plane in the medium outside the wall.

Emittance, Absorptance, and Reflectance

An experiment can be performed as follows: Heat a wall to temperature T_w in a space with cold black walls and observe the spectral radiant intensity of the surface from an area dA_w on the surface and in directions close to θ, ϕ (Fig. 7.18) within a solid angle $d\omega$ subtended by a mirror on lens focusing the power on a detector behind a filter. The filter restricts the photons received by the detector to those with wavenumbers between v and $v + dv$ (wavelengths between $\lambda - d\lambda$ and λ, where $\lambda = 1/v$ and $d\lambda = dv/v^2$). The detector might be a thermistor or thermocouple in the middle of a thin rectangular or circular fin so that the radiant power heats the fin and changes the resistance of emf of the sensor. Compare the signal so obtained with that from a Planckian black body cavity at the same temperature. Since the surrounds are black and cold, all the radiation from the wall must have been emitted and not reflected. We call the ratio of the observed intensities the spectral directional emittance (also called emissivity)

$$\varepsilon(\theta,\phi,v,T_w) = \frac{I_w(\theta,\phi,v)}{I_p(v,T_w)} \qquad ...(7.108)$$

In the same way the total, directional emittance is defined as

$$\varepsilon(\theta,\phi,T_w) = \frac{I_{w,T}(\theta,\phi)}{\sigma T_w^4 / \pi} \qquad ...(7.109)$$

We can also perform an experiment in which we measure how much a wall, at temperature T_w, is heated by directing onto area dA_w a beam from solid angle $d\omega$ about direction, θ, ϕ. A heater is situated behind the wall; its power supply is initially adjusted to balance losses so that the wall is at temperature T_w. When the external radiant source is turned on, the heater power supply is correspondingly reduced to maintain the wall temperature at T_w. The power transfer change at the m-surface is then simply equal to the decrease in power supply to the heater. We define the following :

Spectral directional absorptance

$$\alpha(\theta,\phi,v,T_w) = \frac{\Delta q_m(v)\ dv\ dA_w}{q^-(v)\ dv\ dA_w} \qquad ...(7.110)$$

Total directional absorptance

$$\alpha_T(\theta,\phi,T_e,T_w) = \frac{\Delta q_m dA_w}{q^- dA_w} \qquad ...(7.111)$$

where, if the external source is black,

$$q^-(v)\ dv\ dA_w = I_P\ (v,\ T_e)\ \cos\theta\ dA_w\ d\omega\ dv \qquad ...(7.112)$$

$$q^-\ dA_w = (\sigma T_e^4/\pi)\ \cos\theta\ dA_w\ d\omega \qquad ...(7.113)$$

The remaining power from the source must have been reflected, that is, it must have been delivered to the surroundings which see the s-surface. (Recall that our surface system is opaque, that is, it is not transmitting). It can be arranged to detect the change in power transferred to the surrounds when the source is turned on and find that, within experimental uncertainties, the power not absorbed is indeed reflected so that energy is conserved. Note that when the wall material does not fluoresce, the reflected power has the same wavenumbers as the incident flux. We define the following :

Spectral directional reflectance

$$\rho(\theta,\phi,v,T_w) = \frac{\Delta q^+(v)}{q^-(v)} \qquad ...(7.114)$$

Total directional reflectance

$$\rho_T(\theta,\phi,T_e,T_w) = \frac{\Delta q^+}{q^-} \qquad ...(7.115)$$

where the denominators are given by Eqs. (7.112)) and (7.113). (Note that ρ_T and α_T are T_e-dependent since the v wavenumber distribution is T_e-dependent). For an opaque surface system, conservation of energy requires that

$$\rho\ (\theta,\ \phi,\ v,\ T_w) + \alpha(\theta,\ \phi,\ v,\ T_w) = 1 \qquad ...(7.116)$$

$$\rho_T\ (\theta,\ \phi,\ T_e,\ T_w) + \alpha_T\ (\theta,\ \phi,\ T_e,\ T_w) = 1 \qquad ...(7.117)$$

These relations between reflectance and absorptance are thus a direct consequence of the first law of thermodynamics.

Tables 7.6 and 7.7 show measured spectral reflectances of a few surfaces used in technology. The data were obtained with angles of incidence θ of 25° and with T_w at room temperature.

Table 7.6. Reflectances of some paints and coatings at an angle of incidence of 25° from the normal.

	Reflectance at room temperature					
λ, microns	*3M Black velvet*	*Hard anodised aluminium*	*Anodised titanium*	*White epoxy paint*	*Flame sprayed alumina*	*Aluminium paint*
0.3	0.03	0.05	–	–	0.40	–
0.35	0.03	0.06	–	–	0.52	–
0.4	0.03	0.07	–	0.40	0.66	0.75
0.45	0.03	0.07	–	0.88	0.71	0.75
0.5	0.03	0.07	0.47	0.90	0.73	0.75
0.6	0.03	0.07	0.52	0.85	0.76	0.74
0.7	0.03	0.07	0.52	0.79	0.77	0.71
0.8	0.03	0.07	0.52	0.91	0.77	0.69
1.0	0.03	0.08	0.50	0.92	0.75	0.72
1.5	0.03	0.10	0.50	0.70	0.68	0.75
2.0	0.04	0.15	0.48	0.57	0.49	0.77
3.0	0.04	0.08	0.11	0.07	0.27	0.77
4.0	0.04	0.26	0.24	0.10	0.47	0.77
5.0	0.05	0.30	0.24	0.10	0.38	0.78
6.0	0.04	0.17	0.18	0.09	0.12	0.78
8.0	0.09	0.04	0.17	0.07	0.02	0.74
10.0	0.05	0.02	0.10	0.09	0.02	0.78
12.0	0.05	0.16	0.09	0.07	0.26	0.79
15.0	0.06	0.17	0.12	0.10	0.21	0.80
20.0	0.06	0.20	0.15	0.16	0.25	0.81
30.0	0.03	0.20	–	0.19	–	–
40.0	0.03	–	–	0.20	–	–

A useful relation between the spectral directional absorptance and spectral directional emittance may be derived as follows. The Principle of Detailed Balancing of Statistical Thermodynamics (from which the second law of thermodynamics, in its classical form, results) states that two systems cannot exchange energy when their energy levels are populated with thermodynamic equilibrium populations of the same temperature. Consider two small surfaces; surface 1 is non-black and surface 2 is black. The power emitted by 1 and absorbed by 2 is

$$\varepsilon(\theta, \phi, v, T_1) \, \pi I_P \, (v, T_1) \, A_1 F_{1-2}$$

Table 7.7. Reflectances of some bright metals at an angle of incidence of 25° from the normal.

λ, microns	Reflectance at room temperature					
	Aluminium	Chromium	Copper	Gold	Stainless steel	Titanium
0.3	0.95	0.48	–	0.20		0.29
0.35	0.95	0.52	–	0.22	0.39	0.35
0.4	0.93	0.57	–	0.25	0.43	0.41
0.45	0.93	0.60	–	0.26	0.46	0.44
0.5	0.92	0.61	0.47	0.40	0.47	0.47
0.6	0.89	0.63	0.77	0.83	0.51	0.52
0.7	0.88	0.63	0.86	0.89	0.54	0.53
0.8	0.86	0.63	0.90	0.92	0.56	0.57
1.0	0.92	0.60	0.94	0.957	0.66	0.56
1.5	0.96	0.66	0.968	0.966	0.72	0.60
2.0	0.965	0.74	0.971	0.973	0.75	0.66
3.0	0.971	0.81	0.971	0.975	0.80	0.71
4.0	0.974	0.855	0.978	0.977	0.83	0.76
6.0	0.979	0.912	0.980	0.978	0.86	0.80
8.0	0.981	0.922	0.980	0.978	0.88	0.83
10.0	0.982	0.935	0.982	0.980	0.89	0.85
15.0	0.985	0.950	0.982	0.980	0.91	0.87
20.0	0.986	0.953	0.982	0.980	0.923	0.89
30.0	0.987	0.964	–	–	0.938	0.91
40.0	0.988	0.970	–	–	0.947	0.92

The power emitted by 2 and absorbed by 1 is $\alpha(\theta, \phi, v, T_1) \, \pi I_P \, (v, T_2) A_2 F_{2-1}$. When $T_1 = T_2$ these rates must be equal. Since $A_1 F_{1-2} = A_2 F_{2-1}$, we have

$$\alpha(\theta, \phi, v, T_1) = \varepsilon(\theta, \phi, v, T_1) \qquad \qquad ...(7.118)$$

This relation equating spectral directional absorptance to spectral directional emittance is sometimes called Kirchoff's law.

Total and Hemispherical Characteristics

As we have seen, we can speak of either a spectral characteristic or a total characteristic. It is obvious that the total value can be calculated from spectral values; for example, consider absorptance. From Eqs. (7.110) and (7.112)

$$\Delta q_m (v) = \alpha(\theta, \phi, v, T_w)I_p(v, T_e) \cos\theta \, d\omega \qquad ...(7.119)$$

If wavenumbers of all possible values are considered, the total Δq_m is obtained by integration

$$\Delta q_m = \int_0^\infty \Delta q_m(v) \, dv = \int_0^\infty \alpha(\theta,\phi,v,T_w)I_P(v,T_e) \, dv \cos\theta \, d\omega \qquad ...(7.120)$$

But from Eqs. (7.111) and (7.113)

$$\Delta q_m = \alpha_T (\theta, \phi, Te, T_w)\sigma T_e^4 \cos\theta \, \frac{d\omega}{\pi} \qquad ...(7.121)$$

Comparison of Eqs. (7.121) and (7.120) gives

$$\alpha_T(\theta,\phi,T_e,T_w) = \frac{\int_0^\infty \alpha(\theta,\phi,v,T_w)\pi I_P(v,T_e) \, dv}{\sigma T_e^4} \qquad ...(7.122)$$

If we introduce the quantity f defined by Eq. (7.98a), we can write

$$\alpha_T(\theta,\phi,T_e,T_w) = \int_0^\infty \alpha(\theta,\phi,v,T_w)\left[\frac{-df(hcv/kT_e)}{dv}\right] dv = \int_0^1 \alpha \, df \qquad ...(7.123)$$

A procedure for calculating $\alpha_T (\theta, \phi, T_e, T_w)$ is then as follows: Graph $\alpha(\theta, \phi, v, T_w)$ versus $f (hcv/kT_e)$ and integrate graphically over the interval 0 to 1. Of course numerical integration is to be preferred over graphical integration, but graphical integration has been described to make the procedure clear. In numerical integration a list of absorptances would be read into the computer, perhaps at preselected wavenumbers for which f had desired values. Then a Simpson or Gaussian summing would be performed.

Just as one likes to know the total emittance for all wavelengths, one likes to know the hemispherical emittance for all directions. A black body emits a radiant flux σT_w^4 of all wavelengths and in all directions in a hemisphere. Using this measuring stick, we define the total hemispherical emittance of a real body as

$$\varepsilon_{TH}(T_w) = \frac{q^+}{\sigma T_w^4}, \quad (q^- = 0) \qquad ...(7.124a)$$

We can relate the total hemispherical emittance to the total directional emittance by an integration similar to that in Eq. (7.93). First we return to Eq. (7.91),

$$dq^+ \, dA_w = I_w \cos\theta \, dA_w \, d_\omega.$$

Then we introduce the directional emittance from Eq. (7.109),

$$dq^+ = \varepsilon_T (\theta, \phi, T_w) (\sigma T_w^4/\pi) \cos\theta \, d_\omega.$$

Finally we integrate over the hemisphere

$$\frac{q^+}{\sigma T_w^4} = \frac{1}{\pi} \int_0^{2\pi} \int_0^{\pi/2} \varepsilon_T(\theta, \phi, T_w) \cos\theta \sin\theta \, d\theta \, d\phi$$

$$\varepsilon_{TH}(T_w) = \frac{1}{\pi} \int_0^{2\pi} \int_0^{\pi/2} \varepsilon_T(\theta, \phi, T_w) \cos\theta \sin\theta \, d\theta \, d\phi \qquad ...(7.124b)$$

For a surface with azimuthally random roughness or pigmentation ε is not a function of ϕ. In such a case Eq. (7.123b) can be written

$$\varepsilon_{TH}(T_w) = \int_0^1 \varepsilon_T(\theta, T_w) d(\sin^2\theta) \qquad ...(7.125)$$

In this case the integration can be readily accomplished by regarding ε_T (θ, T_w) to be a function of $\sin^2\theta$.

Hemispherical absorptance and reflectance refer to the fraction of the irradiation absorbed and reflected, respectively when the irradiation is perfectly diffuse, that is, when the incoming intensity I^- is independent of direction θ, ϕ. The same kind of averaging indicated by Eq. (7.123b) then applies.

To conclude this section we note that, for an opaque surface system, there are three system characteristics with which the engineer is often concerned: emittance, absorptance, and reflectance. The two types of averages which are of interest are the total average

$$\varepsilon_T(T_w) = \frac{1}{\sigma T_w^4} \int_0^\infty \varepsilon(v, T_w) \pi I_P(v, T_w) \, dv$$

$$\alpha_T(T_w, T_e) = \frac{1}{\sigma T_w^4} \int_0^\infty p(v, T_w) \pi I_P(v, T_e) \, dv$$

$$\rho_T(T_w, T_e) = \frac{1}{\sigma T_e^4} \int_0^\infty p(v, T_w) \pi I_p(v, T_e)\, dv$$

and the hemispherical average

$$\varepsilon_H = \frac{1}{\pi} \int_0^{2\pi} \int_0^{\pi/2} \varepsilon(\theta, \phi) \cos\theta \sin\theta\, d\theta\, d\phi$$

$$\alpha_H = \frac{1}{\pi} \int_0^{2\pi} \int_0^{\pi/2} \alpha(\theta, \phi) \cos\theta \sin\theta\, d\theta\, d\phi$$

$$\rho_H = \frac{1}{\pi} \int_0^{2\pi} \int_0^{\pi/2} p(\theta, \phi) \cos\theta \sin\theta\, d\theta\, d\phi$$

The two types of average can be taken independently or be combined. For example, the total directional emittance is the ratio of the total intensity of all wavenumbers emitted in a certain direction divided by the total Planckian intensity, and the total hemispherical emittance is the ratio of the total flux of all wavenumbers and in all directions emitted divided by the total flux for the Planckian radiator σT_w^4.

For molecules the absorptance is often referred to as the accommodation coefficient or sticking coefficient, for it is a measure of the fraction of the molecules striking the wall which stick and accommodate themselves (by exchanging energy) to the wall temperature. The notion of a spectral quantity is the same, that is, a quantity which applies only a energies between E and $E + dE$. The concept of a total characteristic is the same with the averaging being done over a Maxwell-Boltzmann distribution. The concept of a hemispherical characteristic is entirely analogous to the case for photon transport.

Example 7.7. Consider the flame sprayed alumina in Table 7.6. Firebrick is often made of alumina and operates typically at 2500°F. Find the total emittance of firebrick at 2500°F using the data in Table 7.6.

Solution. Since the data in Table 7.6 were obtained at a specimen temperature of approximately 80°F, we cannot calculate $\varepsilon(T)$ for $T = 2500°F$. However, we can calculate $\alpha(T_w, T_e)$, where $T_w = 80°F$, $T_e = 2500°F$. If $\alpha(v, T_w)$ is not a strong function of T_w, the value $\alpha(T_w, T_e)$ is a reasonable approximation to $\varepsilon(T_e)$.

$$\alpha(T_w, T_e) = \int_0^1 \alpha(v, T_w)\, df\left(\frac{hcv}{kT_e}\right)$$

From Table 7.4 and a graph of Table 7.6 ($\alpha = 1 - \rho$, $\lambda = 1/v$) we construct Table 7.8, a table of $\alpha(v, T_w)$ versus f. Note $hc/kT_e = 1.44$ cm °K/1643°K = (1/1141) cm. Integration then yields an approximate value for $\varepsilon(T_e)$ of 0.55.

Table 7.8. Computation table for example 7.7.

f	hcv/kT	v	λ	α	f	hcv/kT	v	λ	α
(1)	(2)	(3)	(4)	(5)	(1)	(2)	(3)	(4)	(5)
		(2)1141	10^4/(3)				(2)1141	10^4/(3)	
0.01	9.90	11,300	0.885	0.24	0.55	3.26	3720	2.69	0.64
0.02	8.96	10,220	0.979	0.24	0.60	3.03	3460	2.89	0.73
0.04	7.97	9090	1.101	0.26	0.65	2.80	3190	3.14	0.70
0.06	7.35	8390	1.192	0.27	0.70	2.58	2945	3.40	0.64
0.10	6.55	74.80	1.338	0.29	0.75	2.34	2670	3.75	0.56
0.15	5.88	6710	1.490	0.32	0.80	2.10	2400	4.16	0.53
0.20	5.37	6130	1.632	0.37	0.85	1.83	2090	4.79	0.58
0.25	4.97	5670	1.763	0.41	0.90	1.53	1746	5.73	0.84
0.30	4.61	5260	1.902	0.48	0.94	1.24	1415	7.06	0.94
0.35	4.30	4910	2.035	0.51	0.96	1.06	1210	8.27	0.98
0.40	4.02	4590	2.18	0.54	0.98	0.811	925	10.81	0.80
0.45	3.75	4280	2.34	0.56	0.99	0.628	716	13.97	0.78
0.50	3.50	3990	2.51	0.60	–	–	–	–	–

RADIANT TRANSPORT TO NON-GRAY WALLS

Radiant Energy Balance on an External Surface

Equation (7.91) is fundamental to radiant transport. It applies to both the power radiated $d\dot{Q}^+$ (to a remote object) and the power received $d\dot{Q}$ (from a remote object) at the s-surface on an element of area dA. In both cases the solid angle $d\omega$ is that of the remote object as seen from dA; from $d\dot{Q}^+$, however, the outgoing intensity I^+ (the intensity of dA) is used, whereas for $d\dot{Q}^-$ the incoming intensity I^- (the intensity of the object) is used.

Consider, for example, an element of an opaque surface on the outside of a space vehicle on an interplanetary mission, far from the sun in solar radii. Suppose that the only significant radiant source "seen" by the surface is the sun. To simplify the problem we take the sun to be a black body at $T_s = 5750°K$. In this case Eq. (7.91) yields

$$d\dot{Q}^- = I_p\,(v,\,T_s)\,dA\,\cos\theta_s\,d\omega_s\,dv$$

$$dq^- = d\dot{Q}^-/dA = I_p\,(v,\,T_s)\,\cos\theta_s\,d\omega_s\,dv$$

where $\cos\theta_s$ is the angle from a ray coming in from the sun to the normal of surface dA on the space vehicle, and $d\omega_s$ is the solid angle subtended by the sun when viewed from the vehicle.

The fraction of the irradiation absorbed is dq^- times the absorptance $\alpha(\theta,\,\phi,\,v,\,T_w)$. The area element itself radiates to all of space. The net loss of radiation by the element is then

$$q_r = \int_0^\infty \int_0^{2\pi} \int_0^{\pi/2} \varepsilon(\theta,\phi,v,T_w) I_P(v,T_w) \cos\theta \sin\theta\; d\theta\; d\phi\; dv$$

$$- \int_0^\infty \alpha(\theta_s,\phi_s,v,T_w) I_P(v,T_w) \cos\theta_s\; d\omega_s\; dv$$

$$q_r = \varepsilon_{TH}\sigma T_w^{\,4} - \alpha_T(\theta_s,\phi_s,T_w,T_s)(\sigma\,T_s^{\,4}\,/\,\pi)\cos\theta_s\; d\omega_s$$

The quantity $\sigma T_s^{\,4}\,d\omega_s/\pi$ is just $I_s\,d\omega_s$, so we can write

$$q_r = \varepsilon_{TH}\sigma\,T_w^{\,4} - \alpha_T(\theta_s,\,\phi_s,\,T_w,\,T_s)\,I_s\,d\omega_s\,\cos\theta_s \qquad \text{...(7.126)}$$

At the earth's distance from the sun r_0 (one astronomical unit $= 93 \times 10^6$ miles), $I_s\,d\omega_s$ is approximately 435 Btu/hr ft^2 or 1380 w/m^2. Since $d\omega_s$ is the projected area of the sun divided by distance squared, at a distance r from the sun,

$$d\omega_s = d\omega_0\,(r_0^{\,2}/r^2) \qquad \text{...(7.127)}$$

which allows $I_s\,d\omega_s$ to be conveniently scaled to other values of r from the values given above.

If the vehicle is spherical of radius R, isothermal, uniformly finished, and at steady-state dissipating internal power $\dot{Q}_i$ (say from a radioisotope source), the integral of the flux over the area (only one-half is exposed to solar radiation) yields

$$\dot{Q}_i = \varepsilon_{TH}\,(T_w)\sigma T_w^{\,4}\,(4\pi R^2) - \alpha_{TH}\,(T_w,\,T_s)\,(I_s d\omega_s)\,(\pi R^2) \qquad \text{...(7.128)}$$

This relation may be used to solve for T_w or, from a design point of view, to solve for suitable pairs of ε_{TH} and α_{TH} as a function of distance from the sun and internal power, since the designer chooses surfaces to fix ε_{TH} and α_{TH}.

Internal Surface

Consider an opaque surface on the inside of a space vehicle or other enclosure. We first consider the exchange between an element dA_1 and another, dA_2. Eq. (7.91) gives

$$d\dot{Q}_{1-2}^+ = I_1^+ dA_1 \cos\theta_1 d\omega_{1-2} dv$$

$$d\dot{Q}_{1-2}^+ = I_1^+ dA_1 \cos\theta_1 \frac{dA_2 \cos\theta_2}{r_{1-2}^2} dv$$

Similarly

$$d\dot{Q}^-_{1-2} = I_2^+ dA_1 \cos\theta_1 \frac{dA_2 \cos\theta_2}{r_{1-2}^2}$$

The net power transfer is then

$$d\dot{Q}_{1-2} = (I_1^+ - I_2^+) \frac{\cos\theta_1 \cos\theta_2 dA_1 dA_2}{r_{1-2}^2} dv$$

Multiplying and dividing by π allows us to write

$$d\dot{Q}_{1-2} = (\pi I_1^+ - \pi I_2^+) d(A_1 F_{1-2}) dv \qquad ...(7.129)$$

where

$$d(A_1 F_{1-2}) = \frac{\cos\theta_1 \cos\theta_2 dA_1 dA_2}{\pi r_{1-2}^2} \qquad ...(7.130)$$

The shape factor introduced in Eq. (7.35) is thus seen to arise in a perfectly natural way from Eq. (7.91).

To proceed we must find πI_1^+ and πI_2^+. If all the surfaces are perfectly diffuse, then I^+ is not a function of θ, ϕ, and we may write

$$\pi I_i^+ = \alpha_{H,i}(T_i, v)\pi I_P(v, T_i) + \rho_{H,i}(T_{i,v}) q_{\bar{F},i}(T_i, v) q_{\bar{F},i}(v)$$

$$q_{\bar{F},i} = \sum_{j=1}^{m} F_{i-j}\pi I_j^+ \qquad ...(7.131)$$

Equation (7.131) may be seen to be in the very same form as Eq. (7.43). Therefore, all that follows Eq. (7.43) holds for spectral values provided the total black body radiosity σT_j^4 is replaced with the spectral value πI_P (v, T_j). For example, Eq. (7.56a) becomes

$$\dot{Q}_{i-k}(v) = A_i F_{i-k} [\pi I_P(v, T_i) - \pi I_P(v, T_k)] \frac{\alpha_i(v)\alpha_k(v)}{\alpha_{avg}(v)}$$

Thus total transfer for a spherical enclosure is just

$$\dot{Q}_{i-k} = A_i F_{i-k} \int_0^\infty \frac{\alpha_i(v)\alpha_k(v)}{\alpha_{avg}(v)} [\pi I_P(v,T_i) - \pi I_P(v,T_k)]dv \qquad ...(7.132)$$

To perform the integration when T_i and T_k are not close to the same temperature, Eq. (7.122b) is used,

$$\left(\frac{\alpha_i \alpha_k}{\alpha_{avg}}\right)_{T_k} = \int_0^1 \frac{\alpha_i(v)\alpha_k(v)}{\alpha_{avg}(v)}\left[\frac{-df(hcv/kTk)}{dv}\right]dv \qquad ...(7.133)$$

$$\dot{Q}_{i-k} = A_i F_{i-k}\left\{\left(\frac{\alpha_i \alpha_k}{\alpha_{avg}}\right)_{T_i} \sigma T_i^4 - \left(\frac{\alpha_i \alpha_k}{\alpha_{avg}}\right)_{T_k} \sigma T_k^4\right\} \qquad ...(7.134)$$

But when T_i and T_k are nearly the same, it is best to write Eq. (7.132) in the form

$$\dot{Q}_{i-k} = A_i F_{i-k}\left\{\int_0^\infty \frac{\alpha_i(v)\alpha_k(v)}{\alpha_{avg}(v)}\frac{\partial}{\partial T}(\pi I_P(v,T))\,dv\right\}(T_i - T_k) \qquad ...(7.135)$$

$$\dot{Q}_{i-k} = A_i F_{i-k}\left(\frac{\alpha_i \alpha_k}{\alpha_{avg}}\right)_{TI} 4\sigma T^3(T_i - T_k) \qquad ...(7.136)$$

where the subscript *TI* denotes an *internal* average

$$\left(\frac{\alpha_i \alpha_k}{\alpha_{avg}}\right)_{TI} = \int_0^\infty \frac{\alpha_i(v)\alpha_k(v)}{\alpha_{avg}(v)}\left[\frac{-df_i(hcv/kT)}{dv}\right]dv \qquad ...(7.137)$$

and

$$f_i\left(\frac{hcv}{kT}\right) = \frac{1}{4\sigma T^3}\int_v^\infty \frac{\partial}{\partial T}(\pi I_P(v,T))\,dv \qquad ...(7.138)$$

The adjective *internal* is employed, because the situation where T_i is nearly equal to T_k occurs commonly inside enclosures such as rooms or space vehicles. This internal fraction is tabulated in Table 7.9.

Table 7.9. Internal fraction for the Planck distribution.

$\dfrac{hcv}{kT}$	f_i	$\dfrac{hcv}{kT}$	f_i	$\dfrac{hcv}{kT}$	f_i
0.0	1.00	3.39	0.70	6.11	0.25
0.934	0.99	3.66	0.65	6.56	0.20
1.19	0.98	3.93	0.60	7.11	.0.15
1.52	0.96	4.19	0.55	7.85	0.10
1.76	0.94	4.46	0.50	8.72	0.06
2.13	0.90	4.75	0.45	9.39	0.04
2.50	0.85	5.05	0.40	10.45	0.02
2.82	0.80	5.37	0.35	11.48	0.01
3.11	0.75	5.72	0.30	∞	0.0

RADIATION HEAT TRANSFER COEFFICIENT

As described earlier radiation and convection heat transfer from or to a surface are often comparable and it is convenient to make the necessary calculations using the simplified convection equation, $q = hA\,(t_g - t_w)$, where t_w is the bulk fluid temperature. The rate of radiation heat transfer q_r in such cases can usually be computed from $q_r = 0.1714\varepsilon_g A_g[(T_s/100)^4 - (T_R/100)^4]$ where T_R is the temperature of the walls of the enclosure. Instead of this equation, which involves the fourth powers of the absolute temperatures of the surfaces, it is more convenient to be able to compute q_r from an equation of the form

$$q_r = h_r A_s\,(t_s - t_R) \qquad \ldots(7.139)$$

This is because t_R is usually the same as t_w, and q_r is then based on the same temperature difference as the convection heat transfer. h_r is determined by

$$h_r = \frac{q_r}{A_s(t_s - t_R)} = \frac{0.1714\varepsilon_s[T_s/100)^4 - (T_R/100)^4]}{(T_s - T_R)} \qquad \ldots(7.140)$$

This relation involves only the surface and enclosure temperatures and the surface emissivity. Since ε_s enters as a simple multiplier, it is desirable to plot h_r as a function of T_s using T_R as the parameter as shown in Fig. 7.25. Note that the equation is symmetrical in the two temperatures so that the values of the abscissa or on the curves may be associated with either the surfaces or enclosure wall temperatures.

MEASUREMENT OF TOTAL NORMAL EMISSIVITY

The use of a thermopile radiometer developed by Dunkle and Gier for measurement of the total normal emissivities of surfaces at various temperatures provides a very illustrative example of radiant heat exchange. The radiometer consists essentially of a 160-junction silver-constantan thermopile mounted in a cylindrical metal housing (actual dimensions are shown in Fig. 7.20. Aluminium-foil strips, lamp-blacked on the exposed side, are fastened to the two lines of thermopile junctions with shellac. A rear plate (Fig. 7.20) through which a narrow slot is cut allows radiation entering the front opening of the radiometer to impinge on the hot-junction receiver strip, while shielding the cold-junction receiver strip. Radiant energy absorbed by the hot-junction receiver strip results in a temperature rise over the cold-junction temperature which, of course, produces a measurable emf. The instrument is calibrated by exposing it to the radiation from a standard tungsten-filament lamp. The output of the thermopile has been found to be linear, a typical value being 0.1275 millivolt per Btu/hr ft² (the reciprocal of this, $K = 1/0.1275 = 7.84$ Btu/hr ft² per millivolt, is more convenient for calculations).

For emissivity determination, samples are attached to a copper plate heated by electric coils. In order to prevent direct radiation from the heater case from striking the radiometer, a conical shield is used, as shown in Fig. 7.20. The equation for calculating the emissivity of a sample surface is obtained from a consideration of the radiant energy exchange between the heated sample surface and the thermopile hot-junction receiver strip. For this purpose we need to know the geometrical factor of the receiver strip with respect to the sample surface or vice-versa. In this case it is permissible to regard the receiver strip as an incremental area, ΔA.

The radiometer is placed so that the hot-junction receiver strip sees only the heated sample surface. Because of this the area of the hot surface can be effectively taken as the opening of the conical shield. Consequently the geometric factor will not involve the exact location of the heated surface. Calling the receiver strip area 1 and the heated sample area 2, from equation (7.8)

$$F_{12} = \frac{1}{\Delta A_1 \pi} \int_{A_2} \frac{\cos\theta_1 \cos\theta_2 \Delta A_1}{s^2} dA_2 \qquad ...(7.141)$$

Considering A_2 as the circular opening of the shield, dA_2 is conveniently taken as a ring-shaped area equal to $2\pi r dr$ where r is measured from the radiometer axis. Also $\cos\theta_1 = \cos\theta_2 = l/s$ and $s = \sqrt{r^2 + l^2}$. Therefore,

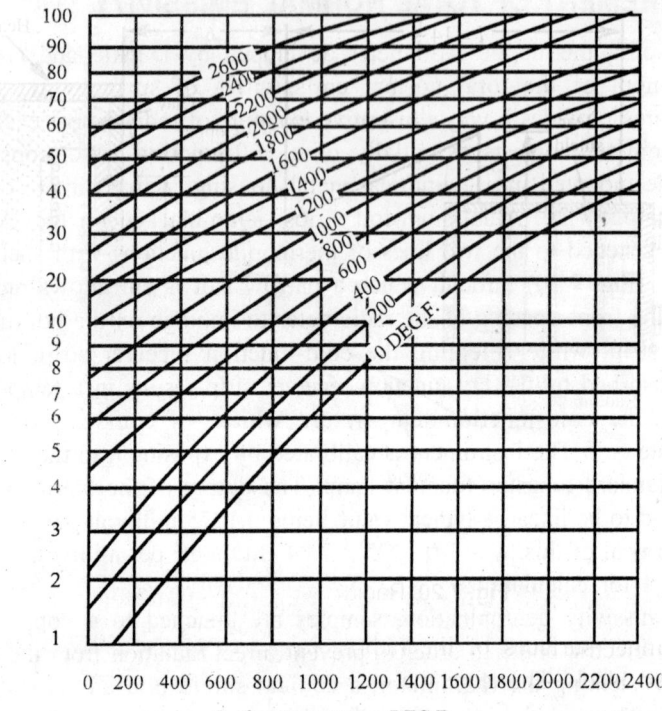

Fig. 7.19. Radiation heat transfer coefficient, hr, for $\varepsilon = 1.0$.

$$F_{12} = \frac{1}{\pi} \int_0^{r_1} \frac{l^2}{(r^2 + l^2)} 2\pi r\, dr = \left. \frac{-l^2}{(r^2 + l^2)} \right|_0^{r_1}$$

$$= \left[\frac{-l^2}{r_1^2 + l^2} + \frac{l^2}{l^2} \right] = \frac{r_1^2}{r_1^2 + l^2} \qquad \text{...(7.142)}$$

The millivolt output of the thermopile, mv, is determined by the net radiation incident upon it. Assuming that the radiometer and hot-junction receiver strip (it will be a negligible amount higher) are at the temperature of the surroundings, T_R, this net radiation will be that from the heated surface plus the radiation from the surroundings reflected from the heated surface to the receiver strip, less the energy emitted by the receiver strip. Assuming the heated sample surface to be gray, and the surroundings to be black, we obtain

$$\Delta A_1 K(\mathrm{mv}) = [T_2^4 + (1 - \varepsilon_2)\, T_R^4] \sigma F_{21} A_2 - F_{12} \Delta A_1 \sigma T_R^4 \qquad \text{...(7.143)}$$

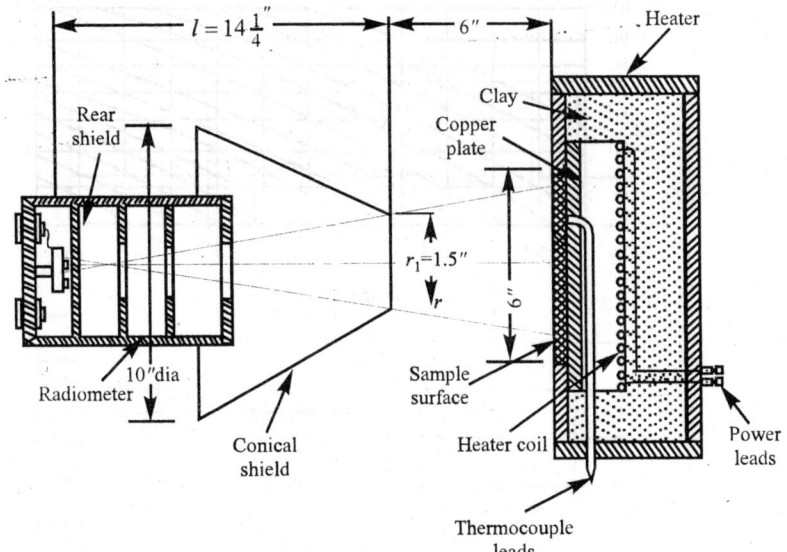

Fig. 7.20. Radiometer test arrangement.

(the inner surfaces of the radiometer seen by the receiver strip are all painted black. Therefore the energy leaving it is proportional to $F_{12}\Delta A_1$). Substituting $F_{12}\Delta A_1 = F_{21}A_2$ gives

$$\varepsilon_2 = \frac{K(mv)}{F_{12}\sigma(T_2^4 - T_R^4)} \qquad ...(7.144)$$

Example 7.9. Calculate the total normal emissivity of an unpolished sample of stainless steel if, when at a temperature of 625°F, a radiometer thermopile output of 0.671 mv was measured. The radiometer was at room temperature of 70°F, the measurements were taken with the equipment set up as in Fig. 7.20, and the instrument constant was 7.84 Btu/hr ft² mv.

Solution. From equation (7.144)

$$F_{12} = \frac{r_1^2}{r_1^2 + l^2} = \frac{(1.5)^2}{(1.5)^2 + (14.25)^2} = 0.011$$

Substituting the observed values in equation (7.146), we have

$$\varepsilon = \frac{7.84(0.671)}{0.011(0.1714)[(10.85)^4 - (5.30)^4]} = 0.23$$

Values of the normal emissivity of unpolished and oxidised stainless steel determined from 100°F to 800°F are shown in Fig. 7.21. As stated before, normal emissivities of such surfaces are satisfactorily representative of the total hemispherical emissivities.

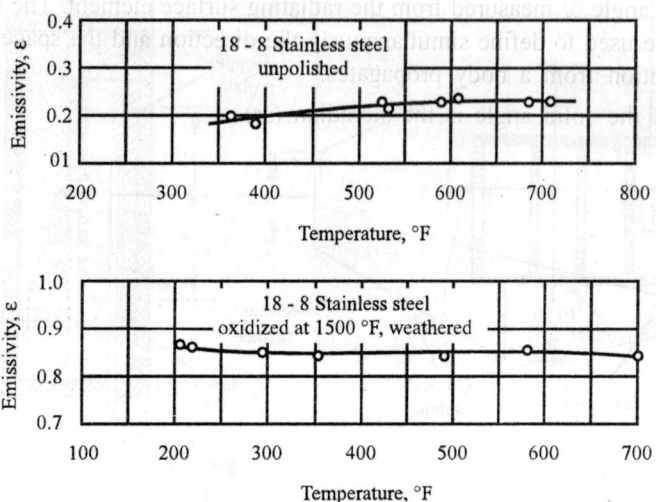

Fig. 7.21. Total normal emissivity measurements for stainless steel.

INTENSITY OF RADIATION

In our discussion so far we have only considered the total amount of radiation leaving a surface, that is, the emissive power. This concept, however, is inadequate for a heat transfer analysis when the amount of radiation passing in a given direction and intercepted by some other body is sought. The amount of radiation passing in a given direction is described in terms of the intensity of radiation, I. Before defining the intensity of radiation, we must have measure of the direction and the space into which a body radiates. As shown in Fig. 7.22a, a differential plane angle $d\alpha$ is defined as the ratio of an element of arc length dl on a circle to the radius r of that circle. Similarly, a differential solid angle $d\omega$ is defined (Fig. 7.22b) as the ratio of the element of area dA_n on a sphere to the square of the radius of sphere, or

$$d\omega = \frac{dA_n}{r^2}$$ ...(7.145)

The rate of radiation heat flow per unit surface area from a body which passes in a given direction can be measured by determining the radiation through an element on the surface of a hemisphere constructed around the radiating surface. If the radius of this hemisphere equals unity, the hemisphere has a surface area of 2π and substends a solid angle of 2π

steradians about a point at the centre of its base. The surface area on such a hemisphere with a radius of unity has the same numerical value as the so-called solid angle ω measured from the radiating surface element. The solid angle can be used to define simultaneously the direction and the space into which radiation from a body propagates.

The unit of the solid angle is the steradian (sr).

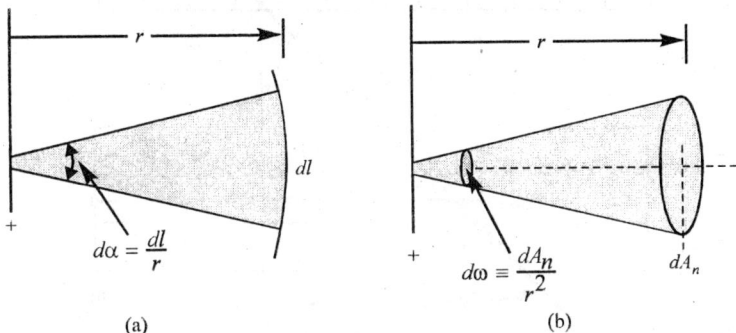

Fig. 7.22. (a) Differential plane angle; and (b) differential solid angle.

The intensity of radiation $I(\theta, \phi)$ is the energy emitted per unit area per unit time into a solid angle $d\omega$ centred around a direction which can be defined in terms of the zenith angle θ and the azimuthal angle ϕ in the spherical coordinate system of Fig. 7.23. The differential area dA_n in Fig. 7.23 is perpendicular to the (θ, ϕ) direction. But for a spherical surface $dA_n = rd\theta\, r \sin \theta\, d\phi$ and therefore

$$d\omega = \sin\theta\, d\theta\, d\phi \qquad ...(7.146)$$

With the above definitions the intensity of radiation $I(\theta, \phi)$ is the rate at which radiation is emitted in the direction (θ, ϕ) per unit area of the emitting surface normal to this direction, per unit solid angle centred about (θ, ϕ).

Since the projected area of emission from (Fig. 7.23) is $dA_1 \cos \theta$, we obtain for the intensity of a black surface, $I_b(\theta, \phi)$

$$I_b(\theta,\phi) = \frac{dq_r}{dA_1 \cos\theta\, d\omega}\,(\text{W}\,/\,\text{m}^2 \text{ sr}) \qquad ...(7.147)$$

where dq_r is the rate at which radiation emitted from dA_1 passes through dA_n.

Example 7.10. A flat black surface of area $A_1 = 10$ cm^2 emits 1000 W/ m^2 sr in the normal direction. A small surface A_2 having the same area as A_1 is placed as shown in Fig. 7.24 relative to A_1 at a distance of 0.5 m^2. Determine the solid angle subtended by A_2 and the rate at which A_2 is irradiated by A_1.

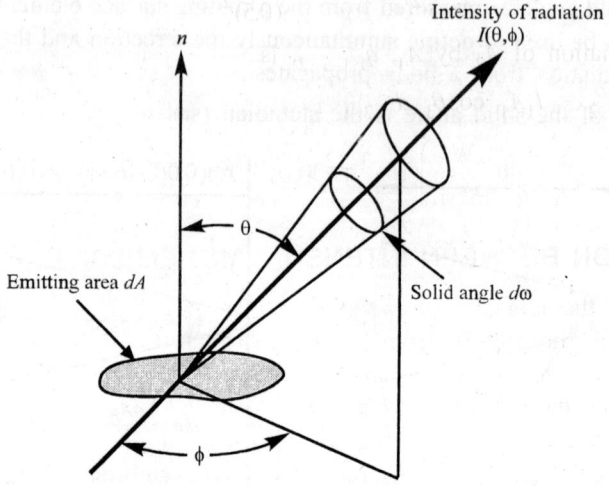

Fig. 7.23. Schematic diagram illustrating intensity of radiation.

Solution. Since A_1 is black, it is a diffuse emitter and its intensity I_b is independent of direction. Moreover, since both areas are quite small, they may be approximated as differential surface areas and the solid angle can be calculated from Eq. (7.145) or $d\omega_{2-1} = dA_{n,2}/r^2$.

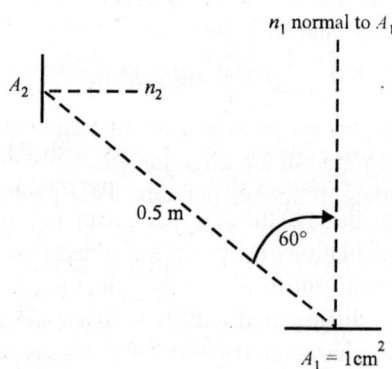

Fig. 7.24. Sketch showing relation between A_1 and A_2.

The area $dA_{n,2}$ is the projection of A_2 in the direction normal to the incident radiation for dA_1, or $dA_{n2} = dA_2 \cos \theta_2$, where θ is the angle between the normal n_2^{-a} and the radiation ray connecting dA_1 and dA_2, that is, $\theta = 30°$. Thus

$$d\omega_{2-1} = \frac{A_2 \cos\theta_2}{r^2} = \frac{10^{-3} \text{ m}^2 \cos 30°}{(0.5)^2} = 0.00346 \text{ sr}$$

The irradiation of A_2 by A_1, $q_{r,1\to 2}$, is

$$q_{r,1 \to 2} = I_1 A_1 \cos\theta_1 \, d\omega_{2-1}$$

$$= \left(1000 \frac{\text{W}}{\text{m}^2\text{sr}}\right)(10^{-3}\text{m}^2)(\cos 60°)(0.00346 \text{ sr}) = 0.00173 \text{ W}$$

RELATION BETWEEN INTENSITY AND EMISSIVE POWER

To relate the intensity of radiation to the emissive power, one simply determines the energy from a surface radiating into a hemispherical enclosure placed above it, as shown in Fig. 7.25. Since the hemisphere will intercept all the radiant rays emanating from the surface, the total amount of radiation passing through the hemispherical surface equals the emissive power. From Eq. (7.147), the rate of radiation emitted from dA_1 passing through dA_n is

$$\frac{dq_r}{dA_1} = I_b(\theta,\phi)\cos\theta \, d\omega \qquad ..(7.148)$$

Substituting for the solid angle $d\omega$ and integrating over the entire hemisphere yields the total rate of radiant emission per unit area, called the emissive power :

$$\left(\frac{q}{A}\right)_r = \int_0^{2\pi}\int_0^{\pi/2} I_b(\theta,\phi)\cos\theta \, \sin\theta \, d\theta \, d\phi \qquad ...(7.149)$$

In order to integrate Eq. (7.149), the intensity with θ and ϕ must be known. The intensity of real surfaces exhibits no appreciable variation with ϕ but does vary with θ. Although this variation can be taken into account, for most engineering calculations it may be assumed that the surface is diffuse and the intensity is uniform in all angular directions. Black body radiation is actually perfectly diffuse and radiation from industrial rough surfaces approaches diffuse characteristics. If the intensity from a surface is independent of direction, it is said to conform to Lambert's cosine law. For a black surface, integration of Eq. (7.149) yields the black body emissive power E_b:

$$\left(\frac{q}{A}\right)_r = E_b = \pi I_b \qquad ...(7.150)$$

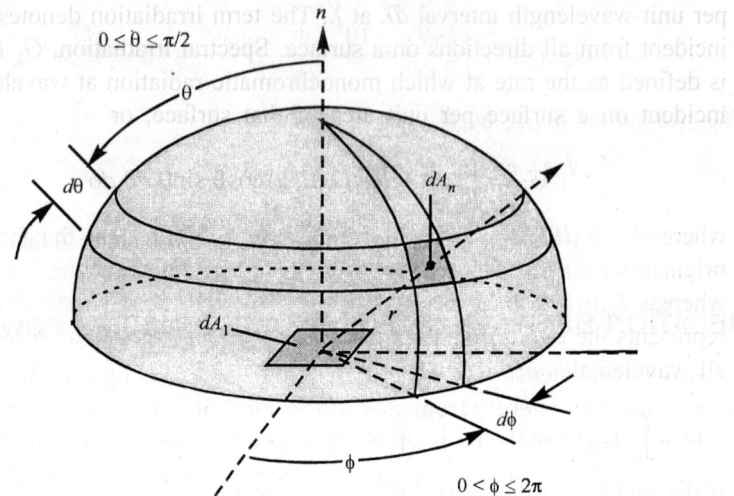

Fig. 7.25. Radiation from a differential area dA into surrounding hemisphere centred at dA.

Thus for a black surface, the emissive power equals π times the intensity. The same relation between emissive power and intensity obtains for any surface that conforms to Lambert's cosine law.

The concept of intensity can be applied to the total radiation over the entire wavelength spectrum, as well as to monochromatic radiation. The relation between the total and the monochromatic intensity I_λ is simply

$$I(\phi,\theta) = \int_0^\infty I_\lambda(\phi,\theta)d\lambda \qquad ...(7.151)$$

If a surface radiates diffusely, it is apparent that also

$$E_\lambda = \pi I_\lambda \qquad ...(7.152)$$

since I_λ is uniform in all directions.

IRRADIATION

To make a heat balance on a body we need to know not only the radiation leaving but also the radiation incident on the surface. The radiation originates from emission and reflection occurring at other surfaces and will in general have a specific directional and spectral distribution. As shown in Fig. 7.26, the incident radiation can be characterised in terms of the incident spectral intensity, $I_{\lambda,i}$, defined as the rate at which radiant energy at wavelength λ from direction (θ, ϕ) per unit area of the intercepting surface normal to this direction, per unit solid angle about the direction (θ, ϕ), and per unit wavelength interval $d\lambda$ at λ. The term irradiation denotes radiation incident from all directions on a surface. Spectral irradiation, G_λ (W/m^2 m) is defined as the rate at which monochromatic radiation at wavelength λ is incident on a surface per unit area of that surface, or

$$G_\lambda = \int_0^{2\pi} \int_0^{\pi/2} I_{\lambda,i}(\lambda,\theta,\phi)\cos\theta \, \sin\theta \, d\theta \, d\phi \qquad ...(7.153a)$$

where $\sin\theta \, d\theta \, d\phi$ is the unit solid angle. Observe that the factor $\cos\theta$ originates from the fact that G_λ is a flux based on the actual surface area, whereas I_λ,i is defined in terms of the projected area. The total irradiation represents the rate of radiation incident per unit area from all direction over all wavelengths and is given by

$$G = \int_0^\infty G_\lambda(\lambda)d\lambda = \int_0^\infty \int_0^{2\pi} \int_0^{\pi/2} I_{\lambda,i}(\lambda,\theta,\phi)\cos\theta\sin\theta \, d\theta \, d\phi \, d\lambda \quad ...(7.153b)$$

If the incident radiation is diffuse—that is the intercepting area is diffusely irradiated and $I_{\lambda,\,i}$ is independent of direction—it follows that

$$G = \pi I_i \qquad ...(7.154)$$

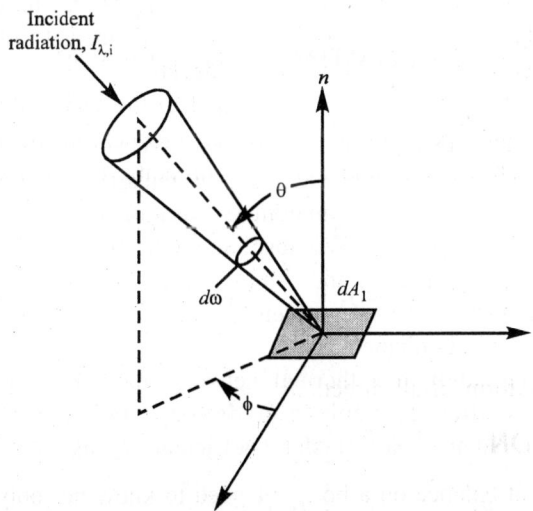

Fig. 7.26. Radiation incident on a differential area dA_1 in a spherical coordinate system.

RADIATION PROPERTIES

Most surfaces encountered in engineering practice do not behave like black bodies. To characterise the radiation properties of non-black surfaces, dimensionless quantities such as the emittance, absorptance, and transmittance are used to relate the emitting, absorbing, and transmitting capabilities of a real surface to those of a blackbody. Radiation properties of real surfaces are functions of wavelength, temperature, and direction. The properties that describe how a surface behaves as a function of wavelength are called monochromatic or spectral properties, and the properties that describe the distribution of radiation with angular direction are called directional properties. For precise heat transfer calculation, we must know the radiative properties of the emitting surface as well as those of all other surfaces with which radiation exchange occurs.

Taking into account the spectral and directional properties of all surfaces, even if they are known, results in complex and involved analyses which can only be carried out by computer. However, engineering calculations of acceptable accuracy can usually be carried out by a simplified approach, using a single radiation property value averaged over the wavelength range of interest and directions are called total properties. Although we will almost exclusively use total radiative properties here, it is important to be aware of the spectral and directional characteristics of surfaces in order to account for them in problems in which these variations are significant.

RADIATION COMBINED WITH CONVECTION AND CONDUCTION

In the preceding sections of this chapter we have considered radiation as an isolated phenomenon. Energy exchange by radiation is the predominant heat flow mechanism at high temperatures because the rate of heat flow depends on the fourth power of the absolute temperature. In many practical problems, however, convection and conduction cannot be neglected, and in this section we shall consider problems which involve two or all three modes of heat flow simultaneously.

To include radiation in a thermal network involving convection and conduction it is often convenient to define a unit thermal radiative conductance, or radiant heat transfer coefficient, $\bar{h}_r$, as

$$\bar{h}_r = \frac{q_r}{A_1(T_1 - T'_2)} = F_{1-2}\left[\frac{\sigma(T_1^4 - T_2^4)}{T_1 - T'_2}\right] \qquad ...(7.155)$$

where

$A_1 =$ area upon which F_{1-2} is based, m^2

$T_1 - T'_2 =$ a reference temperature difference, in K, in which T'_2 may be chosen equal to T_2 or any other convenient temperature in the system

$\bar{h}_r =$ radiant heat transfer coefficient, $W/m^2\ K$

Once a radiant heat transfer coefficient has been calculated, it can be treated similarly to the convective heat transfer coefficient, because the rate of heat flow becomes linearly dependent on the temperature difference and radiation can be incorporated directly in a thermal network for which the temperature is the driving potential. Knowledge of the value of $\bar{h}_r$ is also essential in determining the overall conductance $\bar{h}$ for a surface to or from which heat flows by convection and radiation, i.e.

$$\bar{h} = \bar{h}_c + \bar{h}_r$$

If $T_2 = T'_2$, bracketed expression in Eq. (7.155) is called the temperature factor F_T, and

$$\bar{h}_r = F_{1-2}F_T \qquad \qquad ...(7.156)$$

SOLVED EXAMPLES

Example 7.1. An instrument used to measure high temperature—an optical pyrometer is based on comparing the radiosity of a body investigated with that of an incandescent filament. The instrument is calibrated based on the emission of a black body and, therefore, it measures the temperature which the black body would have at a radiosity equal to that of the black body would have at a radiosity equal to that of the investigated body. The pyrometer incorporates a red light filter ($\lambda = 0.65$ μm).

What is the true temperature of the body, if the pyrometer records the temperature of 1400°C, and the emissivity of the body at $\lambda = 0.65$ μm is equal to 0.6?

Solution. The radiosity of the investigated body

$$B_\lambda = \frac{J_\lambda}{\pi} = \frac{1}{\pi} \frac{\varepsilon_\lambda c_1 \lambda^{-5}}{e^{c_2/\lambda T} - 1},$$

where T is the absolute temperature of the investigated body.

The radiosity of a black body

$$B_{0\lambda} = \frac{J_{0\lambda}}{\pi} = \frac{1}{\pi} \frac{c_1 \lambda^{-5}}{e^{c_2/\lambda T_0} - 1},$$

where T_0 is the absolute temperature of the black body, and at $B_\lambda = B_{0\lambda}$ it will be the temperature indicated by the pyrometer.

Since in the case considered $c_2/\lambda T_0 = 13.2$, then $e^{c_2/\lambda T_0}$ is considerably greater than unity. That is why the digit 1 in the denominator of the equations can be ignored compared with $e^{c_2/\lambda T}$.

It follows from the condition $B_\lambda = B_{0\lambda}$ that

$$\frac{1}{T} = \frac{1}{T_0} - \frac{\lambda}{c^2} \ln \frac{1}{\varepsilon\lambda},$$

whence

$$T = \frac{1}{\dfrac{1}{T_0} - \dfrac{\lambda}{c_2} \ln \dfrac{1}{\varepsilon\lambda}} = \frac{1}{\dfrac{1}{1673} - \dfrac{0.65\times10^{-6}}{1.439\times10^{-2}} 2.3 \log \dfrac{1}{0.6}} = 1740 \text{ K}.$$

The temperature of the body

$$t = 1740 - 273 = 1467°C.$$

Example 7.2. The temperature of the body is measured with the aid of two optical pyrometers fitted with different light filters. The first pyrometer mounts a red light filter ($\lambda_1 = 0.65$ μm) and the second a green one ($\lambda_2 = 0.50$ μm). The temperature readings of the two pyrometers are respectively $t_{01} = 1400°C$ and $t_{02} = 1420°C$. Find the true temperature of the body and its emissivity, assuming the body to be grey.

Solution. The equations derived in solving example 7.1 permit us to write:

$$\left.\begin{array}{c}\dfrac{1}{T_{01}} - \dfrac{1}{T} = \dfrac{\lambda_1}{c_2} \ln \dfrac{1}{\varepsilon_{\lambda 1}}; \\[3mm] \dfrac{1}{T_{02}} - \dfrac{1}{T} = \dfrac{\lambda_2}{c_2} \ln \dfrac{1}{\varepsilon_{\lambda 2}}.\end{array}\right\} \qquad ...(7.1)$$

For a grey body $\varepsilon_{\lambda 1} = \varepsilon_{\lambda 2} = \varepsilon$. From the system of equations (7.1) we obtain the following expressions for T and ε :

$$T = \frac{\dfrac{\lambda_1}{\lambda_2} - 1}{\dfrac{\lambda_1}{\lambda_2} \dfrac{1}{T_{02}} - \dfrac{1}{T_{01}}}; \qquad ...(7.2)$$

$$\ln \varepsilon = \frac{c_2(T_{01} - T_{02})}{T_{01}T_{02}(\lambda_1 - \lambda_2)}. \qquad ...(7.3)$$

Substituting the found values into relationship (7.2) and (7.3), we obtain :

$$T = 1765 \text{ K}; \quad t = 1492°C; \quad \varepsilon = 0.71.$$

Example 7.3. Determine the intensity of solar radiation that is incident upon a plane normal to sun rays and arranged outside the Earth's atmosphere. It is known that solar radiation is close to that of a black body at a temperature $t_0 = 5700°C$. The diameter of the sun $D = 1.391 \times 10^6$ km, the distance from the earth to the sun $l = 149.5 \times 10^6$ km.

Solution. The density of incident solar radiation is determined by formula:

$$E_{in} = B_{d\omega},$$

where B = radiosity of solar radiation;

$d\omega$ = solid angle at which a unit area "sees" the sun.

The radiosity of solar radiation

$$B = \frac{E_0}{\pi} = \frac{C_0(T_0 / 100)^4}{\pi}.$$

The solid angle

$$d\omega = \frac{\pi \dfrac{D^2}{4}}{l^2}.$$

With account taken of these relationships

$$E_{in} = \frac{C_o{}'\left(\dfrac{T_0}{100}\right)^4 \pi D^2}{\pi 4l^2} = \frac{5.67(59.73)^4 \times 1.391^2}{4 \times 149.4^2} = 1550 \text{ W} / \text{m}^2.$$

Example 7.4. An artificial spherical satellite (Sputnik) flies round the earth being launched on its day's light side. The absorptivity of the Sputnik's surface in respect to incident solar radiation is A, and the emissivity of this surface is ε.

Determine the temperature of the Sputnik's surface, assuming that there are no inner heat sources inside the Sputnik and that the surface temperature is the same all over it. The solar radiation reflected from the earth and the radiation emitted from the earth should be ignored.

Solution. Under steady-state conditions the amount of radiant energy absorbed by the Sputnik and the amount of energy emitted by it into space are equal to each other, i.e.

$$AE_{in}F_N = \varepsilon C_0 E\left(\frac{T}{100}\right)^4,$$

where F_N = projection of the Sputnik's irradiated surface on the plane normal to incident radiation;

F = surface area of the Sputnik.

For a sphere

$$\frac{F_N}{F} = \frac{\pi d^2}{4\pi d^2} = \frac{1}{4}.$$

The density of incident radiation E_{in} = 1550 W/m². Finally, the temperature of the Sputnik

$$T = 100\sqrt[4]{\frac{AE_{in}F_N}{\varepsilon C_0 F}} = 100\sqrt[4]{\frac{A \times 1550}{\varepsilon 5.67 \times 4}} = 288\sqrt[4]{\frac{A}{\varepsilon}}.$$

Example 7.5. A spaceship launched from the earth runs to Venus. The distance from Venus to the Sun is 108.1 × 10⁶ km and from the Earth to the Sun, 149.5 × 10⁶ km. The temperature of the spaceship surface near the Earth was equal to $t_1°$C.

How shall change the temperature of the spaceship surface when it approaches the Venus, assuming that the emissivity of the surface does not vary with spaceship temperature?

Solution. The surface temperatures of the spaceship near the Earth and Venus are determined from the following equations :

$$\varepsilon C_0 F \left(\frac{T_1}{100}\right)^4 = AE_{in1}F_N;$$

$$\varepsilon C_0 F \left(\frac{T_2}{100}\right)^4 = AE_{in2}F_N;$$

whence

$$\left(\frac{T_2}{T_1}\right)^4 = \frac{E_{in2}}{E_{in1}} = \left(\frac{l_1}{l_2}\right)^2.$$

Consequently,

$$\frac{T_2}{T_1} = \sqrt{\frac{149.5}{108.1}} = 1.18;$$

$$t_2 + 273 = (t_1 + 273)\,1.18;$$

$$t_2 = (1.18 t_1 + 48),\ °C.$$

Example 7.6. The setting of a boiler furnace is laid from fireclay (chamotte) brick and the outside lagging—from plate steel. The distance between the lagging and setting brick is equal to 30 mm and it may be assumed small, compared with the size of the furnace.

Under steady-state conditions calculate the loss of heat to surroundings by radiation from unit surface area per unit time between the lagging and setting. The temperature of the outer side of the brick setting $t_1 = 127°C$ and the temperature of the steel lagging $t_2 = 50°C$. The emissivity of fireclay $\varepsilon_f = 0.8$ and of plate steel $\varepsilon_s = 0.6$.

Solution. The lagging and setting can be considered as two infinite flat parallel surfaces separated by a transparent participating medium. For a system of this kind net radiation is calculated by formula

$$E_{r1} = q_{1,2} = \varepsilon_{red} C_0 \left[\left(\frac{T_1}{100} \right)^4 - \left(\frac{T_2}{100} \right)^4 \right], \qquad ...(7.4)$$

where the reduced emissivity

$$\varepsilon_{red} = \frac{1}{\dfrac{1}{\varepsilon_1} + \dfrac{1}{\varepsilon_2} - 1} = \frac{1}{\dfrac{1}{0.8} + \dfrac{1}{0.6} - 1} = 0.522.$$

Then,

$$E_{r1} = 0.522 \times 5.67 \left[\left(\frac{127 + 237}{100} \right)^4 - \left(\frac{50 + 237}{100} \right)^4 \right] = 435 \text{ W} / \text{m}^2.$$

Example 7.7. In a channel carrying a hot gas the temperature of the gas is measured with the aid of a thermocouple. Under steady-state conditions the thermocouple indicates a temperature $t_1 = 300°C$ and the wall temperature $t_2 = 200°C$.

Calculate the error admitted in measuring the gas temperature, due to radiation heat transfer between the bead of the thermocouple and the wall of the channel, and also determine the real temperature of the gas. Assume for the thermocouple bead an emissivity $\varepsilon_1 = 0.8$, and a coefficient of heat transfer from the gas to the surface of the bead $\alpha = 58$ W/m^2.°C.

Solution. Let us compose a heat balance equation for the thermocouple bead.

The thermocouple loses heat by radiation

$$Q = F_1 \varepsilon_1 C_0 \left[\left(\frac{T_1}{100} \right)^4 - \left(\frac{T_2}{100} \right)^4 \right]$$

and heat is transferred to it by convection, in the amount

$$Q = F_1 \alpha (t_f - t_1).$$

Under steady-state conditions

$$\alpha(t_f - t_1) = \varepsilon_1 C_0 \left[\left(\frac{T_1}{100} \right)^4 - \left(\frac{T_2}{100} \right)^4 \right].$$

Substituting into the above equation the known quantities, we obtain :

$$t_f - t_1 = \frac{0.8 \times 5.67}{58} \left[\left(\frac{300 + 273}{100} \right)^4 - \left(\frac{200 + 273}{100} \right)^4 \right] = 45.5°C.$$

The actual temperature of the gas
$$t_f = 300 + 45.5 = 345.5°C.$$

Example 7.8. Calculate the radiant heat flux between two black disks arranged in two parallel planes, opposite each other. The temperature of the first disk $t_1 = 500°C$ and of the second disk $t_2 = 200°C$. The disks are of the same size, $d_1 = d_2 = 200$ mm, and they are arranged at a distance $h = 400$ mm from each other.

Solution. The view factor for the system of bodies concerned is calculated from equation :

$$\overline{\varphi}_{1,2} = 1 + 2 \left(\frac{h}{d} \right)^2 - 2 \frac{h}{d} \sqrt{1 + \left(\frac{h^2}{d} \right)}.$$

Substituting into the above equation the numerical values of the quantities present in it, we obtain :

$$\overline{\varphi}_{1,2} = 1 + 2 \left(\frac{400}{200} \right)^2 - 2 \frac{400}{200} \sqrt{1 + \left(\frac{400}{200} \right)^2} = 0.055.$$

Since the disks are of the same size, then $\overline{\varphi}_{1,2} = \overline{\varphi}_{2,1}$. The surface-surface direct exchange area is found to be

$$\overline{H}_{1,2} = \varphi_{1,2} F_1 = 0.055 \times 0.785 \ (0.2)^2 = 1.73 \times 10^{-3} \ m^2.$$

The heat transferred by radiation from the first disk to the second

$$Q_{rad} = C_0 \left[\left(\frac{T_1}{100} \right)^4 - \left(\frac{T_2}{100} \right)^4 \right] \overline{H}_{1,2}$$

$$= 5.67 \left[\left(\frac{500 + 273}{100} \right)^4 - \left(\frac{200 + 273}{100} \right)^4 \right] 1.73 \times 10^{-3} = 30 \ W.$$

Example 7.9. The walls of a boiler furnace are covered with two rows of waterwall tubes having an outside diameter $d = 80$ mm. The waterwall tubes are arranged in the two rows with an equal pitch (in the plane parallel to the wall) s = 400 mm.

Calculate the mean view factor between the surface of the furnace wall and the waterwall tubes.

Solution. The view factor between the wall and one row of waterwall tubes is calculated by the following equation :

$$\bar\varphi_{1,2} = 1 - \sqrt{1 - \left(\frac{d}{s}\right)^2} + \frac{d}{s}\arctan\sqrt{\left(\frac{s}{d}\right)^2 - 1} =$$

$$= 1 - \sqrt{1 - \left(\frac{80}{400}\right)^2} + \frac{80}{400}\arctan\sqrt{\left(\frac{400}{80}\right)^2 - 1} = 0.294.$$

For a multirow bank of tubes $\bar\varphi_{1,2} = 1 - (1 - \bar\varphi_{1,2})^n = 1 - (1 - 0.294)^2$ = 0.502, where n is the number of rows in the bank and in our case $n = 2$.

Example 7.10. Determine the attenuation coefficient for a ray passing through a layer of carbon dioxide 30 mm thick, if it is known that after passing through that layer the spectral intensity of this ray diminished by 90 per cent.

Solution. In an absorbing medium the ray attenuation factor x_λ can be found from the Bouguer's law :

$$J_{\lambda,x} = J_{\lambda,x=0}e^{-x_\lambda x}, \qquad ...(7.5)$$

whence

$$x_\lambda = -\frac{1}{x}\ln\frac{J_{\lambda,x}}{J_{\lambda,x=0}}. \qquad ...(7.6)$$

The conditions of the problem give

$$\frac{J_{\lambda,x}}{J_{\lambda,x=0}} = 0.1.$$

Substituting the numerical values of the quantities given in the problem into equation (7.6), we obtain :

$$x_\lambda = -\frac{1}{3 \times 10^{-2}}2.3\log 0.1 = 76.6 \; 1/m.$$

Example 7.11. The absorptivity of a layer of gas of thickness l_1 at a partial pressure p_1 is $A_{\lambda 1}$.

Determine the absorptivity of the gas after a simultaneous change of the thickness of the layer and of the partial pressure respectively to l_2 and p_2. Bouguer's law is assumed to be valid for the given gas, and the gas temperature is the same in both cases.

Solution. In accordance with Bouguer's law, the absorptivity of gases at a constant temperature is a function of the quantity pl:

$$A_\lambda = 1 - e^{-k(pl)}$$

Let us write down the above equality as applied to the conditions of the problem considered :

$$A_{\lambda 1} = 1 - e^{-k(p_1 l_1)};$$

$$A_{\lambda 2} = 1 - e^{-k(p_2 l_2)}.$$

Eliminating k from the equations, we obtain :

$$A_{\lambda 2} = 1 - (1 - A_{\lambda 1})^{p_2 l_2 / p_1 l_1}.$$

PROBLEMS

Problem 7.1. The surface of a steel product has a temperature $t_w = 727°C$ and its emissivity is $\varepsilon_w = 0.7$. The emitting surface can be considered to be a grey one.

Calculate the density of emitted radiation and the wavelength corresponding to the maximum spectral intensity of radiation.

Answer. $E = 3.97 \times 10^4$ W/m^2; $\lambda_{max} = 2.898$ μm.

Problem 7.2. An optical pyrometer with a red light filter (see example 7.1) records a temperature $t_0 = 1600°C$.

Find the emissivity of the investigated body at $\lambda = 65$ μm, if it is known that its true temperature $t = 1700°C$.

Answer. $\varepsilon_\lambda = 0.55$.

Problem 7.3. A surface covered with a layer of lamp (jet) black emits in the normal direction per unit solid angle the radiant energy $J_{\varphi = 0} = 1.87 \times 10^3$ W/m^2.°C. For black body radiation the absorptivity of lamp black is equal to 0.96. Determine the temperature of this surface, assuming that Lambert's cosine law is valid for lamp black.

Solution. $t = 300°C$.

Problem 7.4. What should be the emissivity of the screen so that the loss of heat to surroundings due to radiation would not exceed 60 W/m^2 in the presence of a protective screen between the furnace setting and steel lagging? All other conditions remain as in example 7.6.

Answer. ε_{sc} = 0.143.

Problem 7.5. A tube 200 mm in diameter and 1 m long has closed ends, and contains a mixture of dry air and carbon dioxide. The total pressure and temperature of the mixture are equal respectively to 98.1 kPa and 800°C. The partial pressure of the carbon dioxide gas is equal to 9 kPa.

Find the emissivity of the mixture of gases in the tube.

Answer. ε = 0.06.

CHAPTER 8

Dimensional Analysis

INTRODUCTION

The subject of dimensional analysis deals with the process whereby all the important variables involved in a physical phenomenon are systematically organised into dimensionless groups which are less numerous than the original variables. For example, a law expressing the relation between fluid density ρ, viscosity μ, velocity V and a length parameter l is designated by a dimensionless number Re known as Reynolds number ($Re = V l \rho/\mu$). The number of unknown quantities is thus, reduced, the problem is generalised and the need for specifying a particular system of units is eliminated. Such dimensionless grouping facilitates the interpretation and extend the range of application of experimental data. Correlation of experimental data with the help of dimensional analysis is fruitfully utilised to develop empirical relations describing a particular phenomenon.

This chapter presents the general method of dimensional analysis and illustrates its application to various problems of fluid mechanics and convective heat transfer.

Dimensional analysis differs from other methods of approach in that it does not yield equations that can be solved. Instead, it combines several variables into dimensionless groups, such as the Nusselt number, which facilitate the interpretation and extend the range of application of experimental data. In practice, convective heat transfer coefficients are generally calculated from empirical equations obtained by correlating experimental data with the aid of dimensional analysis.

The most serious limitation of dimensional analysis is that it gives no information about the nature of a phenomenon. In fact, to apply dimensional analysis it is necessary to know beforehand what variables influence the phenomenon, and the success or failure of the method depends on the proper selection of these variables. It is therefore important to have at least a preliminary theory or a thorough physical understanding of a phenomenon before a dimensional analysis can be performed. However, once the pertinent variables are known, dimensional analysis can be applied to most problems by a routine procedure.

A dimensionally homogeneous equation is one in which all the terms have the same units. These units can be the base units or derived ones (for example, $kg/s^2.m$ or Pa). Such an equation can be used with any system of units provided that the same base or derived units are used throughout the equation. No conversion factors are needed when consistent units are used.

One should be careful in using any equation and always check it for dimensional homogeneity. To do this, a system of units (SI, English, etc.) is first selected. Then units are substituted for each term in the equation and like units in each term cancelled out.

PRIMARY DIMENSIONS AND DIMENSIONAL FORMULAE

The first step is to select a system of primary dimensions. The choice of the primary dimensions is arbitrary, but the dimensional formulae of all pertinent variables must be expressible in terms of them. In the SI system the primary dimensions of length L, time t, temperature T, and mass M are used.

The dimensional formula of a physical quantity follows from definitions or physical laws. For instance, the dimensional formula for the length of a bar is $[L]$ by definition. The average velocity of a fluid particle is equal to a distance divided by the time interval taken to traverse it. The dimensional formula of velocity is therefore $[L/t]$ or $[Lt^{-1}]$ (i.e., a distance or length divided by a time). The units of velocity could be expressed in metres per second, feet per second, or miles per hour, since they all are a length divided by a time. The dimensional formulae and the symbols of physical quantities occurring frequently in heat transfer problems are given in Table 8.1.

Buckingham π Theorem

To determine the number of independent dimensionless groups required to obtain a relation describing a physical phenomenon, the Buckingham π theorem may be used. According to this rule, the required number of independent dimensionless groups that can be formed by combining the physical variables pertinent to a problem is equal to the total number of these physical quantities n (e.g., density, viscosity, heat transfer coefficient) minus the number of primary dimensions m required to express the dimensional formulae of the n physical quantities. If we call these groups

π_1, π_2, etc., the equation expressing the relationship among the variables has a solution of the form

$$F(\pi_1, \pi_2, \pi_3, \ldots) = 0 \qquad \ldots (8.1)$$

Table 8.1. Important heat and mass transfer physical quantities and their dimensions.

Quantity	Symbol	Dimensions in MLtT system
Length	L, x	L
Time	t	t
Mass	M	M
Force	F	ML/t^2
Temperature	T	T
Heat	Q	ML^2/t^2
Velocity	u, U_∞	L/t
Acceleration	a, g	L/t^2
Work	W	ML^2/t^2
Pressure	p	M/t^2L
Density	f	M/L^3
Internal energy	e	L^2/t^2
Enthalpy	h	L^2/t^2
Specific heat	c	L^2/t^2T
Absolute viscosity	μ	M/Lt
Kinematic viscosity	$v = \mu/\rho$	L^2/t
Thermal conductivity	k	ML/t^3T
Thermal diffusivity	α	L^2/t
Thermal resistance	R	Tt^3/ML^2
Coefficient of expansion	β	$1/T$
Surface tension	σ	M/t^2
Shear per unit area	t	M/Lt^2
Heat transfer coefficient	h	M/t^3T
Mass flow rate	m	M/t

In a problem involving five physical quantities and three primary dimensions, $n - m$ is equal to two and the solution either has the form

$$F(\pi_1, \pi_2) = 0 \qquad \ldots (8.2)$$

or the form

$$\pi_1 = f(\pi_2) \qquad \ldots (8.3)$$

Experimental data for such a case can be presented conveniently by plotting π_1 against π_2. The resulting empirical curve reveals the functional relationship between π_1 and π_2, which cannot be deduced from dimensional analysis.

For a phenomenon that can be described in terms of three dimensionless groups (i.e., if $n - m = 3$), Eq. (8.1) has the form

$$F(\pi_1, \pi_2, \pi_3) = 0 \qquad \qquad ...\ (8.4)$$

but can also be written as

$$\pi_1 = f(\pi_2, \pi_3)$$

For such a case, experimental data can be correlated by plotting π_1 against π_2 for various values of π_3. Sometimes it is possible to combine two of the π's in some manner and to plot this parameter against the remaining π on a single curve.

Determination of Dimensionless Groups

A simple method for determining dimensionless groups will now be illustrated by applying it to the problem of correlating experimental convection heat transfer data for a fluid flowing across a heated tube. Exactly the same approach could be used for heat transfer in flow through a tube or over a plate.

From the description of the convective heat transfer process, it is reasonable to expect that the physical quantities listed in Table 8.2 are pertinent to the problem.

Table 8.2. Pertinent physical quantities in convective heat transfer.

Variable	Symbol	Dimensions
Tube diameter	D	$[L]$
Thermal conductivity of fluid	k	$[ML/t^3 T]$
Free-stream velocity of fluid	U_∞	$[L/t]$
Density of fluid	ρ	$[M/L^3]$
Viscosity of fluid	μ	$[M/Lt]$
Specific heat at constant pressure	c_p	$[L^2/t^{2-}T]$
Heat transfer coefficient	h_c	$[M/t^2 T]$

There are seven physical quantities and four primary dimensions. We therefore expect that three dimensionless groups will be required to correlate the data. To find these dimensionless groups, we write π as a product of the variables, each raised to an unknown power

$$\pi = D^a k^b U_\infty^c \rho^d \mu^e c_p^f \bar{h}_c^g \qquad \text{... (8.5)}$$

and substitute the dimensional formulae

$$\pi = [L]^a \left[\frac{ML}{t^3 T}\right]^b \left[\frac{L}{t}\right]^c \left[\frac{M}{L^3}\right]^d \left[\frac{M}{Lt}\right]^e \left[\frac{L^2}{t^2 T}\right]^f \left[\frac{M}{t^3 T}\right]^g \qquad \text{... (8.6)}$$

For π to be dimensionless, the exponents of each primary dimension must separately add up to zero. Equating the sum of the exponents of each primary dimension to zero, we obtain the set of equations

$$b + d + e + g = 0 \text{ for } M$$

$$a + b + c - 3d - e + 2f = 0 \text{ for } L$$

$$-3d - c - e - 2f - 3g = 0 \text{ for } t$$

$$- b - f - g = 0 \text{ for } T$$

Evidently any set of values of a, b, c, d, and e that simultaneously satisfies this set of equations will make π dimensionless. There are seven unknowns, but only four equations. We can therefore choose values for three of the exponents in each of the dimensionless groups. The only restrictions on the choice of the exponents is that each of the selected exponents be independent of the others. An exponent is independent if the determinant formed with the coefficients of the remaining terms does not vanish (i.e., is not equal to zero).

Since $\bar{h}_c$, the convective heat transfer coefficient, is the variable we eventually want to evaluate, it is convenient to set its exponent g equal to unity. At the same time we let $c = d = 0$ to simplify the algebraic manipulations. Solving the equations simultaneously, we obtain $a = 1$, $b = -1$, $e = f = 0$, and the first dimensionless group is

$$\pi_1 = \frac{\bar{h}_c D}{k}$$

which we recognise as the Nusselt number $\overline{Nu}_D$.

For π_2 we select g equal to zero, so that $\bar{h}_c$ will not appear again, and let $a = 1$ and $f = 0$. Simultaneous solution of the equations with these choices yields $b = 0$, $c = d = 1$, $e = -1$ and

$$\pi_2 = \frac{U_\infty D \rho}{\mu}$$

This dimensionless group is a Reynolds number Re_D, with the tube diameter as the length parameter.

If we let $e = 1$ and $c = g = 0$, we obtain the third dimensionless group

$$\pi_3 = \frac{c_p \mu}{k}$$

which is the Prandtl number Pr.

We observe that, although the heat transfer coefficient is a function of six variables, with the aid of dimensional analysis the seven original variables have been combined into three dimensionless groups. According to Eq. (8.4), the functional relationship can be written

$$\overline{Nu}_D = f(Re_D, Pr)$$

and experimental data can now be correlated in terms of three variables instead of the original seven. The importance of this reduction in the variables becomes apparent when we attempt to correlate experimental data.

Correlation of Experimental Data

Suppose that, in a series of tests with air flowing over a 1-in-OD pipe, the heat transfer coefficient has been measured experimentally at velocities ranging from 0.5 to 100 ft/s. This range of velocities corresponds to Reynolds numbers based on the diameter $D\rho U_\infty/\mu$ ranging from 250 to 50,000. Since the velocity was the only variable in these test, the results are correlated in Fig. 8.1(a) by plotting the heat transfer coefficient h_c against the velocity U_∞. The resulting curve permits a direct determination of h_c at any velocity for the system used in the tests, but it cannot be used to determine the heat transfer coefficients for cylinders that are larger or smaller than the one used in the tests. Neither could the heat transfer coefficient be evaluated if the air were under pressure and its density were different from that used in the tests. Unless experimental data could be correlated more effectively, it would be necessary to perform separate experiments for every cylinder diameter, every density, etc. The amount of labour would obviously be enormous.

With the aid of dimensional analysis, however, the results of one series of tests can be applied to a variety of other problems. This is illustrated by Fig. 8.1(b), where the data of Fig. 8.1(a) are replotted in terms of pertinent dimensionless groups. The abscissa in Fig. 8.1(b) is the Reynolds number $U_\infty D\rho/\mu$, and the ordinate is the Nusselt number $\bar{h}_c D/k$. This correlation of the data permits the evaluation of the heat transfer coefficient for air flowing over any size of pipe or wire as long as the Reynolds number of the system falls within the range covered in the experiment.

Experimental data obtained with air alone do not reveal the dependence of the Nusselt number on the Prandtl number, since the Prandtl number is a combination of physical properties whose value does not vary appreciably for gases. To determine the influence of the Prandtl number it is necessary to use different fluids. According to the preceding analysis, experimental data with several fluids whose physical properties yield a wide range of Prandtl numbers are necessary to complete the correlation.

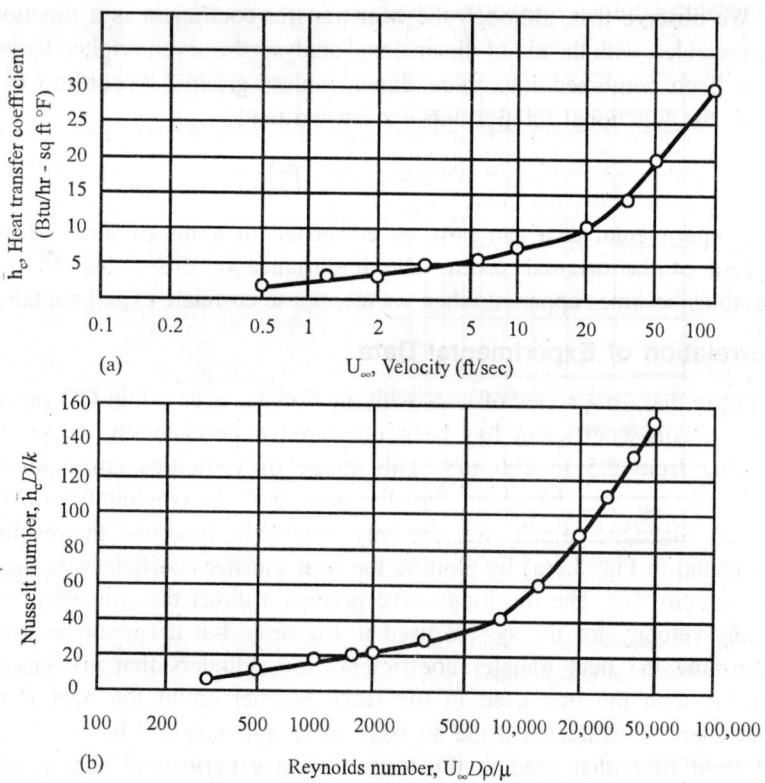

Fig. 8.1. Variation of Nusselt number with Reynolds number for cross-flow of air over a pipe or a long cylinder.

In Fig 8.2 the experimental results of several independent investigations for heat transfer between air, water, and oils in cross-flow over a tube or a wire are plotted for a wide range of temperatures, cylinder sizes, and velocities.

The ordinate in Fig. 8.2 is the dimensionless quantity $\overline{Nu}_D / \mathrm{Pr}^{0.3}$ and the abscissa is Re_D. An inspection of the results shows that all of the data follow a single line reasonably well, so that they can be correlated empirically.

Principle of Similarity

The remarkable result of Fig. 8.2 can be explained by the principle of similarity. According to this principle, often called the model law, the behaviour of two systems will be similar if the ratios of their linear dimensions, forces, velocities, etc., are the same.

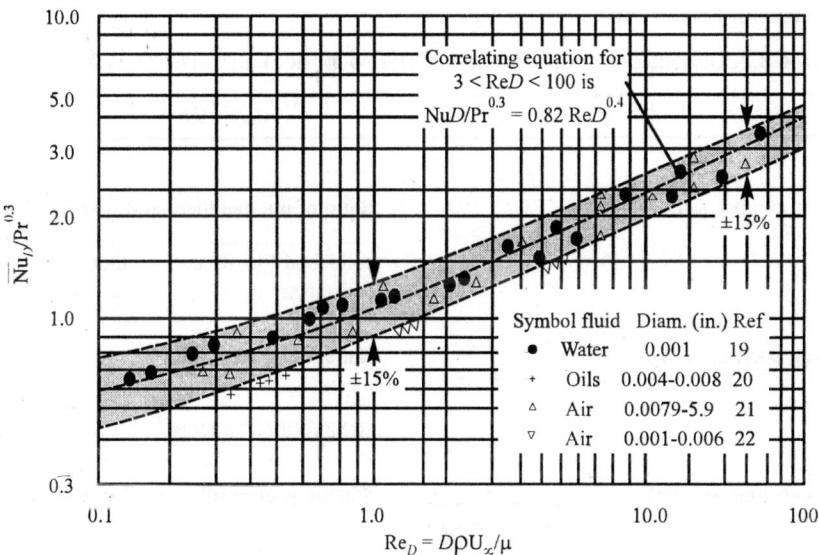

Fig. 8.2. Correlation of experimental heat transfer data for various fluids in cross-flow over pipes and cylinders.

Under conditions of forced convection in geometrically similar systems, the velocity fields will be similar provided the ratio of inertia forces to viscous forces is the same in both fluids.

The Reynolds number is the ratio of these forces, and consequently we expect similar flow conditions in forced convection for a given value of the Reynolds number.

The Prandtl number is the ratio of two molecular transport properties, the kinematic viscosity $v = \mu/\rho$, which affects the velocity distribution, and the thermal diffusivity $k/\rho c_p$, which affects the temperature profile. In other words, it is a dimensionless group that relates the temperature distribution to the velocity distribution.

Hence, in geometrically similar systems having the same Prandtl and Reynolds numbers, the temperature distribution will be similar. The Nusselt number is equal to the ratio of the temperature gradient at a fluid-to-surface interface to a reference-temperature gradient. We expect therefore that, in systems having similar geometries and similar temperature fields, the numerical values of the Nusselt numbers will be equal. This fact is borne out by the experimental results in Fig. 8.2.

Dimensional analyses have been performed for numerous heat and mass transfer systems, and Table 8.3 summarises the most important dimensionless groups used in design.

Table 8.3. Dimensionless groups of importance for heat and mass transfer.

Group	Definition	Interpretation
Biot number (Bi)	$\dfrac{hL}{k_s}$	Ratio of internal thermal resistance of a solid body to its surface thermal resistance
Drag coefficient (Cf)	$\dfrac{\tau_s}{\rho U_\infty^2 / 2}$	Ratio of surface shear stress to free-stream kinetic energy
Eckert number (Ec)	$\dfrac{U_\infty^2}{c_p(T_s - T_\alpha)}$	Kinetic energy of flow relative to boundary-layer enthalphy difference
Fourier number (Fo)	$\dfrac{\alpha t}{L^2}$	Dimensionless time; ratio of rate of heat conduction to rate of internal energy storage in a solid
Friction factor (f)	$\dfrac{\Delta p}{(L / D)(\rho U_m^2 / 2)}$	Dimensionless pressure drop for internal flow
Grashof number (Gr$_L$)	$\dfrac{g\beta(T_s - T_\infty)L^3}{v^2}$	Ratio of buoyancy to viscous forces
Colburn j factor (j$_H$)	$StPr^{2/3}$	Dimensionless heat transfer coefficient

(Table contd...)

Group	Definition	Interpretation
Lewis number (Le)	$\dfrac{\alpha}{D_{AB}}$	Ratio of molecular thermal diffusivity to mass diffusivity
Nusselt number (Nu_L)	$\dfrac{h_c L}{k_f}$	Dimensionless heat transfer coefficient; ratio of convection heat transfer to conduction in a fluid layer of thickness L
Peclet number (Pe_L)	$Re_L Pr$	Product of Grashof and Prandtl numbers
Prandtl number (Pr)	$\dfrac{c_p \mu}{k} = \dfrac{v}{\alpha}$	Ratio of molecular momentum diffusivity to thermal diffusivity
Rayleigh number (Ra)	$GR_L Pr$	Product of Grashof and Prandtl numbers
Reynolds number (Re_L)	$\dfrac{U_\infty L}{v}$	Ratio of inertia and viscous forces
Schmidt number (Sc)	$\dfrac{v}{D_{AB}}$	Ratio of molecular momentum diffusivity to mass diffusivity
Sherwood number (Sh_L)	$\dfrac{h_m L}{D_{AB}}$	Ratio of convection mass transfer to diffusion in a slab of thickness L
Stanton number	$\dfrac{h_c}{\rho U_\infty c_p} = \dfrac{Nu_L}{Re_L Pr}$	Dimensionless heat transfer coefficient

ANALYTIC SOLUTION FOR LAMINAR BOUNDARY-LAYER FLOW OVER A FLAT PLATE

In the preceding section we determined dimesionless groups for correlating experimental data of heat transfer by forced convection. We found that the Nusselt number depends on the Reynolds number and the Prandtl number , i.e.,

$$Nu = \phi \, (Re) \, \psi(Pr) \qquad \qquad \dots (8.7)$$

In the next few sections we shall consider analytical methods for determining the functional relations in Eq. (8.7) for low-speed flow over a flat plate. This system has been selected primarily because it is the simplest to analyse. But the results have many practical applications; for instance, they are good approximations for flow over the surfaces of streamlined bodies such as airplane wings or turbine blades.

In view of the differences in the flow characteristics, the frictional forces as well as the heat and mass transfer are governed by different relations for laminar and turbulent flow. We will first treat the laminar boundary layer, which is amenable to an exact and an approximate method of solution.

To determine the forced-convection heat transfer coefficient and the friction coefficient for incompressible flow over a flat surface we must satisfy the continuity, momentum, and energy equations simultaneously. These relations for convenience are given below.

Continuity:

$$\frac{\partial u}{\partial x} + \frac{\partial v}{\partial y} = 0 \qquad \qquad \dots (8.8)$$

Momentum:

$$\rho \left(u \frac{\partial u}{\partial x} + v \frac{\partial u}{\partial y} \right) = \mu \frac{\partial^2 u}{\partial y^2} - \frac{\partial p}{\partial x} \qquad \qquad \dots (8.9)$$

Energy:

$$u \frac{\partial T}{\partial x} + v \frac{\partial T}{\partial y} = \alpha \frac{\partial^2 T}{\partial y^2} \qquad \qquad \dots (8.10)$$

Boundary-Layer Thickness and Skin Friction

Equation (8.9) must be solved simultaneously with the continuity equation, Eq. (8.8), in order to determine the velocity distribution, boundary-layer thickness, and friction force at the wall. These equations are solved by first defining a stream function $\psi(x,y)$ which automatically satisfies the continuity equation, or

$$u = \frac{\partial \psi}{\partial y} \quad \text{and} \quad v = \frac{\partial \psi}{\partial x}$$

Introducing the new variable

$$\eta = y\sqrt{\frac{U_\infty}{vx}}$$

we can let

$$\psi = \sqrt{vxU_\infty}\,f(\eta)$$

where $f(\eta)$ denotes a dimensionless stream function. In terms of $f(\eta)$, the velocity components are

$$u = \frac{\partial\psi}{\partial y} = \frac{\partial\psi}{\partial\eta}\frac{\partial\eta}{\partial y} = U_\infty\frac{d[f(\eta)]}{d\eta}$$

and

$$v = -\frac{\partial\psi}{\partial x} = \frac{1}{2}\sqrt{\frac{vU_\infty}{x}}\left\{\frac{d[f(\eta)]}{d\eta}\eta - f(\eta)\right\}$$

APPLICATION OF DIMENSIONAL ANALYSIS TO PRESSURE DROP ALONG A PIPE

Pressure drop measurements by Reynolds verified the linear variation with velocity below the critical point. For velocities greater then 1.2 times the critical, the pressure drop varied as the 1.722 power of the mean velocity for smooth pipes and approached the 2.0 power for rough pipes.

It has been possible to present these quite different results in terms of one equation by what might be called a semi-empirical approach. To develop this, we shall employ (and illustrate) another similarity procedure known as dimensional analysis. Although this method is usually attributed to Rayleigh, more complete discussion of the theory and application can be found in works by Buckingham and Bridgman.

In applying the method of dimensional analysis to a problem, it is necessary to know or guess the independent quantities or variables upon which the quantity in question depends. These can usually be determined by physical reasoning about the system under consideration. What can then be determined may not be the final answer or solution, but only physically significant groups of variables for use in guiding experimental work and in correlating experimental data. Although it is possible to set up rules by which to proceed, we can handle the problem of pressure drop along a pipe as follows:

The pressure drop is assumed to be proportional to the length l and is supposed to depend on the inner diameter d of the pipe, the mean velocity

u_m, the mass density ρ, and the dynamic viscosity μ of the fluid. This is expressed in equation form as

$$\frac{\Delta p}{l} = Cd^a u_m^{\ b} \rho^c \mu^e \qquad \qquad \cdots (8.11)$$

where C is a dimensionless constant and a, b, c, and e are exponents to be determined. In order for this equation to be physically correct, the sum of the exponents of the basic units (mass, length, and time) must be the same on the right as on the left side. For finding the relationship between the exponents, we may assume a case where all quantities except the exponents are equal to unity. Taking unit length to be denoted by L, and mass by M, and unit time by S, equation (8.11) could be written

$$\frac{m}{L^2 S^2} = CL^a \left(\frac{L}{S}\right)^b \left(\frac{M}{L^3}\right)^c \left(\frac{M}{LS}\right)^c \qquad \qquad \cdots (8.12)$$

Equating the sum of the exponents on both sides, for

$L: \quad -2 = a + b - 3c \quad - e$

$M: \quad 1 = \qquad \qquad c + e$

$S: \quad -2 = -b \qquad \qquad - e$

Solving in terms of a gives

$b = a + 3$

$c = a + 2$

$e = -a - 1$

Substituting in equation (8.11) shows

$$\frac{\Delta p}{l} = Cd^a u_m^{\ a+3} \rho^{a+2} \mu^{-a-1} \qquad \qquad \cdots (8.13)$$

Multiplying the right side by d/d and combining quantities with the same exponent yields

$$\frac{\Delta p}{l} = \frac{C}{d} \left(\frac{du_m \rho}{\mu}\right)^{a+1} \rho u_m^{\ 2} \qquad \qquad \cdots (8.14)$$

To determine the final form, divide both sides by $w = \rho g$ and multiply the right by 2/2. Transferring l to the right then gives

$$\frac{\Delta p}{w} = 2C \frac{l}{d} \left(\frac{du_m \rho}{\mu}\right)^{a+1} \left(\frac{u_m^{\ 2}}{2g}\right) \qquad \qquad \cdots (8.15)$$

The term $\Delta p/w$ has the net dimension of a length; it should be regarded, however, as an energy or head loss per unit weight of fluid flowing (e.g., ft lb/lb). Noting that $u_m^2/2g$ is the kinetic energy per unit weight of fluid flowing, the final result can be written in terms of dimensionless groups as

$$\frac{\Delta p \,/\, w}{u_m^{\,2} \,/\, 2g} = \frac{\Delta p}{\rho u_m^{\,2} \,/\, 2} = 2C \frac{l}{d} \left(\frac{d u_m \rho}{\mu} \right)^{a+1} \qquad \dots (8.16)$$

This is the general equation for the pressure drop in a smooth pipe.

DIMENSIONAL ANALYSIS IN MASS TRANSFER

The use of dimensional analysis enables us to predict the various dimensional groups which are very helpful in correlating experimental mass-transfer data. As we saw in fluid flow and in heat transfer, the Reynolds number, the Prandtl number, the Grashof number, and the Nusselt number were often used in correlating experimental data. The Buckingham theorem as already discussed, states that the functional relationship among q quantities or variables whose units may be given in terms of u fundamental units or dimensions may be written as $(q - u)$ dimensionless groups.

Dimensional Analysis for Convective Mass Transfer

We consider a case of convective mass transfer where a fluid is flowing by forced convection in a pipe and mass transfer is occurring from the wall to the fluid. The fluid flows at a velocity v inside a pipe of diameter D and we wish to relate the mass-transfer coefficient k'_c to the variables D, ρ, $\mu, v,$ and D_{AB}. The total number of variables is $q = 6$. The fundamental units or dimensions are $u = 3$ and are mass M, length L, and time t.

$$k'_c = \frac{L}{t} \quad \rho = \frac{M}{L^3} \quad \mu = \frac{M}{Lt} \quad v = \frac{L}{t} \quad D_{AB} = \frac{L^2}{t} \quad D = L$$

The number of dimensionless groups or π' are then $6 - 3$, or 3. Then,

$$\pi_1 = f(\pi_2, \pi_3) \qquad \dots (8.17)$$

We choose the variables D_{AB}, ρ, and D to be the variables common to all the dimensionless groups, which are

$$\pi_1 = D^a_{AB} \rho^b D^c k'_c \qquad \dots (8.18)$$

$$\pi_2 = D^d_{AB} \rho^e D^f v \qquad \dots (8.19)$$

$$\pi_3 = D^g_{AB} \, \rho^h D^i \mu \qquad \dots (8.20)$$

For π_1 we substitute the actual dimensions as follows:

$$1 = \left(\frac{L^2}{t}\right)^a \left(\frac{M}{L^3}\right)^b (L)^c \left(\frac{L}{t}\right) \qquad \dots (8.21)$$

Summing for each exponent,

(L) $0 = 2a - 3b + c + 1$

(M) $0 = b$

(t) $0 = a - 1$ $\dots (8.22)$

Solving these equation simultaneously, $a = -1$, $b = 0$, $c = 1$. Substituting these values into Eq. (8.18),

$$\pi_1 = \frac{k'_c D}{D_{AB}} = N_{Sh} \qquad \dots (8.23)$$

$$\pi_2 = \frac{vD}{D_{AB}} \qquad \dots (8.24)$$

$$\pi_3 = \frac{\mu}{\rho D_{AB}} = N_{Sc} \qquad \dots (8.25)$$

If we divide π_2 by π_3 we obtain the Reynolds number.

$$\frac{\pi_2}{\pi_3} = \frac{vD}{D_{AB}} \bigg/ \left(\frac{\mu}{\rho D_{AB}}\right) = \frac{Dv\rho}{\mu} = N_{Re} \qquad \dots (8.26)$$

Hence, substituting into Eq. (8.17),

$$N_{Sh} = f(N_{Re}, N_{Sc}) \qquad \dots (8.27)$$

DIMENSIONAL ANALYSIS IN HEAT TRANSFER

As seen in many of the correlations for fluid flow and heat transfer, many dimensionless groups, such as the Reynolds number and Prandtl number, occur in these correlations. Dimensional analysis is often used to group the variables in a given physical situation into dimensionless parameters or numbers which can be useful in experimentation and correlating data.

An important method of obtaining these dimensionless groups is the Buckingham method, in which the listing of the significant variables in the particular physical problem is done first. Then we determine the number of dimensionless parameters into which the variables may be combined.

Buckingham Method

Heat transfer inside a pipe

The Buckingham theorem states that the function relationship among q quantities or variables whose units may be given in terms of u fundamental units or dimensions may be written as $(q - u)$ dimensionless groups.

As an additional example to illustrate the use of this method, let us consider a fluid flowing in turbulent flow at velocity v inside a pipe of diameter D and undergoing heat transfer to the wall. We wish to predict the dimensionless groups relating the heat transfer coefficient h to the variables D, ρ, μ, c_p, k, and v. The total number of variables is $q = 7$.

The fundamental units or dimensions are $u = 4$ and are mass M, length L, time t, and temperature T. The units of the variables in terms of these fundamental units are as follows:

$$h = \frac{M}{t^3 T} \quad D = L \quad \rho = \frac{M}{L^3} \quad \mu = \frac{M}{Lt} \quad c_p = \frac{L^2}{t^2 T} \quad k = \frac{ML}{t^3 T} \quad v = \frac{T}{t}$$

Hence, the number of dimensionless groups or π's can be assumed to be $7 - 4$, or 3. Then

$$\pi_1 = f(\pi_2, \pi_3) \qquad \dots (8.28)$$

We will choose the four variables D, k, μ, and v to be common to all the dimensionless groups. Then the three dimensionless groups are

$$\pi_1 = D^a k^b \mu^c v^d \rho \qquad \dots (8.29)$$

$$\pi_2 = D^e k^f \mu^g v^h c_p \qquad \dots (8.30)$$

$$\pi_3 = D^i k^j \mu^k v^l h \qquad \dots (8.31)$$

For π_1, substituting the actual dimensions,

$$M^0 L^0 t^0 T^0 = 1 = L^a \left(\frac{ML}{t^3 T}\right)^b \left(\frac{M}{LT}\right)^c \left(\frac{L}{T}\right)^d \left(\frac{M}{L^3}\right) \qquad \dots (8.32)$$

Summing for each exponent,

$$(L) \qquad 0 = a + b - c + d - 3$$

$$(M) \qquad 0 = b + c + 1 \qquad \dots(8.33)$$

$$(t) \qquad 0 = -3b - c - d$$

$$(T) \qquad 0 = -b$$

Solving these equations simultaneously, $a = 1$, $b = 0$, $c = -1$, $d = 1$.

Substituting these values into Eq. (8.29),

$$\pi_1 = \frac{Dv\rho}{\mu} = N_{Re} \qquad \ldots (8.34)$$

Repeating for π_2 and π_3 and substituting the actual dimensions,

$$\pi_2 = \frac{c_p\mu}{K} = N_{Pr} \qquad \ldots (8.35)$$

$$\pi_3 = \frac{hD}{k} = N_{Nu} \qquad \ldots (8.36)$$

Substituting π_1, π_2, and π_3 into Eq. (8.28) and rearranging,

$$\frac{hD}{k} = f\left(\frac{Dv\rho}{\mu}, \frac{c_p\mu}{k}\right) \qquad \ldots (8.37)$$

This is in the form of the familiar equation for heat transfer inside pipes.

This type of analysis is useful in empirical correlations of heat-transfer data. The importance of each dimensionless group, however, must be determined by experimentation.

Natural convection heat transfer outside a vertical plane

In the case of natural-convection heat transfer from a vertical plane wall of length L to an adjacent fluid, different dimensionless groups should be expected when compared to forced convection inside a pipe since velocity is not a variable. The buoyant force due to the difference in density between the cold and heated fluid should be a factor. As already seen the buoyant force depends upon the variables β, g, ρ, and ΔT. Hence, the list of variables to be considered and their fundamental units are as follows:

$$L = L \quad \rho = \frac{M}{L^3} \quad \mu = \frac{M}{Lt} \quad c_p = \frac{L^2}{t^2 T} \quad \beta = \frac{1}{T}$$

$$g = \frac{L}{t^2} \quad \Delta T = T \quad h = \frac{M}{t^3 T} \quad k = \frac{ML}{t^3 T}$$

The number of variables is $q = 9$. Since $u = 4$, the number of dimensionless groups or π's is $9 - 4$, or 5. Then $\pi_1 = f(\pi_2, \pi_3, \pi_4, \pi_5)$.

EMPIRICAL CORRELATION FOR VARIOUS SHAPES

After experimental data have been correlated by dimensional analysis, it is general practice to write an equation for the line that best fits the data. It is also useful to compare the experimental results with those obtained by analytic means, if they are available. This comparison allows one to determine whether the analytic method adequately describes the experimental results. If the two agree, one can describe the physical mechanisms that are important for the problem with confidence.

In this section the results of some experimental studies on natural convention for a number of geometric shapes of practical interest are presented. Each shape has been associated with a characteristic dimension, such as its distance from the leading edge x, length L, diameter D, and so on. The characteristic dimension is attached as a subscript to the dimensionless parameters Nu and Gr. Average values of the Nusselt number for a given surface are identified by a bar, that is, $\overline{Nu}$; local values are without a bar. All physical properties are to be evaluated at the arithmetic mean between the surface temperature T_s and the temperature of the undisturbed fluid T_∞. The temperature difference in the Grashof number ΔT represents the absolute value of the difference between the temperatures T_s and T_∞. The accuracy with which in practice the unit-surface conductance can be predicted from any of the equations is generally no better than 20 per cent, because most experimental data scatter by as much as ± 15 per cent or more, and in a majority of engineering applications stray currents due to some interaction with surfaces other than the one transferring the heat are unavoidable.

In the following sub-sections we present correlation equations for several important geometries.

Vertical Plates and Cylinders

For a flat vertical surface, it is possible to find analysitcal and approximate solutions to the momentum and energy equations. Details of the method for natural convection may indicate that the local value of the heat transfer coefficient for laminar natural convection from an isothermal vertical plate or cylinder at a distance x from the leading edge is

$$h_{cx} = 0.508 \, Pr^{1/2} \frac{Gr_x^{1/4}}{(0.952 + Pr)^{1/4}} \frac{k}{x} \qquad \ldots (8.38)$$

Since $Gr_x \sim x^3$, Eq. (8.38) shows that the heat transfer coefficient decreases with the distance from the leading edge to the 1/4 power. The leading edge is the lower edge for a heated surface and the upper edge for a surface cooler than the surrounding fluid. The average value of the heat transfer coefficient for a height L is obtained by integrating Eq. (8.38) and dividing by L, or

$$\bar{h}_c = \frac{1}{L}\int_0^L h_{cx} = 0.68\,\mathrm{Pr}^{1/2}\,\frac{\mathrm{Gr}_L^{1/4}}{(0.952 + \mathrm{Pr})^{1/4}} \qquad \cdots \ (8.39a)$$

In dimensionless form, the average Nusselt number is

$$\overline{\mathrm{Nu}} = \frac{\bar{h}_c L}{k} = 0.68\,\mathrm{Pr}^{1/2}\,\frac{\mathrm{Gr}_L^{1/4}}{(0.952 + \mathrm{Pr})^{1/4}} \qquad \cdots \ (8.39b)$$

Gryzagorids has shown experimentally that Eq. (8.39b) adequately represents the data in the regime $10 < \mathrm{Gr}_L\mathrm{Pr} < 10^8$.

For a vertical plate submerged in a liquid metal (Pr < 0.03), the average Nusselt number in laminar flow is

$$\overline{\mathrm{Nu}}_L = \frac{\bar{h}_c L}{k} = 0.68(\mathrm{Gr}_L\,\mathrm{Pr}^2)^{1/4} \qquad \cdots \ (8.39c)$$

In the turbulent region, the value of h_{cx}, the local heat transfer coefficient, is nearly constant over the surface. In fact, McAdams recommends for $\mathrm{Gr}_L > 10^9$ the equation

$$\overline{Nu}_L = \frac{\bar{h}_c L}{k} = 0.13(Gr_L\,\mathrm{Pr})^{1/3} \qquad \cdots \ (8.40)$$

according to which the heat transfer coefficient is independent of the length L.

A theoretical analysis by Sparrow and Gregg, supported by experimental data from Dotson, indicates that the equations for laminar natural convection from a vertical flat plate apply to a constant surface temperature as well as to a uniform heat flux over the surface. In the latter case the surface temperature T_s is to be taken at one-half of the total height of the plate.

For a vertical plate, or a plate tilted at an angle θ from the vertical, with the heated surface facing downward (or cooled surface facing upward), Fujii and Imura found that the equation

$$\overline{Nu}_L = 0.56(Gr_L\,\mathrm{Pr}\cos\theta)^{1/4} \qquad \cdots \ (8.41)$$

applies in the range

$$10^5 < \mathrm{Gr}_L\,\mathrm{Pr}\,\cos\theta < 10^{11} \text{ and } 0 \le \theta \le 89°$$

In Eq. (8.41) the plate length L is the dimension which rotates in a vertical plane as θ increases. If the heated surface is facing upward (or cooled surface facing downward), Eq. (8.41) is only recommended up to the critical Rayleigh number, because at higher Ra the data deviate significantly from Eq. (8.41) and boundary-layer separation was observed. For natural convection to or from spheres of diameter D the empirical equation

$$\overline{Nu}_D = 2 + 0.39(Gr_D)^{1/4} \text{ for } 1 < Gr_D < 10^5 \qquad \dots (8.42)$$

is recommended. For very small spheres, as the Grashof number approaches zero, the Nusselt number approaches a value of 2, that is, $\overline{h}_c \, D/k \rightarrow 2$. This condition corresponds to pure conduction through a stagnant layer of fluid surrounding the sphere.

Experimental data for natural convection from vertical cones pointing downward with vertex angles between 3 and 12 degrees have been correlated by

$$\overline{Nu}_L = 0.63(1 + 0.72 \in)Gr_L^{1/4} \qquad \dots (8.43)$$

where $3 < \phi < 12°$, $7.5 < \log Gr_L < 8.7$, $0.2 \le \in \le 0.8$, and

$$\in = \frac{2}{Gr_L^{1/4} \tan(\phi/2)}$$

ϕ = vertex angle

L = slant height of the cone

Enclosed Spaces

Natural-convection heat transfer across enclosures such as shown in Fig. 8.3 is important for determining heat loss through windows, from flat-plate solar collectors, through building walls, and in many other applications. The enclosure consists of two isothermal vertical surfaces at temperature T_1 and T_2 spaced a distance δ apart and of height L. The top and bottom of the enclosure are insulated. The Grashof number is defined by

$$Gr_\delta = \frac{g\beta(T_1 - T_2)\delta^3}{v^2} \qquad \dots (8.44)$$

Any temperature difference will produce flow in the enclosure. Hollands and Konicek found that for $Gr_\delta \gtrsim 8000$ the flow consists of one large cell rotating in the enclosure. The heat transfer mechanism is equivalent to conduction across the enclosure for $Gr_\delta < 8000$. As the Grashof numbr is increased beyond this value, the flow becomes more of a boundary-layer type with fluid rising in a layer near the heated surface, turning the corner at the top, and flowing downward in a layer near the cooled surface. The boundary-layer thickness decreases with $Gr_\delta^{1/4}$ and the core region is more or less inactive and thermally stratified.

For the geometry in Fig. 8.3 Catton recommends the correlation of Berkovsky and Polevikov:

$$\overline{Nu}_\delta = 0.22 \left(\frac{L}{\delta}\right)^{-1/4} \left(\frac{Pr}{0.2 + Pr} Ra_\delta\right)^{0.28} \qquad \text{... (8.45)}$$

in the range

$$2 < L/\delta < 10, \ Pr < 10, \ \text{and} \ Ra_\delta < 10^{10}$$

and

$$\overline{Nu}_\delta = 0.18 \left(\frac{Pr}{0.2 + Pr} Ra_\delta\right)^{0.29} \qquad \text{... (8.46)}$$

in the range

$$1 < L/\delta < 2, \ 10^{-3} < Pr < 10^5, \ \text{and} \ 10^3 < \frac{Ra_\delta \, Pr}{0.2 + Pr}$$

Data are lacking for select ratios L/δ less than one. Imberger found that as $Ra_\delta \to \infty$, $Nu_\delta \to (L/\delta)Ra_\delta^{1/4}$ for $L/\delta = 0.01$ and 0.02. Bejan et al. found that $Nu_\delta = 0.014 \, Ra_\delta^{0.38}$ for $L/\delta = 0.0625$ and $2 \times 10^8 < Ra_\delta < 2 \times 10^9$. Nansteel and Greif found $Nu_\delta = 0.748 \, Ra_\delta^{0.226}$ for $L/\delta = 0.5$, $2 \times 10^{10} < Ra_\delta \leq 10^{11}$, and $3.0 \leq Pr \leq 4.3$.

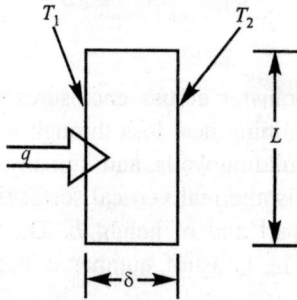

Fig. 8.3. Nomenclature for natural convection in enclosed vertical spaces.

In a horizontal fluid layer with heating from above, heat transfer is by conduction only. Heating from below results in conduction heat transfer only if $Ra_\delta < 1700$, where the length scale is the spacing enclosing the layer. Above this value of Ra_δ the fluid motion is in the form of multiple cells rotating with a horizontal axis, which are known as Benard cells. The flow begins to become turbulent for $Ra_\delta \sim 5500$ for $Pr = 0.7$ and for Ra_δ 55,000 for $Pr = 8500$ and becomes fully turbulent for $Ra_\delta \sim 10^6$.

Hollands et al. correlated data for horizontal air layers heated from below over a very wide range of Rayleigh numbers with

$$\overline{\mathrm{Nu}}_\delta = 1 + 1.44\left[1 - \frac{1708}{\mathrm{Ra}_\delta}\right] + \left[\left(\frac{\mathrm{Ra}_\delta}{5830}\right)^{1/3} - 1\right]^{\cdot} \qquad \dots (8.47)$$

where notation [ ]· indicates that if the quantity inside the bracket is negative the quantity is to be taken as zero. This equation closely represented data for air from the critical Rayleigh number (Ra_δ = 1700) to Ra_δ = 10^8. To closely match data for water, it was necessary to add a term to the above equation

$$\overline{\mathrm{Nu}}_\delta = 1 + 1.44\left[1 - \frac{1708}{\mathrm{Ra}_\delta}\right] + \left[\left(\frac{\mathrm{Ra}_\delta}{5830}\right)^{1/3} - 1\right] + 2.0\left[\frac{\mathrm{Ra}_\delta^{1/3}}{140}\right]^{[1-\ln(\mathrm{Ra}_\delta^{1/3}/140)]} \qquad \dots(9.48)$$

which is then valid from the critical Rayleigh number ($\sim$ 1700) to Ra_δ = 3.5 × 10^9. These two correlation equations with experimental data are shown in Figs. 8.4 and 8.5.

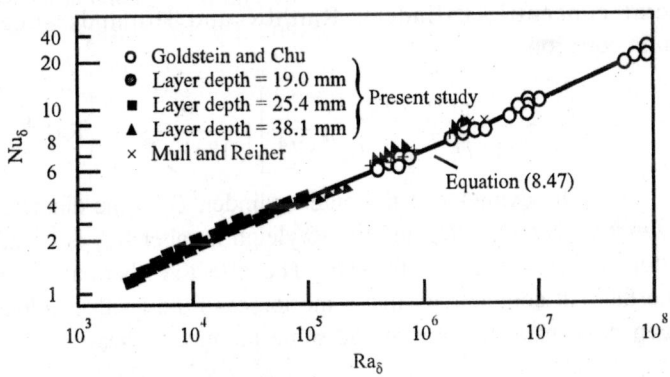

Fig. 8.4. Correlation of data for natural-convection heat transfer across a horizontal layer of air heated from below.

For natural convection inside spherical cavities of diameter D the relation

$$\frac{D\overline{h}_c}{k} = C(\mathrm{Gr}_D\,\mathrm{Pr})^n \qquad \dots(8.49)$$

is recommended with the constants C and n selected from the tabulation below :

$\mathrm{Gr}_D\mathrm{Pr}$	C	n
$10^4 - 10^9$	0.59	1/4
$10^9 - 10^{12}$	0.13	1/3

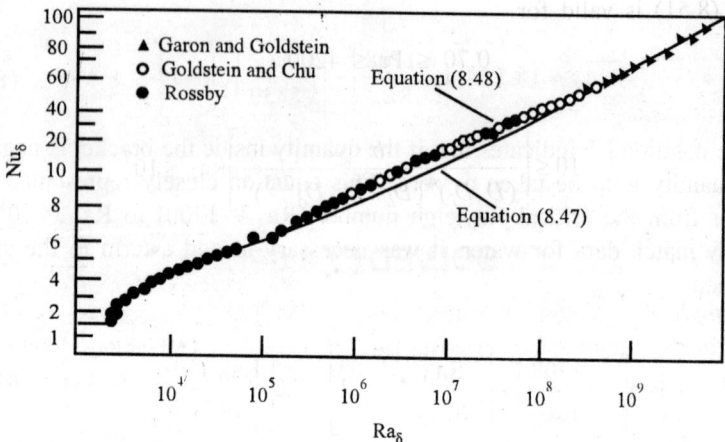

Fig. 8.5. Correlation of data for natural-convection heat transfer across a horizontal layer of water heated from below.

For natural-convection heat transfer across the gap between two horizontal concentric cylinders, Raithby and Hollands suggest the correlation equation

$$\frac{K_{eff}}{k} = 0.386 \left[\frac{\ln(D_o / D_i)}{b^{3/4}(1/D_i^{3/5} + 1/D_o^{3/5})^{5/4}} \right] \left(\frac{Pr}{0.861 + Pr} \right)^{1/4} Ra_b^{1/4} \quad ...(8.50)$$

Here, D_o is the diameter of the outer cylinder, D_i is the diameter of the inner cylinder, $2b = D_o - D_i$, and the Rayleigh number Ra_b is based on the temperature difference across the gap. The effective thermal conductivity k_{eff} is the thermal conductivity that a motionless fluid (with conductivity k) in the gap must have to transfer the same amount of heat as the moving fluid.

The correlation Eq. (8.50) is valid over the following range of parameters:

$$0.70 \le Pr \le 6000$$

$$10 \le \left[\frac{\ln(D_o / D_i)}{b^{3/4}(1/D_i^{3/5} + 1/D_o^{3/5})^{5/4}} \right]^4 Ra_b \le 10^7$$

For concentric spheres, Raithby and Hollands recommend

$$\frac{k_{eff}}{k} = 0.74 \left[\frac{b^{1/4}}{D_o D_i (D_i^{-7/5} + D_o^{-7/5})^{5/4}} \right] Ra_b^{1/4} \left(\frac{Pr}{0.861 + Pr} \right)^{1/4} \quad ... (8.51)$$

Eq. (8.51) is valid for

$$0.70 \leq Pr \leq 4200$$

and

$$10 \leq \left[\frac{b}{(D_o D_i)^4 (D_i^{-7/5} + D_o^{-7/5})^5} \right] Ra_b \leq 10^7$$

SOLVED EXAMPLE

Example 8.1. A pan of water 8 cm deep is placed on a stove-top burner. The burner element is thermostatically controlled and maintains the bottom of the pan at 100°C. Assuming the top surface of the water is initially at room temperature, 20°C, what is the initial rate of heat transfer from the burner to the water? The pan is circular and is 15 cm in diameter.

Solution. For the properties of water at 60°C

$$Ra_\infty = \frac{(9.8\ m/s^2)(5.18 \times 10^{-4} K^{-1})(80K)(0.08m)^3(3.02)}{(0.478 \times 10^{-6} m^2/s)^2}$$

$$Ra_\delta = 2.75 \times 109$$

From Eq. (8.48) we find

$$Nu_\delta = 1 + 1.44 + 76.8 + 0.1$$

$$= 79.3$$

$$\bar{h}_c = \overline{Nu}_\delta \frac{k}{\delta} = \frac{(79.3)(0.657\ W/mK)}{0.08\ m}$$

$$= 651\ W/m^2\ K$$

The initial rate of heat transfer is therefore

$$Q = (651 W/m^2 K) \left(\frac{\pi\ 0.15^2\ m^2}{4} \right)(80K)$$

$$= 920\ W$$

CHAPTER 9

Hydrodynamic and Thermal Boundary Layers

INTRODUCTION

When a fluid flows around an object or when the object moves through a body of fluid, their exists a thin layer of fluid close to the solid surface within which shear stresses significantly influence the velocity distribution. The fluid velocity varies from zero at the solid surface to the velocity of free stream flow at a certain distance away from the solid surface. This thin layer of changing velocity has been called the hydrodynamic boundary layer; a concept first suggested by Ludwig Prandtl. Heat transfer occurs due to heat conduction and energy transport by moving fluid within this thin layer. Hence, the value of convection coefficient and heat transfer is highly dependent upon the thickness and characteristics of the boundary layer. For better understanding of the combined fluid-dynamic and heat transfer phenomenon, recourse has to be made to the realms of fluid mechanics with particular emphasis on the development and growth of boundary layer.

Our approach to this study of convection starts with laminar flow of a fluid parallel to a flat plate and in a tube. Next is the consideration of these same systems but with turbulent flow.

HYDRODYNAMIC AND THERMAL BOUNDARY LAYERS

The heat exchange in the laminar boundary laye region of a flat plate parallel to a fluid stream is probably the best example to use in presenting the basic ideas involved in convection heat transfer. The physical concepts are easily visualised and an exact solution has been obtained. Consider a very thin flat plate parallel to an unbounded incompressible viscous fluid flowing with a uniform velocity u_∞. When the fluid meets the leading edge, friction causes it to adhere to the surface and a laminar boundary layer develops. Fig 9.1 shows the boundary layer region along the upper surface of the plate. At any distance x from the leading edge the velocity u increases from zero at the plate surface to u_∞ at the edge of the boundary layer ($y = \delta$).

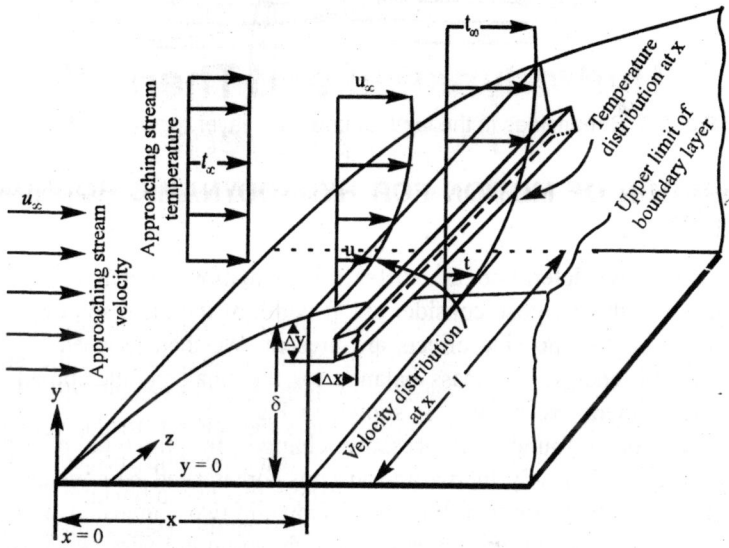

Fig. 9.1. Velocity and temperature distributions in a laminar boundary layer along a flat plate (temperature considered as datum).

If the approaching free stream temperature is t_∞ degrees above the plate temperature t_s, a thermal boundary layer will exist as well as the velocity or hydrodynamic boundary layer. Assuming that these two layers are of the same thickness, the temperature distribution would be similar to the velocity distribution, both increasing from zero to free stream conditions (considering the plate temperature as the zero point) at the upper limit of the boundary layer.

As the boundary layer thickness increases with x, more and more fluid is included with an attendant decrease in velocity. In order to satisfy flow continuity, it is apparent that the initially parallel streamlines of the oncoming stream must diverge away from the plate as shown in Fig. 9.2.

It is evident from this divergence that a velocity component (even though very small) will exist in the direction normal to the plate. This component is denoted by v and is clearly proportional to the rate of divergence of the streamlines.

$$v = -\frac{\partial \psi}{\partial x} \qquad \qquad \dots (9.1)$$

where ψ is the stream function of flow.

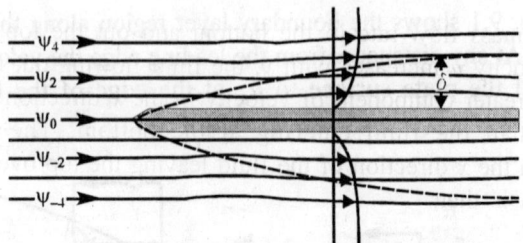

Fig. 9.2. Streamlines in the laminar boundary layer region of a flat plate.

EQUATION OF MOTION FOR HYDRODYNAMIC BOUNDARY LAYER

We shall now determine the differential equations which apply to this system. For this purpose consider a 1 ft width of the plate somewhere near its centre so that no side effects are present. We then proceed to make a momentum, energy, and mass balance for the small parallelepiped-shaped, volume element, $\Delta x \times \Delta y \times 1$, shown in Fig. 9.1.

It has been found that pressure changes in this type of flow are negligible, so that only inertia and viscous forces need be considered. The pertinent quantities are therefore those shown in two dimensions in Fig. 9.3. (Both viscous and inertia forces in the y direction are very small and have been neglected). Now apply Newton's second law in terms of the change in momentum per unit time being equal t the force acting to the x direction.

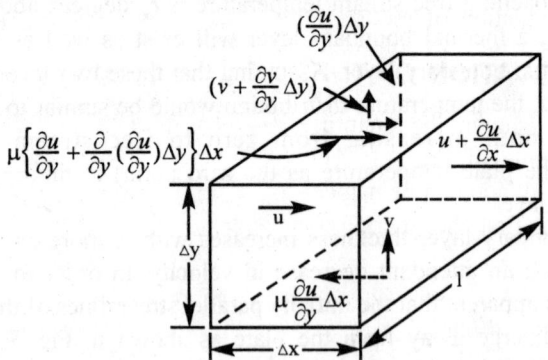

Fig 9.3. Velocity changes and viscous forces acting on elemental volume of Fig. 9.1.

The average mass flow rate in the left face and out the right is $\rho u \Delta y$. The change in momentum per unit time is, therefore.

$$\rho u \Delta y \left(u + \frac{\partial u}{\partial x} \Delta x - u \right) = \rho u \frac{\partial u}{\partial x} \Delta x \Delta y \qquad ...(9.2)$$

The average mass flow rate in the bottom and out the top will be $\rho v \Delta x$. Since the velocity u increases with y, the fluid flowing out though the top will have a greater component of velocity in the x direction by an amount $(\partial u/\partial y \Delta y)$ than the fluid entering at the bottom. The increment of momentum in the x direction of the fluid leaving the top over that entering at the bottom is then

$$\rho v \Delta x \frac{\partial u}{\partial y} \Delta y \qquad \qquad ...\ (9.3)$$

The net force acting is simply the difference in the viscous drag on the upper and lower surfaces; this is

$$\mu \Delta x \left\{ \frac{\partial u}{\partial y} + \frac{\partial}{\partial y}\left(\frac{\partial u}{\partial y}\right) \Delta y - \frac{\partial u}{\partial y} \right\} = \mu \frac{\partial^2 u}{\partial y^2} \Delta x \Delta y \qquad ...\ (9.4)$$

Equating the sum of equations (9.2) and (9.3) to (9.4) gives

$$\rho u \frac{\partial u}{\partial x} + \rho v \frac{\partial u}{\partial y} = \mu \frac{\partial^2 u}{\partial y^2}$$

or

$$u \frac{\partial u}{\partial x} + v \frac{\partial u}{\partial y} = v \frac{\partial^2 u}{\partial y^2} \qquad \qquad ...\ (9.5)$$

which is the equation of motion for this flow.

ENERGY EQUATION OF THE THERMAL BOUNDARY LAYER

In fig 9.4 are shown the quantities of energy entering and leaving the elemental volume of Fig 9.1. The rate of temperature change in the x direction is small and can be neglected without introducing significant error.

Considering first the energy convected in for the x direction, the net amount is

$$u \Delta y w c_p t - \left(u + \frac{\partial u}{\partial x}\Delta x\right) \Delta y w c_p \left(t + \frac{\partial t}{\partial x}\Delta x\right)$$

which reduces to

$$-w c_p \left(u \frac{\partial t}{\partial x} + t \frac{\partial u}{\partial y}\right) \Delta x \Delta y \qquad \qquad ...\ (9.6)$$

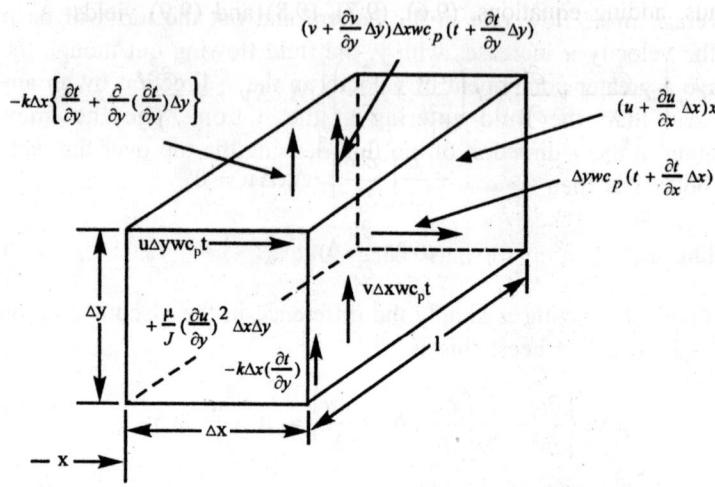

Fig 9.4. Energy quantities entering and leaving elemental volume in Fig. 9.1.

if the term involving the product $(\partial u/\partial x)(\partial t/\Delta x)$ is neglected. A similar expression obtained for the y direction is

$$-wc_p\left(v\frac{\partial t}{\partial y} + t\frac{\partial v}{\partial y}\right)\Delta x\Delta y \qquad \dots (9.7)$$

Since temperature changes in the x direction are negligible, conduction is considered in the y direction only. The net heat flow in by conduction is then

$$-k\Delta x\frac{\partial t}{\partial y} - (-k\Delta x)\left\{\frac{\partial t}{\partial y} + \frac{\partial}{\partial y}\left(\frac{\partial t}{\partial y}\right)\Delta y\right\} = k\frac{\partial^2 t}{\partial y^2}\Delta x\Delta y \qquad \dots (9.8)$$

Finally we must account for the heat generated by the fluid friction. The average viscous force is $\mu(\partial u/\partial y)\Delta x$. The distance through which this force can be considered to act is determined by the relative movement of the fluid at the upper surface over that along the lower; in unit time this is equivalent to $(\partial u/\partial y)\Delta y$. The rate of work is then $\mu (\partial u/\partial y)2\,\Delta x\,\Delta y$, which, divided by the mechanical equivalent of heat J, gives the net rate of energy added to the element as

$$\frac{\mu}{J}\left(\frac{\partial u}{\partial y}\right)^2\Delta x\Delta y \qquad \dots (9.9)$$

The temperature varies only with location and not time in this system. The sum of the various energy quantities entering the elemental volume must, therefore, be equal to zero.

Thus, adding equations, (9.6), (9.7), (9.8) and (9.9) yields

$$-wc_p \left\{ u \frac{\partial t}{\partial x} + v \frac{\partial t}{\partial y} + t \left(\frac{\partial u}{\partial x} + \frac{\partial v}{\partial y} \right) \right\} \Delta x \Delta y + k \frac{\partial^2 t}{\partial y^2} \Delta x \Delta y$$

$$+ \frac{\mu}{J} \left(\frac{\partial u}{\partial y} \right)^2 \Delta x \Delta y = 0 \qquad \dots (9.10)$$

Recalling the equation for mass balance shows

$$\frac{\partial u}{\partial x} + \frac{\partial v}{\partial y} = 0 \qquad \dots (9.11)$$

equation (9.10) can be simplified to

$$u \frac{\partial t}{\partial x} + v \frac{\partial t}{\partial y} = \frac{k}{wc_p} \frac{\partial^2 t}{\partial y^2} + \frac{\mu}{Jwc_p} \left(\frac{\partial u}{\partial y} \right)^2 \qquad \dots (9.12)$$

which is the differential energy equation for this plate system.

VELOCITY DISTRIBUTION IN THE BOUNDARY LAYER

The next step is to find solutions for the velocity and temperature at any point x,y, in the boundary layer. The boundary conditions to be satisfied are

(a) $u = 0$ and $v = 0$ at $y = 0$

$u = u_\infty$ at $y = \infty$

and

(b) $t = t_s$ (taken as zero) at $y = 0$

$t = t_\infty$ at $y = \infty$ ... (9.13)

In the actual course of events the velocity distribution was determined first by solving the hydrodynamic equation (9.5) in conjuction with the continuity equation (9.11). This velocity distribution was then used in the solution of the energy equation (9.12) to obtain the temperature distribution.

The key to solving equation (9.5) was provided by Prandtl when he pointed out that the number of variables could be reduced by considering the local velocity u to be a function of a new variable combining x and y in the form $y / \sqrt{x}$. The reasoning which led to this started with an order of magnitude analysis of the boundary layer thickness. In equation (9.5) the terms on the left represent the inertia forces and must be balanced by viscous forces given by the term on the right. Since u is large compared to v, we may consider as a first approximation the inertia forces proportional to $\rho u \partial u / \partial x$. Along a plate of length l, u will be proportional to the free stream velocity u_∞ and $\partial u / \partial x$ to u_∞ / l. Hence the inertia force is of the order of $\rho u_\infty^2 / l$. The velocity gradient $\partial u / \partial y$ in the direction normal to the wall will be proportional to u_∞ / δ so that the viscous or friction force $\mu \partial^2 u / \partial y^2$ is of the order of $\mu u_\infty / \delta^2$. Since the inertia and viscous forces are of the same order of magnitude

$$\frac{\rho u_\infty^2}{l} \sim \mu \frac{u_\infty}{\delta^2} \qquad \text{... (9.14)}$$

or solving for the boundary layer thickness δ,

$$\delta \sim \sqrt{\frac{\mu l}{\rho u_\infty}} = \sqrt{\frac{\nu l}{u_\infty}} \quad \text{or} \quad \sqrt{\frac{\nu x}{u_\infty}} \qquad \text{... (9.15)}$$

It is now reasoned that velocity profiles at various distances from the leading edge of the plate are similar to each other. This means if we can express u at any location x as a function of the dimensionless distance from the wall y/δ, i.e.,

$$\frac{u}{u_\infty} = f\left(\frac{y}{\delta}\right) \qquad \text{... (9.16)}$$

it must change along the plate according to some scale factor involving only x. If this scale factor involved y, the velocity profiles would not be the same. The desired scale factor quite obviously must involve the increase in δ with x, which is expressed by equation (9.15). Substituting for δ in equation (9.16) from (9.15) gives

$$\frac{u}{u_\infty} = f\left(\frac{y}{\sqrt{x}} \sqrt{\frac{u_\infty}{\nu}}\right) \qquad \text{... (9.17)}$$

Equation (9.17) then suggests it is possible to express the velocity ratio u/u_∞ at any point x and y in the boundary layer in terms of one variable η defined as

$$\eta = y \sqrt{\frac{u_\infty}{\nu x}} \qquad \text{... (9.18)}$$

So far we have overlooked the fact that the vertical component of velocity v occurs in equation (9.5). In order to account for this it is necessary to define a stream function

$$\frac{\psi}{u_\infty} = \sqrt{\frac{vx}{u_\infty}} f(\eta) \qquad \text{... (9.19)}$$

rather than u as in equation (9.17). Then

$$u = \frac{\partial \psi}{\partial y} = \frac{\partial \psi}{\partial \eta} \frac{\partial \eta}{\partial y} = u_\infty \frac{df}{d\eta} \qquad \text{... (9.20)}$$

and

$$v = -\frac{\partial \psi}{\partial y} = \frac{1}{2} \sqrt{\frac{vu_\infty}{x}} \left(\eta \frac{df}{dn} - f \right) \qquad \text{... 9.21)}$$

After substituting from equations (9.20) and (9.21) in (9.5), the following ordinary differential equation obtained:

$$f \frac{d^2 f}{d\mu^2} + 2 \frac{d^3 f}{d\eta^3} = 0 \qquad \text{...(9.22)}$$

The boundary conditions are

$$\text{at } \eta = 0 : f = 0 \text{ and } \frac{df}{d\eta} = 0 ; \text{ at } \eta = \infty : \frac{df}{d\eta} = 1 \qquad \text{...(9.23)}$$

Thus the partial differential equation for the boundary layer velocity distribution has been transformed into an ordinary differential equation of the third order.

The general solution of equation (9.22) cannot be obtained, and so it is necessary to solve it by numerical methods or by a series expansion. A solution was first presented by Blasius. Additional solution have since been obtained by others, the most accurate being the recent work of Howarth. The velocity distribution parallel to the surface can be presented as one curve using the dimensionless coordinates, u/u_∞ and η, as in Fig. 9.5. The variation of the normal velocity, v/u_∞, is shown in Fig. 9.6. It is interesting to note that, at the outer edge of the boundary layer where $\eta \to \infty$, this does not go to zero but approaches the value

$$v = \frac{0.865}{\sqrt{xu_\infty / v}} u_\infty \qquad \text{... (9.24)}$$

Also note that the magnitude of v/u_∞ is of the order of u/u_∞ divided by $\sqrt{xu_\infty / v} = \sqrt{N_{Rex}}$.

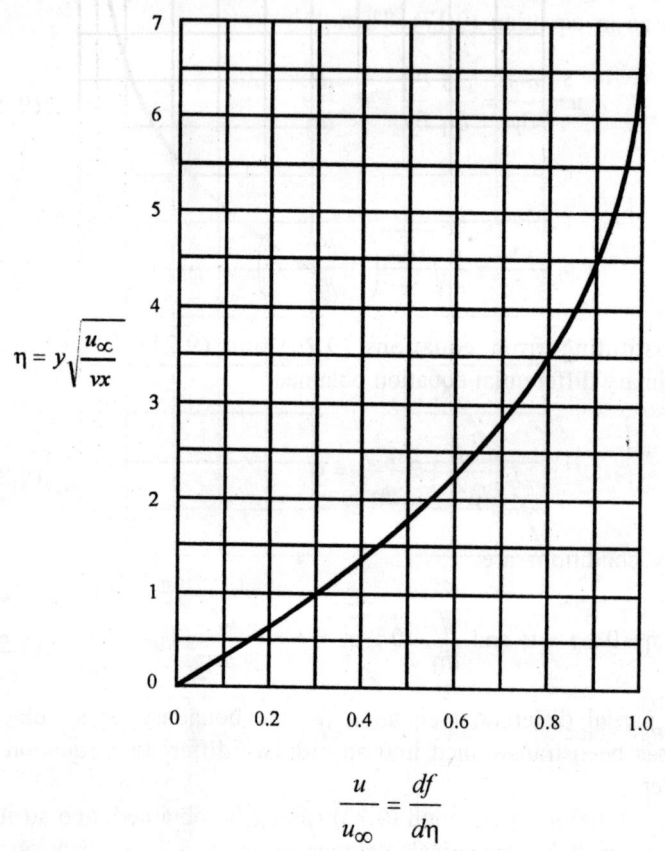

$$\frac{u}{u_\infty} = \frac{df}{d\eta}$$

Fig. 9.5. Dimensionless velocity distribution in the laminar boundary layer along a flat plate.

PRANDTL NUMBER

We are now in a position to evaluate the temperature field in the boundary layer. It will be desirable first to study the problem when the free stream velocity is relatively low and the temperature difference between the free stream and plate is small (of the order of 100°F). Under these conditions it is possible to neglect the effect of frictional heat generation (i.e., the term $\mu(\partial u/\partial y)^2/wc_p J$) and equation (9.12) reduces to

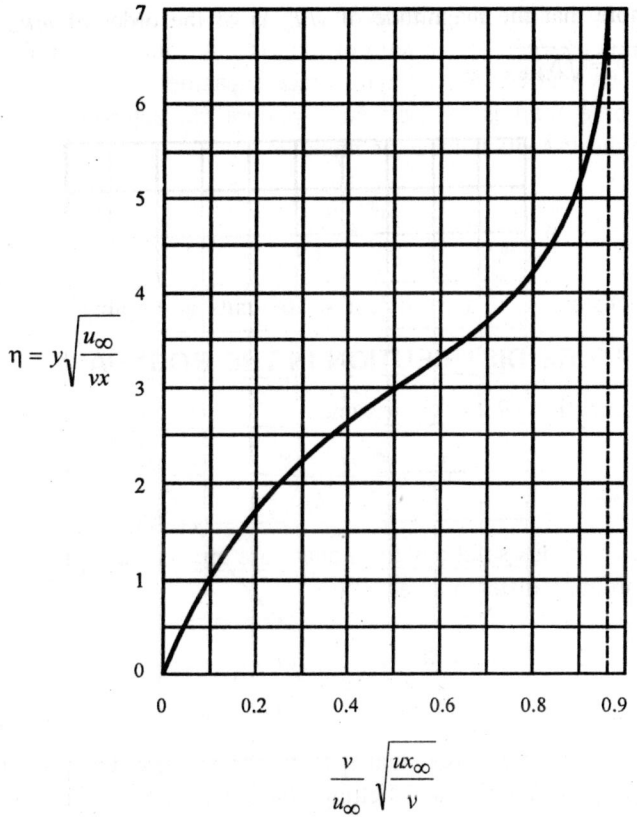

Fig. 9.6. Normal velocity distribution in the laminar boundary layer along a flat plate.

$$u\frac{\partial t}{\partial x} + v\frac{\partial t}{\partial y} = \alpha\frac{\partial^2 t}{\partial y^2} \qquad \text{... (9.25)}$$

where k/wc_p has been replaced by the diffusivity, α. It may be observed that, if $\alpha = v$, this equation is identical to equation (9.5) for the velocity with t replacing u. The temperature and velocity distributions would then be identical or

$$\frac{t - t_s}{t_\infty - t_s} = \frac{u}{u_\infty} = \frac{df}{d\eta} \qquad \text{... (9.26)}$$

This result is of considerable practical importance because for all gases α is approximately equal to v. It also suggests that the more the ratio v/α differs from unity, the greater will be the difference in the temperature and velocity profiles. That this should be true is quite reasonable when v/α is recognised as $\mu/(k/c_p)$, or the ratio of the viscous nature of a fluid and the quotient of its heat conducting and storage capacities. This combination of properties of a fluid is in itself an important characteristic. It has been defined as the Prandtl number or modulus

$$N_{\mathrm{Pr}} = \frac{v}{\alpha} = \frac{\mu c_p}{k} \qquad \ldots (9.27)$$

and is dimensionless if a consistent set of units is employed.

TEMPERATURE DISTRIBUTION IN THE BOUNDARY LAYER

In solving equation (9.25) we assume

$$\theta(\eta) = \frac{t(\eta) - t_s}{t_\infty - t_s} \qquad \ldots (9.28)$$

and that the previous solutions for u and v are applicable. This leads to the following ordinary differential equation:

$$\frac{d^2\theta}{d\eta^2} + \frac{1}{2} N_{\mathrm{Pr}} f \frac{d\theta}{d\eta} = 0 \qquad \ldots (9.29)$$

with $\theta = 0$ at $\eta = 0$ and $\theta = 1$ at $\eta = \infty$. The solution to equation (9.29) was first given by Pohlhausen. Letting $d\theta/d\eta = p$, a first integration gives

$$p(\eta) = \frac{d\theta}{d\eta} = \frac{d\theta(0)}{d\eta} \exp\left[-\frac{N_{\mathrm{Pr}}}{2} \int_0^\eta f d\eta\right] \qquad \ldots (9.30)$$

A second integration then yields

$$\theta(\eta) = \frac{d\theta(0)}{d\eta} \int_0^\eta \exp\left[-\frac{N_{\mathrm{Pr}}}{2} \int_0^\eta f d\eta\right] d\eta \qquad \ldots (9.31)$$

$d\theta(0)/d\eta$ can be evaluated by letting $\eta \to \infty$, giving

$$\frac{d\theta(0)}{d\eta} = \frac{1}{\displaystyle\int_0^\infty \exp\left[-\frac{N_{\mathrm{Pr}}}{2} \int_0^\eta f d\eta\right] d\eta} \qquad \ldots (9.32)$$

Equation (9.32) is the dimensionless slope at the surface of the temperature distribution. It is a function only of N_{Pr} and Pohlhausen found that it could be represented with good accuracy by

$$\frac{d\theta(0)}{d\eta} = 0.332 N_{\mathrm{Pr}}{}^{\frac{1}{3}} \qquad \qquad \text{... (9.33)}$$

Temperature distribution for different values of N_{Pr} are shown plotted in Fig. 9.7. The curve for $N_{\mathrm{Pr}} = 1$ is the same as that in Fig. 9.5. Note that for values of $N_{\mathrm{Pr}} < 1$ the thermal boundary layer is thicker than the velocity boundary layer, while for $N_{\mathrm{Pr}} > 1$ the thermal boundary layer is thinner. An oil with $N_{\mathrm{Pr}} = 1000$, for example, has a thermal boundary layer thickness only about one-tenth that of the velocity boundary layer.

LOCAL HEAT TRANSFER COEFFICIENT

We are now ready to consider the heat transfer to or from a flat plate. where the mechanism of convection was discussed and the concept of a heat transfer coefficient h was introduced.

The temperature distributions in the boundary layer at several locations along a plate have been plotted in Fig. 9.8. The scale for the normal distance from the plate is at the right. The free stream Reynolds number was selected as 3.2×10^5 which would obtain, for example, if the velocity were 50 ft/sec, the fluid air at a temperature of 80°F, and the plate temperature 30°F (the viscosity being evaluated at the arithmetic average of the air and plate temperatures, 55°F). N_{Pr} was taken equal to one.

The rate of heat flow per unit area at any location x along the plate is by definition

$$\frac{q(x)}{A} = -k\left(\frac{\partial t}{\partial y}\right)_{sx} = h_x(t_8 - t_\infty) \qquad \qquad \text{... (9.34)}$$

where $(\partial t/\partial y)_{sx}$ is the slope of the temperature distribution in the fluid at the plate surface and the location x, k is the fluid thermal conductivity, and h_x is the local or point heat transfer coefficient at the point x. Note that since $(\partial t/\partial y)_{sx}$ as shown in Fig. 9.8 is positive, the heat flow is in the negative y direction and $q(x)/A$ is, therefore, negative. Hence h_x, which is a positive, is multiplied by a negative quantity $(t_s - t_\infty)$. Solving equation (9.34) for h_x gives

$$h_x = \frac{-k\left(\frac{\partial t}{\partial y}\right)_{sx}}{(t_s - t_\infty)} \qquad \qquad \text{... (9.35)}$$

$(t_s - t_\infty)$ being a constant, we see that the local heat transfer coefficient varies along the plate as the slope of the temperature distribution in the fluid right at the wall. It is proportional to the tangent of the angle ϕ shown in Fig. 9.8.

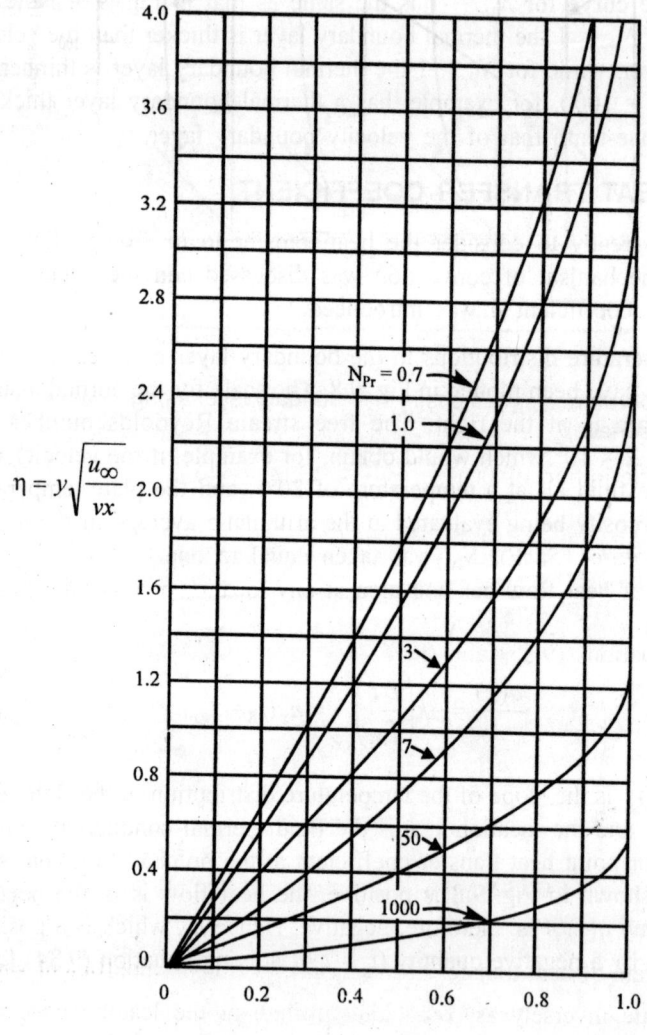

$$\theta(\eta) = t(\eta) - t_s \, / \, (t_\infty - t_s)$$

Fig. 9.7. Temperature distributions in the laminar boundary layer of a flat plate (frictional heat neglected).

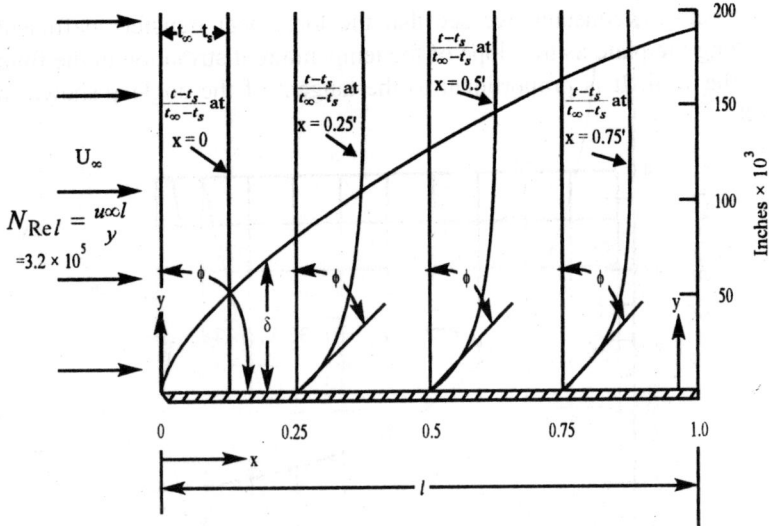

Fig. 9.8. Temperature distributions in the boundary layer along a flat plate (tan $\phi = \partial t/\partial y$ at the surface).

From the definition of $\theta(\eta) = (t(\eta) - t_s)/(t_\infty - t_s)$

$$\tan\phi = \left(\frac{\partial t}{\partial y}\right)_{sx} = (t_\infty - t_s)\left(\frac{\partial\theta}{\partial y}\right)_s = (t_\infty - t_s)\left[\frac{d\theta}{d\eta}\frac{\partial n}{\partial y}\right]_{y=0} \qquad \dots (9.36)$$

and from equations (9.33) and (9.18)

$$\left(\frac{\partial t}{\partial y}\right)_{sx} = 0.332\,N_{Pr}^{\frac{1}{3}}\sqrt{\frac{u_\infty}{vx}}(t_\infty - t_s) \qquad \dots (9.37)$$

$$_x = 0.332kN_{Pr}^{\frac{1}{3}}\sqrt{\frac{u_\infty}{vx}} \qquad \dots (9.38)$$

us the rate of heat transfer or the local heat transfer coefficient varies along the plate inversely as $x^{\frac{1}{2}}$, being infinite at the leading edge and decreasing in the direction of flow, as shown in Fig. 9.9.

AVERAGE HEAT TRANSFER COEFFICIENT

The total heat transfer from a plate of unit width can be determined by integrating the local rate over the entire length l, i.e.,

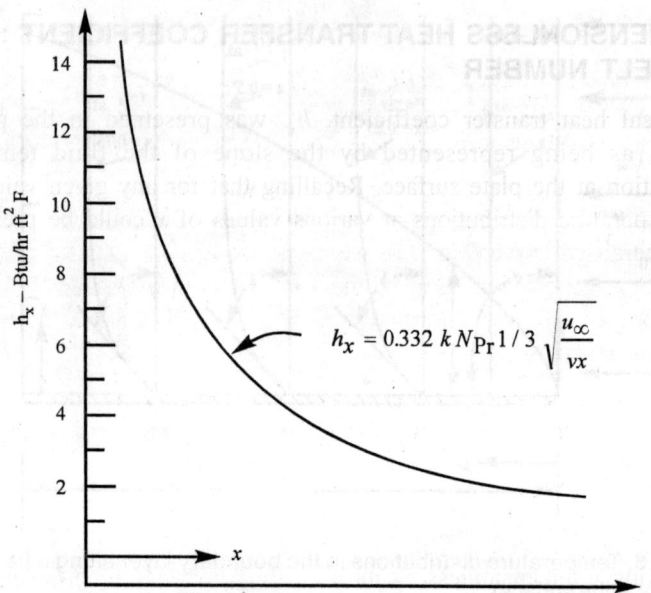

Fig. 9.9. Variation of local heat transfer coefficient along flat plate in Fig. 9.8 ($N_{Pr} = 1$ and $k = 0.0145$ Btu/hr ft F).

$$q = \int_0^l - k \left(\frac{\partial t}{\partial y} \right)_{sx} dx = \int_0^l h_x (t_\infty - t_s) dx \qquad \ldots (9.39)$$

Substituting from equation (9.38) we obtain

$$q = 0.664 k N_{Pr}^{\frac{1}{3}} \sqrt{\frac{u_\infty l}{v}} (t_s - t_\infty)$$

$$= 0.664 k N_{Pr}^{\frac{1}{3}} \sqrt{N_{Rel}} (t_s - t_\infty) \qquad \ldots (9.40)$$

For practical calculations it is desirable to define an average heat transfer coefficient, h, such that

$$q = hl \, (t_s - t_\infty)$$

Equating (9.40) and (9.41) and solving for h gives

$$h = 0.664 k N_{Pr^{\frac{1}{3}}} \sqrt{\frac{u_\infty}{vl}} \qquad \ldots (9.42)$$

It may be noted from equations (9.38) and (9.42) that the average coefficient for a certain length l is just twice the local coefficient at the point l.

A DIMENSIONLESS HEAT TRANSFER COEFFICIENT : NUSSELT NUMBER

The local heat transfer coefficient, h_x, was presented in the preceding section as being represented by the slope of the fluid temperature distribution at the plate surface. Recalling that for any given value of N_{Pr} the temperature distributions at various values of x could be presented as one generalised curve (Fig 9.7), we might expect the slope of this curve to represent a generalised form of the heat transfer coefficient.

Before considering this directly, let us rewrite equation (9.35) in the following form:

$$h_x = k \left[\frac{\partial \left(\frac{t-t_s}{t_\infty -t_s} \right)}{\partial y} \right]_{y=0} \qquad \text{... (9.43)}$$

Making y dimensionless by introducing the distance x from the leading edge gives

$$h_x = \frac{k}{x} \left[\frac{\partial \left(\frac{t-t_s}{t_\infty -t_s} \right)}{\partial (y/x)} \right]_{y=0}$$

or

$$\frac{h_x x}{k} = \left[\frac{\partial \left(\frac{t-t_s}{t_\infty -t_s} \right)}{\partial \left(\frac{y}{x} \right)} \right]_{y=0} \qquad \text{... (9.44)}$$

Thus we see that the product of the local heat transfer coefficient and the distance from the leading edge at which it applies, divided by the thermal conductivity of the fluid, is equal to the dimensionless slope of the fluid temperature distribution at the surface.

The group $(h_x x/k)$ must, therefore, be dimensionless. Although it is sometimes referred to as the dimensionless heat transfer coefficient, it is better known as the *Nusselt number*, N_{Nu}, named in honour of Willelm Nusselt, a German engineer and teacher who made important contributions to the study of conduction and convection early in the present century.

Returning now to the anticipated general relationship between h_x and the slope of the generalised temperature distribution in the boundary layer along a flat plate, we see from equations (9.35) and (9.36) that

$$h_x = k \left(\frac{d\theta}{d\eta} \right)_{\eta=0} \sqrt{\frac{u_\infty}{vx}} \qquad \text{... (9.45)}$$

Again introducing x/x on the right, we find

$$N_{Nux} = \frac{h_x x}{k} = \left(\frac{d\theta}{d\eta} \right)_{\eta=0} \sqrt{\frac{xu_\infty}{v}} = \left(\frac{d\theta}{d\eta} \right)_{\eta=0} \sqrt{N_{Rex}} \qquad \text{... (9.46)}$$

Consequently we see that the slope at the surface of the generalised temperature curve, $(d\theta/d\eta)_{\eta=0}$, is equal to the local Nusselt number divided by the square root of the local Reynolds number.

Since $(d\theta/d\eta)_{\eta=0}$ varies only with N_{Pr} according to equation (9.33) equation (9.46) becomes

$$N_{Nu_x} = 0.332 N_{Pr}^{\frac{1}{3}} N_{Re_x}^{\frac{1}{2}} \qquad \text{... (9.47)}$$

Note that this could also have been obtained from equation (9.38). The Nusselt number, based on the average coefficient for a plate of unit width and length l, would be

$$N_{Nu} = \frac{hl}{K} = 0.664 N_{Pr}^{\frac{1}{3}} N_{Re_x} 1/3 \qquad \text{...(9.48)}$$

VELOCITY DISTRIBUTION IN THE ENTRANCE REGION OF A TUBE

Similar to the flow over a flat plate, a fluid of uniform velocity entering a tube is retarded near the walls and a boundary layer begins to develop (Fig. 9.10). A short distance from the entrance the velocity will increase parabolically from zero at the wall to a uniform distribution u_x over the centre region or core.

Assuming that the mass flow rate is constant, the velocity in the core must increase as the boundary layer thickens. When the boundary layer thickness becomes equal to the tube radius, no further change in the velocity distribution will occur. The fully develop profile will be that of a parabola (so-called Poiseuille flow).

If the tube is heated, a thermal boundary layer begins to develop simultaneously with the hydrodynamic layer. The relative rates at which the two layers develop depends on the characteristics of the fluid as expressed by the Prandtl number, $N_{Pr} = \mu c_p/k$. If $N_{Pr} = 1$, the two develop at an equal rate. When $N_{Pr} < 1$ (liquid metals, for example)—signifying a low velocity and high conducting ability—the thermal boundary layer develops faster than the hydrodynamic. Conversely for $N_{Pr} < 1$ (oils may have values of several hundred) the predominating influence of the viscosity causes very rapid development of fully developed parabolic flow, while the temperature distribution changes slowly, attaining a fully developed profile much further along the tube.

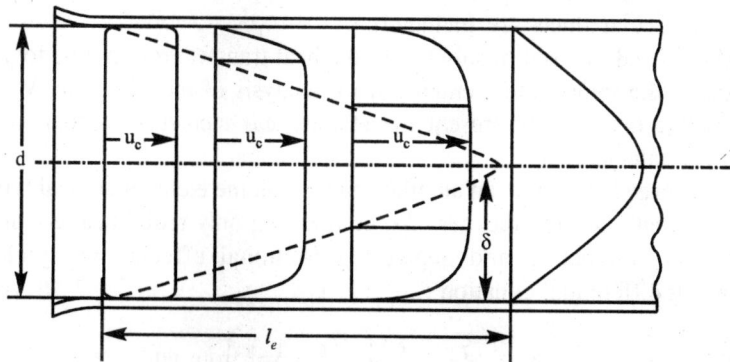

Fig. 9.10. Development of a laminar velocity profile in the intake region of a tube.

The velocity distribution change in the entrance region has been anlysed by Langhaar. The results are expressed in dimensionless form as a function of the distance from the entrance, x/d, and Reynolds number. The length l_e required to develop the parabolic v'   .ty distribution varies with the Reynolds number approximately as

$$\frac{l_e}{d} = 0.05 \frac{du_m}{v} = 0.05 N_{Re} \qquad \text{... (9.49)}$$

where u_m is the mean velocity. Recalling that the critical value of N_{Re} for transition from laminar to turbulent flow is about 2300, equation (9.49) indicates that the velocity profile may be changing for distances of the order of 100 or more tube diameters from the entrance.

FORCED CONVECTION IN LAMINAR FLOW THROUGH A TUBE OR DUCT

A general solution for the temperature field in the entrance region of a heated tube has not yet been obtained because of the complexity of the problem, Solutions for $N_{Pr} = 0.7$ have, however, been worked out numerically be Kays. If the tube wall is not heated until after the velocity profile is fully established, the problem is less complex and has been solved. As early as 1885 Graetz presented a solution for the case of a uniform wall temperature. As for the laminar boundary layer in the previous sections variations of the thermal properties of the fluid with temperature were neglected. Refinements of this work were carried out by Nusselt and Groeber and Jakob. The problem of a non-uniform pipe wall temperature or a specified heat rate distribution along the pipe has also been considered. Just recently a generalised procedure based on a superpositions principle has been described for treating all types of boundary conductions (flat as well as circular ducts are included).

The procedure used in solving for the heat transfer from a tube to a fluid is exactly the same as that employed for analysis of the flat plate. We shall outline the main steps, present the results, and then discuss their use for practical calculations.

An energy balance on an annular shaped volume element coaxial with the tube of length Δx and thickness Δr, considering only radial heat conduction and axial convection and neglecting frictional effects, will yield the following differential equation :

$$u \frac{\partial t}{\partial x} = \frac{k}{w c_p} \left(\frac{\partial^2 t}{\partial r^2} + \frac{1}{r} \frac{\partial t}{\partial r} \right) \qquad \qquad ...(9.50)$$

For the purpose of generalising the solution, equation (9.50) is put into a dimensionless form by introducing the tube radius, r_1, the initial temperature difference between the fluid and wall $t_0 - t_s$, the mean velocity um, and the distance form the entrance x (this procedure is essentially that of requiring differential similarity between two systems). Defining

$$r = \frac{r}{r_1} r_1 = R r_1 ; u = \frac{u}{u_m} u_m = U u_m \text{ and } t = \frac{t - t_s}{t_0 - t_s} (t_0 - t_s) = \theta(t_0 - t_s)$$

and substituting in equation (9.50) gives

$$U_{u_m} (t_0 - t_1) \frac{\partial \theta}{\partial x} = \frac{k}{w c_p} \frac{1}{R r_1} \frac{\partial \left(r_1 R (t_0 - t_s) \frac{\partial \theta}{r_1 \partial R} \right)}{r_1 \partial R} \qquad \qquad ... (9.51)$$

Equation (9.51) can be written as

$$U \frac{\partial \theta}{\partial \left(\dfrac{kx}{w c_p u_m r_1^2} \right)} = \frac{1}{R} \frac{\partial \left(R \frac{\partial \theta}{\partial R} \right)}{\partial R} \qquad \qquad ... (9.52)$$

The group $(kx/w c_p u_m r_1^2)$ is a dimensionless length parameter which must be used to present the results in perfectly general terms. By introducing π we see that this group, expressed in terms of the weight flow rate $W = wu_m \pi r_1^2$, is

$$\pi \frac{kx}{Wc_p} \qquad \qquad ... (9.53)$$

The group Wc_p/kx is known as the Graetz number, N_{Gz}.

A modified form of the Graetz number, which we shall use, is

$$N_{Gz} = \frac{4kx}{4wc_p u_m r_1^2} = \frac{4x}{d(du_m\rho / \mu)(\mu c_p g / k)} = \frac{4x / d}{N_{Re} N_{Pr}} \qquad ... (9.54)$$

Equation (9.52) is then

$$U \frac{\partial \theta}{\partial N_{Gz}} = \frac{1}{R} \frac{\partial \left(R \frac{\partial \theta}{\partial R} \right)}{\partial R} \qquad ... (9.55)$$

Kays integrated equation (9.55) for several different boundary conditions using the velocity distribution given by Langhaar and $N_{Pr} = 0.7$. Fig. 9.11 shows the local Nusselt number N_{Nux} as a function of $(x/d)/N_{Re}N_{Pr}$ for the cases of constant wall temperature, constant temperature difference between wall and fluid, and for constant heat input along the tube.

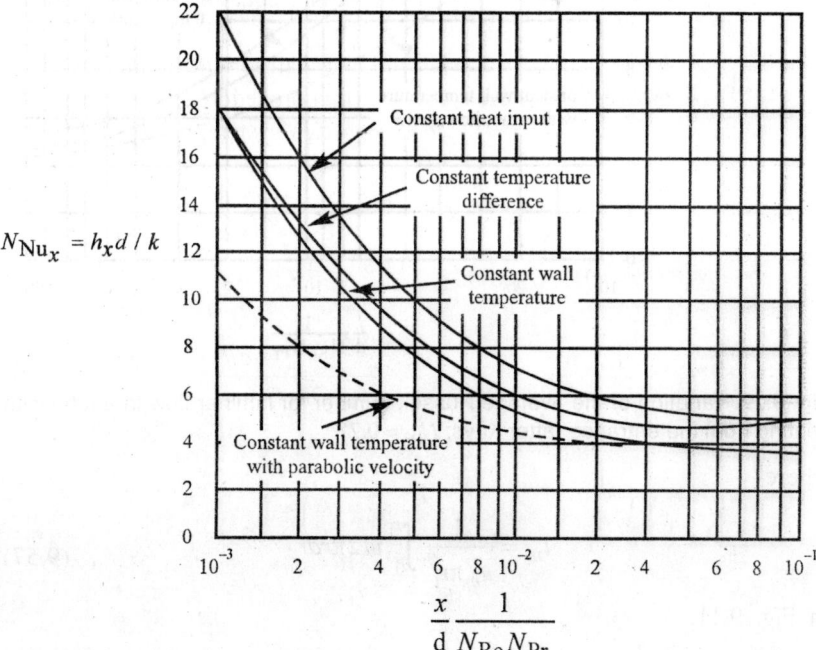

Fig. 9.11. Variation of the local Nusselt number for laminar flow in a tube with heating from the entrance (after Kays; $N_{Pr} = 0.7$).

LOCAL AND AVERAGE HEAT TRANSFER COEFFICIENTS FOR LAMINAR FLOW IN A TUBE

For flow in a pipe the heat transfer coefficient is not specified in terms of a uniform temperature difference between the pipe wall and fluid. The procedure is to define the local coefficient h_x in terms of the difference between the wall temperature and the mean fluid temperature at that point; i.e.,

$$\frac{q(x)}{A} = h_x(t_s - t_m) = -k\left(\frac{\partial t}{\partial r}\right)_{r_1,x} \qquad \ldots (9.56)$$

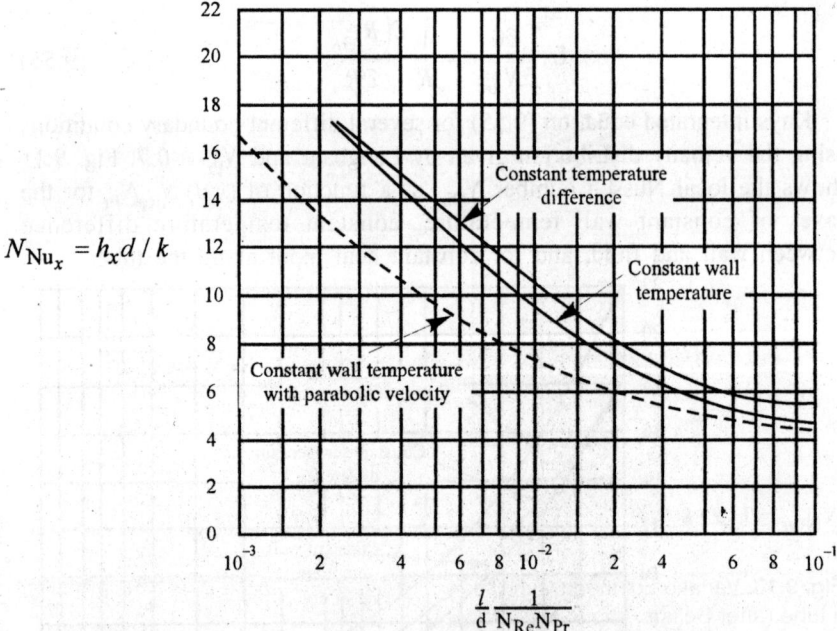

$$N_{Nu_x} = h_x d / k$$

Fig. 9.12. Variation of the average Nusselt number for laminar flow in a tube with heating from the entrance (after Kays; $N_{Pr} = 0.7$).

where

$$t_m = \frac{1}{u_m \pi r_1^2} \int_0^{r_1} ut\, 2\pi r\, dr \qquad \ldots (9.57)$$

In Fig. 9.11,

$$N_{Nux} = \frac{h_x d}{k} = \frac{2\left(\frac{\partial \theta}{\partial R}\right)_{R=1}}{\theta_\pi - \theta_m} \qquad \ldots (9.58)$$

In practice the average coefficient or Nusselt number is often needed. This is obtained from

$$h = \frac{1}{l}\int_0^l h_x dx \qquad \qquad \dots (9.59)$$

where l is the pipe length. The results of Fig. 9.11 in terms of the average Nusselt number are shown in Fig. 9.12.

In the case where heating does not start until the parabolic velocity distribution is fully established, an analytical solution to equation (9.51) is obtained by use of the separation of variable technique. To account for boundary conditions—other than a uniform wall temperature—requires additional calculation. Solutions for the local Nusselt number are shown in Fig. 9.13.

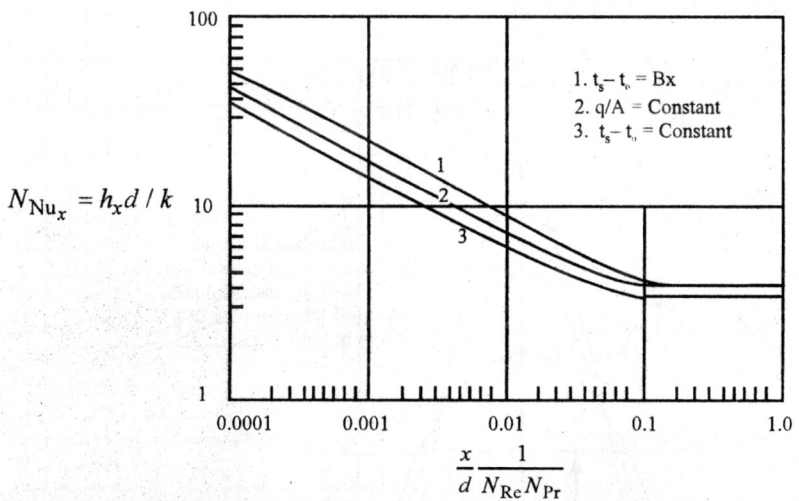

Fig. 9.13. Variation of the local Nusselt number for fully developed laminar flow in a tube (after Sellars, Tribus and Klein).

Although it is not evident in Figs. 9.11 and 9.13 because of the use of a log scale for the abscissa, the local Nusselt number at the beginning of the heated section is infinitely large. It decreases rapidly in the direction of flow asymptotically approaching a constant value as the temperature distribution becomes fully established. This constant value is 3.65 when the wall temperature is uniform and 4.36 when the heat rate is constant or the wall temperature increase is linear. From Fig. 9.13 it is reached when $(x/d)/N_{Re}N_{Pr}$ is approximately 0.05. The length l_t in which the temperature distribution is changing can, therefore, be estimated by

$$\frac{l_t}{d} = 0.05 N_{Re} N_{Pr} \qquad \qquad \ldots (9.60)$$

Comparing equations (9.60) and (9.49), we see that they differ only by N_{Pr}. Thus for fluids of very small N_{Pr} the thermal transition length is very short compared with the velocity transition length; the heat transfer in this case can be estimated by assuming a uniform velocity distribution. For N_{Pr} around unity Figs. 9.11 and 9.12 are applicable. For fluids with large values for N_{Pr} the thermal transition length is mush longer than the hydrodynamic. In this case the solution for a fully established velocity distribution can be used (Fig. 9.13).

For practical calculations the following empirical formulas, representing the average Nusselt number for a uniform wall temperature and fully developed flow, is very useful:

$$N_{Nu} = \frac{hd}{k} = 3.65 + \frac{0.0668(d/l) N_{Re} N_{Pr}}{1 + 0.04\,[(d/l) N_{Re} N_{Pr}]^{2/3}} \qquad \ldots (9.61)$$

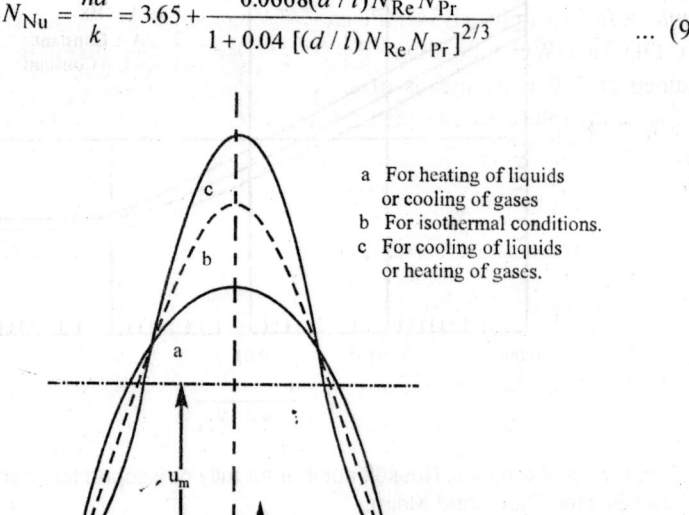

a For heating of liquids
   or cooling of gases
b For isothermal conditions.
c For cooling of liquids
   or heating of gases.

Fig. 9.14. Influence of large temperature differences on velocity distribution in a tube.

Equation (9.61) and the previous solutions are for constant physical properties of the fluid. When temperature differences are small, such an assumption is quite satisfactory. For large temperature differences the fluid velocity distributions will be influenced as indicated in Fig. 9.14.

Curve *b* shows the fully developed parabolic distributions that would obtain for isothermal or very small temperatures difference flow. When the heating of the liquids or cooling of gases is significant, the viscosity is lowest near the wall; as a result, the velocity increases faster (curve *a*) than for the isothermal case (curve *b*). For the cooling of liquids or heating or gases the reverse occurs as indicated by curve *c* of Fig. 9.14. An exact analysis of the effect of these variations on the heat transfer is not practical. It has been found, however, that they can be accounted for by multiplying equation (9.61) by the ratio $(\mu_B/\mu_S)^{0.14}$ where μ_B is the viscosity at the fluid bulk temperature (arithmetic average of mean inlet and outlet temperatures) and μ_S is the viscosity at the wall temperature.

An illustration of heat transfer during laminar flow in a tube is given in the following example:

SOLVED EXAMPLE

Example 9.1. 350 lb/hr of 34°API crude oil at 100 F are flowing through a 1 in. OD 16 BWG tube (ID = 0.87 in.) If a 12 ft length of the tube is maintained at 210 F by means of a steam jacket, estimate the increase in the mean temperature of the oil as it passes through the heated section.

Solution. The following average properties were determined for the oil :

$c = 0.485$ Btu/lb F; $\qquad\qquad\qquad w = 53.5$ lb/ft^3

$v = 5.5 \times 10^{-5}$ ft^2/sec $= 0.198$ ft^2/hr; $\qquad k = 0.0775$ Btu/hr ftF

Since the oil travels through the tube for some distance before entering the heated section and the wall temperature is uniform, equation (9.61) is applicable.

$$N_{Re} = \frac{du_m}{v} = \frac{(0.87/12)(350/53.5)}{0.198(\pi/4)(0.87/12)^2} = 580$$

$$N_{Pr} = \frac{v}{\alpha} = \frac{vwc}{k} = \frac{0.98(53.5)(0.485)}{0.0775} = 66.3$$

Substituting in equation (9.61)

$$N_{Nu} = \frac{hd}{k} = 03.65 + \frac{0.0668(0.00603)580(66.3)}{1 + 0.04[0.00603(580)66.3]^{\frac{2}{3}}} = 3.65 + 6.15 = 980$$

and
$$h = 9.80 \frac{k}{d} = 9.80 \frac{0.0775}{0.87/12} = 10.5 \text{ Btu / hr ft}^2 \text{ F}$$

To calculate the increase of the oil temperature we set
$$hA\Delta t_m = Wc(t_2 - t_1)$$

where Δt_m is an appropriate mean temperature difference between the wall and oil and t_1 and t_2 are the mean temperatures of the oil entering and leaving. Considering h as uniform along the pipe the appropriate Δt_m can be shown to be the logartithmic-mean-temperature difference,

$$\Delta t_{lm} = [(t_s - t_1) - (t_s - t_2)]/\ln\,(t_s - t_1)/(t_s - t_2)$$

In many case, however, computation is simplified and acceptable accuracy obtained by using the arithmetic-mean-temperature difference, which in this case is

$$\Delta t_a = t_5 - (t_1 + t_2)/2$$

Then
$$10.5\pi(0.87/12)12\left(210 - \frac{100 + t_2}{2}\right) = 350(0.485)(t_2 - 100)$$

and
$$t_2 = 117 \text{ F}$$

which differs negligibly from the value determined by using Δt_{lm}.

ASYMPTOTIC VALUE OF h_x IN TUBES

The decrease of the local heat transfer coefficient h_x with distance from the entrance of a tube and its approach to a constant value was already pointed out. In describing this phenomenon briefly, we may visualise fluid flowing between two parallel walls. Also to simplify the picture, assume the fluid is ideal so that it moves as a "slug" between the walls (Fig 9.15).

This will not change the general trend of h_x. If this slug of fluid suddenly enters a section in which the walls are cooled and maintained at a constant temperature lower than that of the fluid, heat begins to flow from the fluid to the walls. Initially the process is identical to the formation of a thermal boundary layer along a flat plate parallel to a fluid stream. The fluid in contact with the wall is quickly cooled to the wall temperature, resulting in a very large temperature gradient in the fluid near the surface (see point (a) in Fig 9.12). As the stream moves along the channel, heat transfer to the wall requires fluid farther and father from the wall to give up some of its energy. The temperature gradient at the surface decreases and the temperature distribution is like the curve at (*b*) in Fig. 9.15. This process continues until the fluid all the way to the centreline has sensed the energy transfer occurring and the temperature distribution is as shown at (*c*). This marks the end of the transition region.

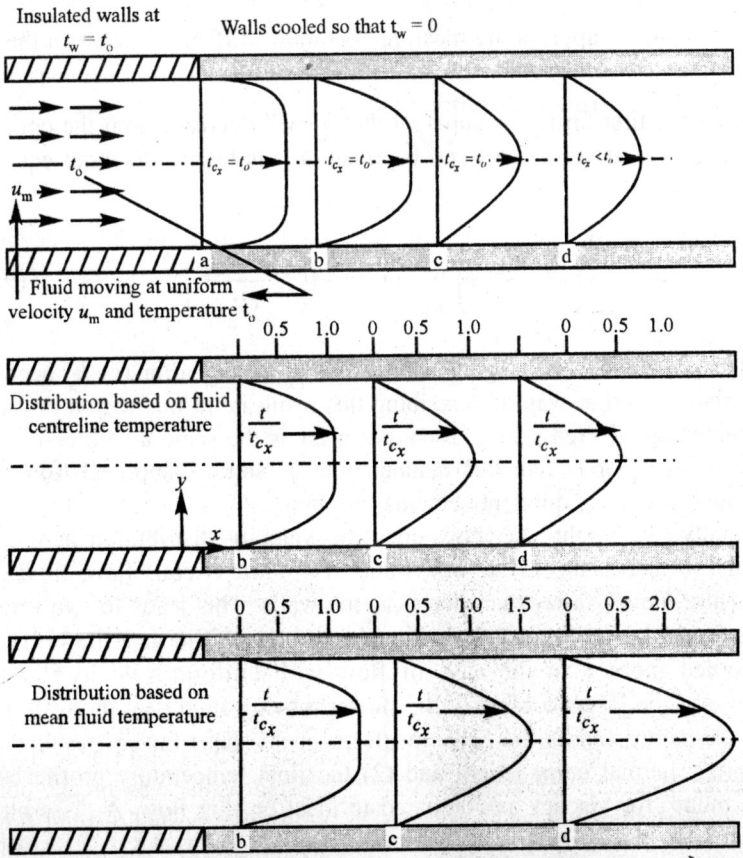

Fig. 9.15. Temperature distributions in a fluid flowing through a channel and losing heat to walls at a uniform temperature.

From this point on, fluid at various distances from the wall gives up energy in proportion to its temperature as it moves along. This means the profile of the temperature distribution will remain the same since the t at any y (assuming constant density and specific heat) will be decreased by a constant factor with increase in distance x along the channel. The unchanging nature of the profile after (*c*) is apparent if the ratio of t at any y to the centreline temperature or the mean temperature curve varies from

0 to 1.0. If the mean temperature t_{mx} at any x is used, the ratio t/t_{mx} exceeds 1.0 at the centre.

What is the significance of this in terms of hx ? Consider the definition of h_x as

$$h_x = \frac{-k\left(\frac{\partial t}{\partial y}\right)_{y=0}}{t_{cx}} \quad \text{or} \quad = \frac{-k\left(\frac{\partial t}{\partial y}\right)_{y=0}}{t_{mx}} \qquad \text{... (9.62)}$$

From the first form it is apparent that h_x will decrease until the point (c) is reached. To observe the trend beyond (c) it is helpful to write equation (9.62) as

$$h_x = -k\left(\frac{\partial(t/t_{cx})}{\partial y}\right)_{y=0} \quad \text{or} \quad = -k\left(\frac{\partial(t/t_{mx})}{\partial y}\right)_{y=0} \qquad \text{... (9.63)}$$

Because of the unchanging nature of t/t_{cx} or t/t_m after (c) h_x becomes a constant. Another way of describing this result is: in this region of a tube the percentage decrease in q_x for a given Δx is the same as the percentage decrease in t_{cx} or t_{mx}. In this manner both q_x and t_{mx} approach zero in a long tube but their quotient remains constant.

Finally, we might ask, how does the velocity distribution across the channel influence the asymptotic value of h_x ? In viscous flow, fluid near the centre moves faster than that near the walls. This leads to two effects. First, for the same distance from the tube entrance, fluid near the walls will be cooled more than the case of flow with uniform velocity. Second, because of the faster cooling of the slower moving fluid near the wall, fluid near the centre senses the heat transfer sooner. These factors result in (1) a shorter thermal entry length and (2) the final temperature profile being "less blunt" for viscous as compared to ideal or slug flow. A "less blunt" temperature profile means a lower value of $\partial(t/t_{mx})/\partial y$ at $y = 0$, which is the explanation for $N_{Nux} \rightarrow 5.75$ for slug flow in a round tube, whereas $N_{Nux} \rightarrow 3.65$ when the velocity distribution is parabolic.

CHAPTER 10

Condensation and Boiling

INTRODUCTION

Condensation and boiling are the convective heat transfer processes that are associated with change in the phase of a fluid. Condensation of a vapour to a liquid and vapourisation of a liquid to a vapour both involve a change of phase of a fluid with large heat-transfer coefficients. Condensation occurs when a saturated vapour such as steam comes in contact with a solid whose surface temperature is below the saturation temperature, to form a liquid such as water.

Normally, when a vapour condenses on a surface such as a vertical or horizontal tube or other surfaces, a film of condensate is formed on the surface and flows over the surface by the action of gravity. It is thin film of liquid between the surface and the vapour that forms the main resistance to heat transfer. This is called film-type condensation.

Another type of condensation, dropwise condensation, can occur where small drops are formed on the surface. These drops grow, coalesce, and the liquid flows from the surface. During this condensation, large areas of tube are devoid of any liquid and are exposed directly to the vapour. Very high rates of heat transfer occur on these bare areas. The average coefficient can be as high as 110000 $W/m^2.K$ (20000 $btu/h.ft^2.°F$), which is 5 to 10 times larger than film-type coefficients. Condensing film coefficients are normally much greater than those in forced convection and are of the order of magnitude of several thousand $W/m^2.K$ or more.

Dropwise condensation occurs on contaminated surfaces and when impurities are present. Film-type condensation is more dependable and more common. Hence, for normal design purposes, film-type condensation is assumed.

Many of our most important industrial processes, however, are concerned with energy transfers to accomplish phase changes or control chemical reactions. In this chapter we shall briefly describe the phenomena associated with phase changes and the procedures used in determining energy transfers involved.

FILMWISE CONDENSATION ON A VERTICAL SURFACE

Visualise a wide, cooled, vertical surface on which condensing vapour forms a continuous film. As shown in Fig. 10.1 the film thickness is zero at the upper edge and increases with distance down the vertical surface. This results from the continuous addition of condensate to that flowing downward due to the action of gravity. Nusselt showed that at any location x along the surface, the velocity profile through the film was parabolic and could be represented by

$$u = u_\infty \left(\frac{2y}{\delta} - \frac{y^2}{\delta^2} \right) \qquad \qquad ...(10.1)$$

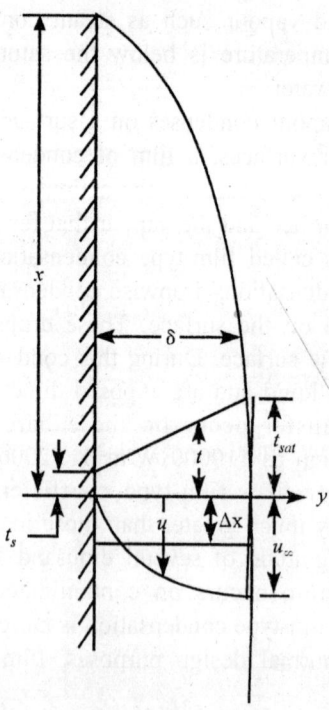

Fig. 10.1. Film condensation on a vertical wall showing film thickness, velocity and temperature profiles.

The heat given up by the condensing vapour to the wall will consist of the latent heat of condensation and sensible heat of the fluid in the film as it is cooled below the saturation temperature. The differences between saturation and surface temperatures are usually small so that the amount of sensible heat given up by the condensed vapour is small compared with the latent heat of condensation. Nusselt assumed that this could be neglected; he also assumed a uniform surface temperature and that the temperature at the interface between the film and the vapour was equal to the saturation temperature of the vapour. Under these conditions all of the heat released during condensation of vapour at the outer border of the film must flow by conduction through the film, and the temperature profile will be linear.

The heat flow through an element of the film of length Δx and unit width in the horizontal direction along the surface is then

$$dq = k\Delta x \frac{(t_{sat} - t_s)}{\delta} \qquad ...(10.2)$$

Introducing the local heat transfer coefficient h_x defined by

$$dq = h_x \Delta x (t_{sat} - t_s),$$

equation (10.2) leads to

$$h_x = \frac{k}{\delta} \qquad ...(10.3)$$

To evaluate δ it is first necessary to determine an expression for the average velocity u_m at any location x. Note if pressure and acceleration forces are neglected, the weight of the fluid in the element Δx must be supported by the shear stress at the wall. Thus

$$\Delta x w \delta = \sigma_s \Delta x = \mu \left(\frac{du}{dy} \right)_s \Delta x$$

Determining $(du/dy)_s$ from equation (10.1) and substituting gives

$$u_\infty = \frac{w\delta^2}{2\mu}$$

and

$$u = \frac{w}{2\mu}\delta^2 \left(\frac{2y}{\delta} - \frac{y^2}{\delta^2} \right) \qquad ...(10.4)$$

The average value of u from $y = 0$ to $y = \delta$ is then found to be

$$u_m = \frac{w\delta^2}{3\mu} \qquad \text{...(10.5)}$$

We can now write for the weight rate of fluid flow W in the film per unit width at the section x

$$W = \frac{w^2\delta^2}{3\mu}$$

The rate of increase of W in the distance Δx is then given by

$$\Delta W = \frac{w^2\delta^2}{\mu}\Delta\delta$$

ΔW may also be expressed in terms of the latent heat of condensation h_{fg} and the heat flow rate

$$\Delta W = \frac{k\Delta x}{h_{fg}\delta}(t_{sat} - t_s)$$

Equating these two expressions and replacing the increments with differentials gives

$$\delta^3 d\delta = \frac{\mu k}{w^2 h_{fg}}(t_{sat} - t_s)dx$$

which when integrated from 0 to x yields

$$\delta^4 = \frac{4\mu k}{w^2 h_{fg}}(t_{sat} - t_s)x \qquad \text{...(10.6)}$$

CONDENSATION HEAT TRANSFER COEFFICIENTS

Substituting δ from this equation in (10.3) the local heat transfer coefficient is

$$h_x = \frac{k}{\delta} = \left(\frac{k^3 w^2 h_{fg}}{4\mu(t_{sat} - t_s)x}\right)^{\frac{1}{4}} \qquad \text{...(10.7)}$$

In the form of a Nusselt number it becomes

$$N_{Nu_x} = \frac{h_x x}{k} = \frac{x}{\delta} = \left(\frac{h_{fg} w^2 x^3}{4\mu k(t_{sat} - t_s)}\right)^{\frac{1}{4}} \qquad \text{...(10.8)}$$

The average coefficient over a length l is from integration of equation (10.7)

$$h = \frac{4}{3} h_{x=l} \qquad \qquad ...(10.9)$$

Property values are introduced at the average temperature of the film.

Nusselt also investigated the heat transfer by condensation on circular horizontal tubes. For conditions similar to those for a vertical wall he found that the average heat transfer coefficients would be the same if $l = 2.5d$, where d is the tube diameter.

FILM REYNOLDS NUMBER AND EQUATIONS FOR CONDENSATION ON VERTICAL SURFACES

With film condensation of vapour on tall vertical surfaces or tubes, it is easy to obtain condensation rates such that turbulent flow occurs in the film. This has been found to begin at a critical value of a Reynolds number of the film. If this Reynolds number is defined as

$$N_{Re_\delta} = \frac{u_m \delta}{v} \qquad \qquad ...(10.10)$$

the critical value is around 400 to 450 (cf. the value of 300 for free convection). An alternate definition uses the equivalent diameter, d_e, of the film. For a film of unit width and thickness δ, the film diameters, d_e is simply 4δ, and N_{Re_d} four times the value given by equation (10.10). For practical purposes the condensate rate, W, rather than δ and u_m, is used so that the alternative N_{Re_d} becomes

$$N_{Re_d} = \left(\frac{4W}{\mu g}\right) \qquad \qquad ...(10.11)$$

Substituting $h_{fg} = hl(t_{sat} - t_s)/W$ in equation (10.9) and rearranging leads to

$$h\left(\frac{\mu^2}{k^3 \rho^2 g}\right)^{\frac{1}{2}} = 1.47 \left(\frac{4W}{\mu g}\right)^{\frac{1}{2}} = 1.47(N_{Re_d})^{-\frac{1}{2}} \qquad ...(10.12)$$

The form of equation (10.12) has been used extensively for the correlation of experimental data. The left-hand side involves only the properties of the fluid film and the average coefficient, which can be evaluated from the temperature and heat rate measurements. The right side involves only the total condensate rate. For values of $(4W/\mu g)$ less than 1800, McAdams recommends multiplying values of h from equation (10.12) by 1.28. For $(4W/\mu g)$ greater than 1800 the equation recommended is

$$h\left(\frac{\mu^2}{k^3\rho^2 g}\right)^{\frac{1}{2}} = 0.0077 \left(\frac{4W}{\mu g}\right)^{0.4} \qquad ...(10.13)$$

Solved Example

Example 10.1. What length of vertical 2 in. nominal tube at 138 F is required to condense 34 lb/hr of steam at 142 F ?

Solution. Calculating first the Reynolds number of the film, $W = 34/\pi d =$ 34/3.14 (2.375/12) = 54.8 lb/hr ft of tube circumference, and

$$N_{Re_d} = \frac{4(54.8)}{9.5 \times 10^{-6}(32.2)3600} = 200$$

We see that equation (10.12) is to be used.

$$h\left(\frac{(9.5\times 10^{-6})^2}{(0.38)^3(61.3/32.2)^2 32.2}\right)^{\frac{1}{2}} = 1.47(200)^{-\frac{1}{2}}$$

$$h(14.1 \times 10^{-12})^{\frac{1}{2}} = 0.251$$

$$h = \frac{0.251}{2.48 \times 10^{-4}} = 1010 \text{ Btu / hr ft}^2 \text{ F}$$

Multiplying by 1.28 gives h = 1295 Btu/hr ft² F.

Since $q = hA\,(t_{sat} - t_s) = \pi dlh\Delta t = W\pi dh_{fg}$

$$l = \frac{Wh_{fg}}{h\Delta t} = \frac{54.8(1013)}{1295(4)} = 10.8 \text{ ft}$$

Note the importance of the temperature difference in condensation energy transfer. An increase of only 2 F would decrease the required tube length by $\frac{1}{3}$.

BOILING LIQUIDS

The conversion of a liquid to vapour is one of the most familiar of processes; it calls to mind the boiler, the distillation column, the refrigerator evaporator, and numerous cooling applications. Although engineers have been successfully designing this type of equipment for a long time, only recently has the actual nature of the process been subjected to careful study. Most of this work has been with the boiling of liquids on the outside of submerged tubes or other surfaces where the circulation is usually natural. The results of several observers for heat transfer from an electrically heated wire indicate the existence of several regimes for this type of boiling. This is illustrated in Fig. 10.2, which shows the heat transfer rate from an electrically heated platinum wire submerged in water as a function of increasing temperature difference Δt between the surface of the wire and the saturation temperature of the water.

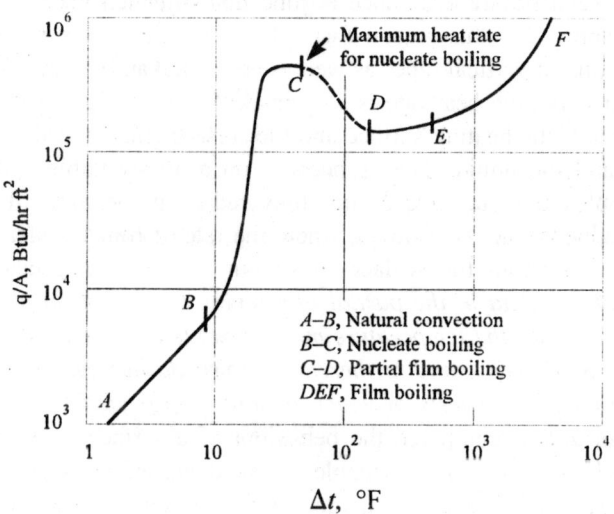

Fig. 10.2. Heat rate versus temperature difference during boiling of water at 212°F on an electrically heated platinum wire.

For values of Δt up to approximately 10 F (A to B on the curve) the liquid water is being superheated by natural convection; the heat transferred is carried away by vapourisation at the surface of the pool, q/A is proportional to $\Delta t^{5/4}$. With further increase in Δt (from B to C) bubbles form at active nuclei on the heated surface. These bubbles break away from the surface and rise through the pool, their sweeping and stirring action causing the heat transfer to be much above that due to natural convection. This phenomenon is called *nucleate boiling*. q/A varies as Δt^n with n between 3 and 4. The rate of bubble formation increases with Δt in the range B to C until the vapour formation becomes so rapid that the bubbles cannot get away before they tend to merge and spread out over the surface. Thus a vapour film forms through which heat has to be transferred by conduction. This type of boiling occurs after a "critical Δt" is reached at which the heat rate passes through a maximum (point C). The heat rate then decrease with Δt until complete film boiling is reached at D. In the film boiling range E–F, heat is transferred through the film by conduction and radiation. Point F corresponds to the melting point of the wire.

A curve of the heat transfer coefficient would be very similar to the heat rate curve of Fig. 10.2. The maximum point is reached, however, when the Δt is a few degrees less than the critical.

Interestingly enough, shape, size, or inclination does not seem to have much effect on the heat rate to a liquid boiling on a submerged surface in the range of temperature difference beyond that in which the laws of free convection apply. For this reason Fig. 10.2 is typical of the results from larger horizontal or vertical tubes as well as horizontal and inclined surfaces. On the other hand, the heat rate is very markedly influenced by the nature and condition of the heating surface and the surface tension at the interface between it and the liquid. This is because on a non-wettable surface the vapour bubbles tend to spread out, thus reducing the area of contact between heating surface and liquid, where the rate of heat transfer is much higher than it is from the surface to vapour. For example, Jakob (*It is difficult to obtain data in the partial film region C–D with an electrically heated wire because the temperature must be made to increase as the heat rate is decreased. The curve is therefore shown dashed and is based on trends determined from tests in which heating is provided by condensing vapour inside tubes*) compared the behaviour of a surface covered with a thin layer of oil, a partially wettable polished chromium surface, and a specially prepared "screen" surface with cubical cavities of linear dimensions and spacings about 0.01 in., which become fully wetted. Three typical shapes of vapour bubbles were observed, as shown in Fig. 10.3. On the unwetted surface (a) the bubbles spread out, forming a wedge between the water and heating surface, thereby permitting hydrostatic forces to resist the action of buoyancy. On the smooth chromium-plated surface (b) the bubbles rose in a steady stream from a few irregularly distributed points; the number of points of origin increased with heat rate. On the specially prepared screen surface (c) the roughness counteracted the surface tension so that the plate became completely wetted. In this case the liquid tended to push away and shear off the bubbles, causing them to become globular or oval and leave the surface while still very small. Addition of agents to reduce the surface tension was found to have the same effect as providing a wettable surface and to give increased rates of heat transfer.

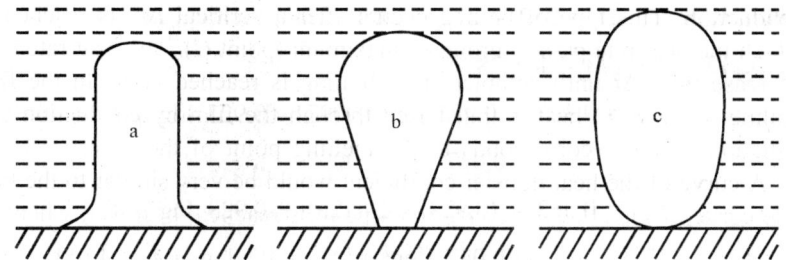

Fig. 10.3. Typical shapes of steam bubbles. (a) surface unwetted; (b) partially wetted; and (c) totally wetted.

The heat transfer coefficient, or heat rate for a liquid boiling on a submerged surface, for any given temperature difference Δt also increases as the pressure and boiling point are increased above atmospheric conditions and decreases as the pressure and boiling point are reduced below atmospheric conditions. Although this has been definitely established at lower pressures, the evidence for pressures above 1/3 of the critical is inconclusive.

Calculations of submerged boiling heat transfer coefficients or heat rates are usually made with equations developed from experimental results. For water at atmospheric pressure a rather complex general correlation proposed by Insinger and Bliss reduces to

$$\log (0.112q/A\Delta t) = 0.363 + 0.923 \log (0.0097q/A)$$
$$- 0.047 \, [\log (0.0097q/A)]^2 \qquad ...(10.14)$$

When Δt is given, this equation can be used to determine q/A. h can then be calculated as $q/A\Delta t$. Because of possible variations due to surface conditions, however, actual values may differ widely from calculated ones.

When a saturated liquid is pumped through a heated tube some evaporation will occur. This is usually only a small percentage of the circulated liquid, and the heat transfer can be most conveniently described by comparison with the familiar expression for forced convection in tubes, i.e., equation (8.26) $N_{Nu} = 0.023 N_{Re}^{0.8} N_{Pr}^{0.4}$. For values of N_{Re} exceeding 65,000 average coefficients for tubes in which evaporation occurs may be expected to be about 25 per cent higher than predicted by equation (8.36). At lower values of N_{Re} when a higher percentage of the water is evaporated, values of h may exceed those of equation (8.36) by as much as 40 per cent.

Example 10.2. Determine the rate of heat transfer to boiling at 212°F on a surface which is maintained at a temperature of 250°F.

Solution. Since $\Delta t = 250 - 212 = 38°F$, from equation (10.14)

$\log 0.112 + \log q/A - \log 38$

$\qquad = 0.363 + 0.923 \log 0.0097 + 0.923 \log q/A$

$\qquad\qquad - 0.047 [\log 0.0097 + \log q/A]$

$\qquad = 0.363 - (0.923 \times 2.0125) + 0.923 \log q/A$

$\qquad\qquad - 0.047 [\log^2 q/A - 4.02 \log q/A + 4.05]$

Then

$(1 - 0.923 - 0.189) \log q/A + 0.047 \log^2 q/A$

$\qquad\qquad = 0.363 + 1.58 + 0.951 - 1.857 - 0.190$

$\qquad\qquad = 0.847$

or

$\qquad 0.047 \log^2 q/A - 0.112 \log q/A - 0.847 = 0$

Solving gives

$$\log q/A = 5.58$$

and

$$q/A = 388,000 \text{ Btu/hr ft}^2; \quad h = 10,200 \text{ Btu/hr ft}^2°F$$

DROPWISE CONDENSATION

When a condensing surface material prevents the condensate from wetting the surface, such as is the case for a metallic (non-oxide) coating, the vapour will condense in drops rather than as a continuous film. This is known as dropwise condensation. A large part of the surface is not covered by an insulating film under these conditions, and the heat transfer coefficients are four to eight times as high as in filmwise condensation. The ratio of condensate mass flux for dropwise condensation, $\dot{m}_D$, from the outside of a horizontal tube of diameter D, to that for film condensation, $\dot{m}_f$, may be estimated from

$$\frac{\dot{m}_D}{\dot{m}_f} = \left(\frac{\rho_l^2 D^2 g}{24.2 \mu_l \dot{m}_f} \right)^{1/9} \qquad \qquad ...(10.15)$$

For steam at atmospheric pressure and $\dot{m}_f = 0.014$ kg/m²s, Eq. (10.15) predicts a ratio of 6.5.

For the purpose of calculating the unit conductance in practice, a conservative approach is to assume filmwise condensation because, even, with steam, dropwise condensation can be expected only under carefully controlled conditions, which cannot always be maintained in practice. Dropwise condensation of steam may, however, be a useful technique in experimental work when it is desirable to reduce the thermal resistance on one side of a surface to a negligible value.

BOILING

Mechanisms of Boiling

Heat transfer to a boiling liquid is very important in evaporation and distillation and also in other kinds of chemical and biological processing, such as petroleum processing, control of the temperature of chemical reactions, evaporation of liquid foods, and so on. The boiling liquid is usually contained in a vessel with a heating surface of tubes or vertical or horizontal plates which supply the heat for boiling. The heating surfaces can be heated electrically or by a hot or condensing fluid on the other side of the heated surface.

In boiling the temperature of the liquid is the boiling point of this liquid at the pressure in the equipment. The heated surface is, of course, at a temperature above the boiling point. Bubbles of vapour are generated at the heated surface and rise through the mass of liquid. The vapour accumulates in a vapour space above the liquid level and is withdrawn.

Boiling is a complex phenomenon. Suppose we consider a small heated horizontal tube or wire immersed in a vessel containing water boiling at 373.2 K (100°C). The heat flux is q/A W/m^2, $\Delta T = T_w - 373.2$ K, where T_w is the tube or wire wall temperature and h is the heat-transfer coefficient in W/m^{2}K. Starting with a low ΔT, the q/A and h values are measured. This is repeated at higher values of ΔT and the data obtained are shown in Fig. 10.4 plotted as q/A versus ΔT.

In the first region A of the plot in Fig. 10.4, at low temperature drops, the mechanism of boiling is essentially that of heat transfer to a liquid in natural convection. The variation of h with $\Delta T^{0.25}$ is approximately the same as that for natural convection to horizontal plates or cylinders. The very few bubbles formed are released from the surface of the metal and rise and do not disturb appreciably the normal natural convection.

In the region B of nucleate boiling for a ΔT of about 5–25 K (9–45°F), the rate of bubble production increases so that the velocity of circulation of the liquid increases. The heat-transfer coefficient h increases rapidly and is proportional to ΔT^2 to ΔT^3 in this region.

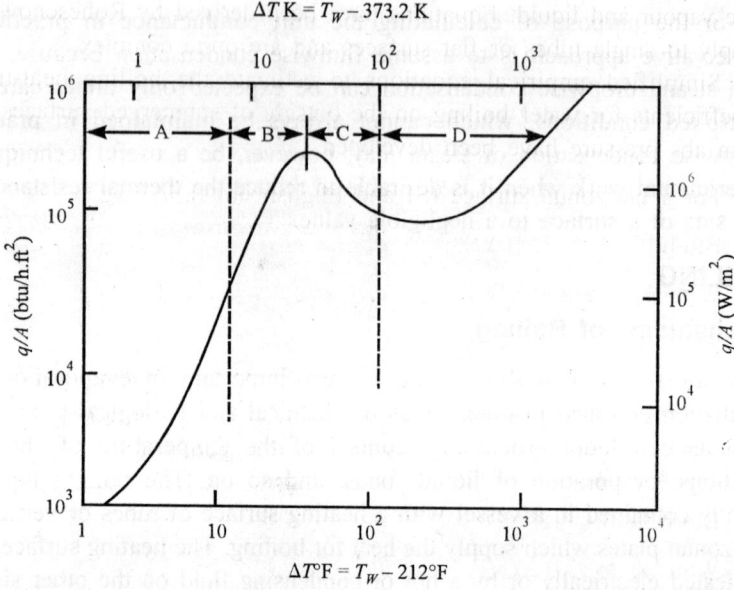

Fig. 10.4. Boiling mechanisms for water at atmospheric pressure, heat flux vs. temperature drop : (A) natural convection; (B) nucleate boiling; (C) transition boiling; and (D) film boiling.

In the region C of transition boiling, many bubbles are formed so quickly that they tend to coalesce and form a layer of insulating vapour. Increasing the ΔT increases the thickness of this layer and the heat flux and h drop as ΔT is increased. In region D or film boiling, bubbles detach themselves regularly and rise upward. At higher ΔT values radiation through the vapour layer next to the surface helps increase the q/A and h.

The curve of h versus ΔT has approximately the same shape as Fig. 10.4. The values of h are quite large. At the beginning of region B in Fig. 10.4 for nucleate boiling, h has a value of about 5700–11400 W/m². K, or 1000–2000 btu/h.ft².°F. These values are quite high, and in most cases the per cent resistance of the boiling film is only a few per cent of the overall resistance to heat transfer.

The regions of commercial interest are the nucleate and film-boiling regions. Nucleate boiling occurs in kettle-type and natural-circulation re-boilers.

NUCLEATE BOILING

In the nucleate boiling region the heat flux is affected by ΔT, pressure, nature and geometry of the surface and system, and physical properties of the vapour and liquid. Equations have been derived by Rohesenow. They apply to single tubes or flat surfaces and are quite complex.

Simplified empirical equations to estimate the boiling heat-transfer coefficients for water boiling on the outside of submerged surfaces at 1.0 atm abs pressure have been developed.

For a horizontal surface (SI and English units),

h, Btu/h.ft^2.°F $= 151(\Delta T°F)^{1/3}$ $\quad q/A$, Btu/h.ft^2, < 5000 $\qquad$ ...(10.16)

h, W/m^2.K $= 1043(\Delta TK)^{1/3}$ $\quad q/A$, kW/m^2, < 16

h, Btu/h.ft^2.°F $= 0.168(\Delta T°F)^3$ $\quad 5000 < q/A$,

$\qquad\qquad\qquad\qquad\qquad\qquad\qquad$ Btu/h.ft^2, < 75000 $\qquad$ ...(10.17)

h, W/m^2.K $= 5.56(\Delta TK)^3$ $\qquad 16 < q/A$, kW/m^2, < 240

For a vertical surface,

h, Btu/h.ft^2.°F $= 87(\Delta T°F)^{1/7}$ $\quad q/A$, Btu/h.ft^2, < 1000 $\qquad$ ...(10.18)

h, W/m^2.K $= 537(\Delta TK)^{1/7}$ $\quad q/A$, kW/m^2, < 3

h, Btu/h.ft^2.°F $= 0.240(\Delta T°F)^3$ $\quad 1000 < q/A$,

$\qquad\qquad\qquad\qquad\qquad\qquad\qquad$ Btu/h.ft^2, < 20000 $\qquad$ ...(10.19)

h, W/m^2.K $= 7.95(\Delta TK)^3$ $\qquad 3 < q/A$, kW/m^2, < 63

where $\Delta T = T_w - T_{sat}$ K or °F.

If the pressure is p atm abs, the values of h at 1 atm given above are multiplied by $(p/1)^{0.4}$. Equations (10.16) and (10.18) are in the natural convection region.

For forced convection boiling inside tubes, the following simplified relation can be used.

$$h = 2.55(\Delta TK)^3 e^{p/1551} \text{ W/m}^2.K \qquad \text{(SI)}$$

$$h = 0.077(\Delta T°F)^3 e^{p/225} \text{ Btu/h.ft}^2.°F \quad \text{(English)} \qquad \text{...(10.20)}$$

where p in this case is in kPa (SI units) and psia (English units).

FILM BOILING

In the film-boiling region the heat-transfer rate is low in view of the large temperature drop used, which is not utilised effectively. Film boiling has been subjected to considerable theoretical analysis. Bromley gives the following equation to predict the heat-transfer coefficient in the film-boiling region on a horizontal tube.

$$h = 0.62 \left[\frac{k_v^3 f_v (\rho_l - \rho_v) g (h_{fg} + 0.4 c_{pv} \Delta T)}{D \mu_v \Delta T} \right]^{1/4} \qquad ...(10.21)$$

where k_v is the thermal conductivity of the vapour in W/m.K, ρ_v the density of the vapour in kg/m³, ρ_l the density of the liquid in kg/m³, h_{fg} the latent heat of vapourisation in J/kg, $\Delta T = T_w - T_{sat}$, T_{sat} the temperature of saturated vapour in K, D the outside tube diameter in m, μ_v the viscosity of the vapour in Pa.s and g the acceleration of gravity in m/s². The physical properties of the vapour are evaluated at the film temperature of $T_f = (T_w + T_{sat})/2$ and h_{fg} at the saturation temperature. If the temperature difference is quite high, some additional heat transfer occurs by radiation.

SOLVED EXAMPLES

Example 10.1. Dry saturated steam condenses on the external wall of a horizontal tube, having a diameter $d = 20$ mm and length $l = 2$ m, at a pressure $p = 1 \times 10^5$ Pa. Tube surface temperature $t_w = 94.5°C$.

Determine the mean coefficient of the heat transfer from the condensing steam to the wall and the amount of steam G, kg/h, condensing on the surface of the tube.

Solution. With dry saturated steam in film-type condensation on horizontal tubes, the circumferentially-mean heat-transfer coefficient can be determined by the following equation :

$$Re = 3.25 Z^{0.75}, \qquad ...(10.1)$$

where

$$Re = \alpha \Delta t \pi R \frac{4}{r \rho v};$$

$$Z = \Delta t \pi R \left(\frac{g}{v^2} \right)^{1/3} \frac{\lambda}{r \rho v}$$

is the reduced length of the tube; $\Delta t = t_s - t_w$–temperature difference; R–tube radius; λ, v and ρ–thermal conductivity, dynamic viscosity coefficient and density of the condensate at the saturation temperature t_s; r–heat of vapourisation at t_s.

The formula holds at $d < 20 \ (\sigma/\rho g)^{0.5}$ (σ–surface tension coefficient) and provided the condensate film is in laminar flow which is determined by the condition $Z < 3900$. For commonly encountered practical applications, these two conditions are usually satisfied.

Equation (10.1) can be presented in the following form :

$$\alpha = 3.25 \frac{A^{0.75}}{B} \frac{1}{(\Delta t \pi R)^{0.25}} \qquad ...(10.2)$$

where

$$A = \left(\frac{q}{v^2}\right)^{1/3} \frac{\lambda}{r \rho v}, 1/m. \, °C;$$

$$B = \frac{4}{r \rho v}, \, m/W.$$

The terms A and B depend only on the kind of liquid and saturation temperature. For water these terms are shown as a function of saturation temperature t_s in Table 10.1.

Table 10.1. Values of A and B in equations (10.1) and (10.2) for water.

t_s, °C	A, 1/m.°C	$B \times 10^3$, m/W	t_s, °C	A, 1/m.°C	$B \times 10^3$, m/W
20	5.16	1.62	170	136	12.04
30	7.88	2.06	180	150	12.90
40	11.4	2.54	190	167	14.02
50	15.6	3.06	200	182	15.05
60	20.9	3.62	210	197	16.08
70	27.1	4.22	220	218	17.63
80	34.5	4.88	230	227	18.40
90	42.7	5.57	240	246	19.78
100	51.5	6.28	250	264	21.32
110	60.7	6.95	260	278	22.70
120	70.3	7.65	270	296	24.42
130	82.0	8.47	280	312	26.31
140	94.0	9.29	290	336	28.72
150	107	10.15	300	354	31.21
160	122	11.09	NA	NA	NA

In the problem considered at $p = 1 \times 10^5$ Pa $t_s = 99.6°C$ and from Table 10.1 we find :

$$A = 51.2 \, 1/m.°C; \, B = 6.25 \times 10^{-3} \, m/W.$$

The temperature difference

$$\Delta t = t_s - t_w = 99.6 - 94.5 = 5.1°C.$$

Substituting the found values into equation (10.2), we obtain :

$$\alpha = 3.25 \frac{(51.2)^{0.75}}{6.25 \times 10^{-3}} \frac{1}{(3.14 \times 0.01 \times 5.1)^{0.25}} = 15\,600 \text{ W / m}^2.^{\circ}\text{C}.$$

The amount of condensing steam is found from the heat balance equation

$$\text{Gr} = \alpha \Delta t F,$$

where $F = \pi\, dl$, m^2, is the surface area of the tube.

At $t_s = 99.6^{\circ}\text{C}$ the heat of vapourisation $r = 2258$ kJ/kg, hence,

$$G = \pi\, dl \frac{\alpha \Delta t}{r} = 3.14 \times 0.02 \times 2 \frac{15\,600 \times 5.1}{2258 \times 10^3} = 4.43 \times 10^{-3} \text{ kg / s}$$

or $G = 4.43 \times 10^{-3} \times 3600 = 15.9$ kg/h.

Example 10.2. What wall temperature t_s must be maintained to ensure film-type condensation $G = 6.5 \times 10^{-3}$ kg/s of dry saturated steam on the surface of a horizontal tube having a diameter $d = 16$ mm and length $l = 2.4$ m. The steam pressure $p = 5 \times 10^5$ Pa.

Also, determine the heat-transfer coefficient under these conditions.

Solution. The heat balance equations yield

$$\alpha = \frac{Gr}{\Delta t 2\pi Rl}, \text{ W / m}^2.^{\circ}\text{C}.$$

On the other hand, according to equation (10.2), the heat-transfer coefficient

$$\alpha = 3.25 \frac{A^{0.75}}{B} \frac{1}{(\pi R\Delta t)^{0.25}}.$$

Equating the right-hand sides of the two equations, we obtain the expression for the temperature difference :

$$\Delta t^{0.75} = \frac{Gr}{6.5(\pi R)^{0.75} l} \frac{B}{A^{0.75}}.$$

In the problem considered, at $p = 5 \times 10^5$ Pa, the saturation temperature $t_s = 151.8^{\circ}\text{C}$, at which $r = 2109$ kJ/kg and from Table 10.1 we find $A = 109.7$ 1/m.$^{\circ}$C; $B = 10.3 \times 10^{-3}$ m/W. Consequently,

$$\Delta t^{0.75} = \frac{6.5 \times 10^{-3} \times 2109 \times 10^3}{6.5 \,(3.14 \times 8 \times 10^{-3})^{0.75} \times 2.4 \,(109.7)^{0.75}} \, \frac{10.3 \times 10^{-3}}{} = 4.22$$

whence the temperature difference

$$\Delta t = (4.22)^{4/3} = 6.8°C$$

and the wall temperature required

$$t_w = t_s - \Delta t = 151.8 - 6.8 = 145°C.$$

The heat-transfer coefficient is found by equation (10.4).

$$\alpha = 3.25 \frac{(109.7)^{0.75}}{10.3 \times 10^{-3}} \frac{1}{(3.14 \times 8 \times 10^{-3} \times 6.8)^{0.25}} = 16\,600 \ W/m^2.°C.$$

Example 10.3. Dry saturated steam condenses on the vertical tube of a water heater at a pressure $p = 8.6$ MPa. The temperature of the tube external surface $t_w = 287°C$ and the height of the tube $H = 1.8$ m.

Determine the mean coefficient of heat transfer from the condensing steam to the wall of the tube.

Solution. At $p = 8.6$ MPa $t_s = 300°C$; from Table 10.1 we find: $A = 354$ 1/m,°C, $B = 31.21 \times 10^{-3}$ m/W.

The temperature difference $\Delta t = t_s - t_w = 300 - 287 = 13°C$, consequently, the reduced length of the tube

$$Z = \Delta t HA = 13 \times 1.8 \times 354 = 8380 > 2300.$$

Since the reduced length exceeds the critical, the condensate film is in turbulent flow along the bottom part of the tube.

With film-type condensation of dry saturated steam and the condensate film in a combined pattern of flow the lengthwise mean heat-transfer coefficient can be determined by the following equation :

$$Re = \left[253 + 0.069 \left(\frac{Pr}{Pr_w} \right)^{0.25} Pr^{0.5}(Z - 2300) \right]^{4/3}, \quad ...(10.3)$$

where Pr and Pr_w are Prandtl numbers for condensate respectively at the temperatures t_s and t_w. The remaining notations are as in equation (10.5).

Formula (10.3) is valid provided $Z \geqslant 2300$.

In the problem considered at $t_s = 300°C$ $Pr = 0.97$; at $t_w = 287°C$ $Pr_w = 0.921$. Equation (10.6) gives :

$$Re = \left[253 + 0.069 \left(\frac{0.97}{0.921} \right)^{0.25} \times 0.97^{0.5} (8380 - 2300) \right) \right]^{4/3} = 5930.$$

Allowing for $Re = \alpha \Delta t HB$, we find :

$$\alpha = \frac{Re}{\Delta t HB} = \frac{5930}{13 \times 1.8 \times 31.21 \times 10^{-3}} = 8100 \ W / m^2.^{\circ}C.$$

PROBLEMS

Problem 10.1. Solve example 10.1, assuming a steam pressure $p = 2 \times 10^5$ Pa, with all other conditions remaining unchanged. Compare calculation results with the answer to example 10.1.

Answer. $\alpha = 10800 \ W/m^2.^{\circ}C$; $G = 57$ kg/h.

Problem 10.2. Determine the amount of dry saturated steam G, kg/h, condensing on the surface of a horizontal tube of a diameter $d = 16$ mm and length $l = 1.5$ m, if the steam pressure $p = 1.2$ MPa and the tube surface temperature $t_w = 180^{\circ}C$.

Answer. $G = 99$ kg/h.

Problem 10.3. What temperature difference must be maintained to ensure a rate of heat flow, $q = 5.8 \times 10^4 \ W/m^2$, upon film-type condensation of dry saturated steam on the surface of a horizontal tube of a diameter $d = 34$ mm? The steam pressure $p = 1 \times 10^5$ Pa.

Also, determine the heat-transfer coefficient under these conditions.

Answer. $\Delta t = 4^{\circ}C$; $\alpha = 14500 \ W/m^2.^{\circ}C$.

Problem 10.4. Dry saturated steam condenses on the surface of a horizontal brass tube, having a diameter $d_2/d_1 = 20/18$ mm, at a pressure $p = 2.4 \times 10^5$ Pa. The tube carries cooling water. The rate of flow and mean water temperature are respectively $G_1 = 400$ kg/h, $t_{f_1} = 40^{\circ}C$.

Determine the amount of steam condensing per 1 h on length of 1 m of the brass tube, G_2, kg/m.h.

Answer. $G_2 = 20.8$ kg/m.h.

CHAPTER 11

Heat Exchangers

INTRODUCTION

Heat exchanger is a device used to transfer heat between two or more fluids. The fluids can be single or two phase and, depending on the exchange type, may be separated or in direct contact. Devices involving energy sources such as nuclear fuel pins or fired heaters are not normally regarded as heat exchangers although many of the principles involved in their design are the same.

In order to discuss heat exchangers it is necessary to provide some form of categorisation. There are two approaches that are normally taken. The first considers the flow configuration within the heat exchanger, while the second is based on the classification of equipment type primarily by construction.

CLASSIFICATION OF HEAT EXCHANGERS BY FLOW CONFIGURATION

There are four basic flow configurations : (i) counter flow; (ii) co-current flow; (iii) cross-flow; and (iv) hybrids such as cross counterflow and multi-pass flow.

Fig. 11.1 illustrates an idealised counter flow exchanger in which the two fluids flow parallel to each other but in opposite directions. This type of flow arrangement allows the largest change in temperature of both fluids and is therefore most efficient (where efficiency is the amount of actual heat transferred compared with the theoretical maximum amount of heat that can be transferred).

In co-current flow heat exchangers, the streams flow parallel to each other and in the same direction as shown in Fig. 11.2. This is less efficient than counter current flow but does provide more uniform wall temperatures. Cross flow heat exchangers are intermediate in efficiency between counter current flow and parallel flow exchangers. In these units, the streams flow at right angles to each other as shown in Fig. 11.3.

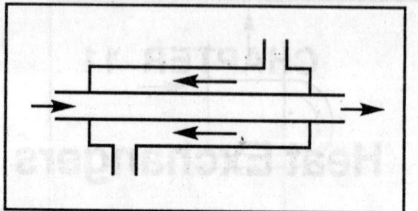

Fig. 11.1. Counter current flow.

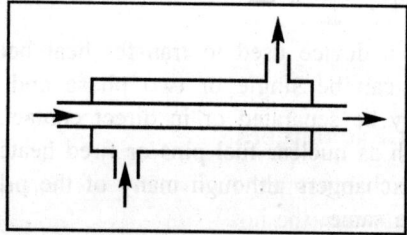

Fig. 11.2. Co-current flow.

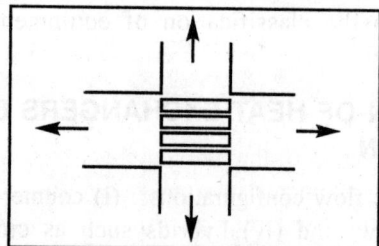

Fig. 11.3. Cross flow.

In industrial heat exchangers, hybrids of the above flow types are often found. Examples of these are combined cross flow/counter flow heat exchangers and multi pass flow heat exchangers (Fig. 11.4).

CLASSIFICATION OF HEAT EXCHANGERS BY CONSTRUCTION

The first level of classification is to divide heat exchanger types into recuperative or regenerative. A recuperative heat exchanger has separate flow paths for each fluid and fluids flow simultaneously through the exchanger exchanging heat across the wall separating the flow paths. A regenerative heat exchanger has a single flow path which the hot and cold fluids alternately pass through.

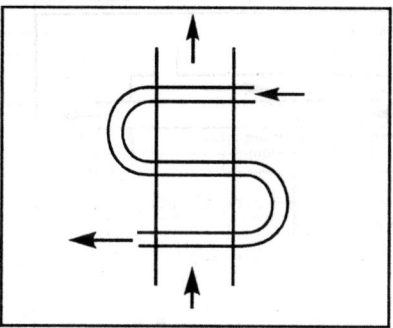

Fig. 11.4. Cross/counter flow.

Regenerative Heat Exchangers

In a regenerative heat exchanger, the flow path normally consists of a matrix which is heated when the hot fluid passes through it (this is known as the "hot blow"). This heat is then released to the cold fluid when this flows through the matrix (the "cold blow"). Regenerative heat exchangers are sometimes known as capacitive heat exchangers.

Regenerators are mainly used in gas/gas heat recovery applications in power stations and other energy intensive industries. The two main types of regenerator are static and dynamic. Both types of regenerator are transient in operation and unless great care is taken in their design there is normally cross contamination of the hot and cold streams. However, the use of regenerators is likely to increase in the future as attempts are made to improve energy efficiency and recover more low grade heat. However, because regenerative heat exchangers tend to be used for specialist applications recuperative heat exchangers are more common.

Recuperative Heat Exchangers

There are many types of recuperative exchangers which can broadly be grouped into indirect contact, direct contact and specials. Indirect contact heat exchangers keep the fluids exchanging heat separate by the use of tubes or plates etc. Direct contact exchangers do not separate the fluids exchanging heat and in fact rely on the fluids being in close contact.

Types of Heat Exchanger

This section briefly describes some of the more common types of heat exchanger and is arranged according to the classification given in Fig. 11.5.

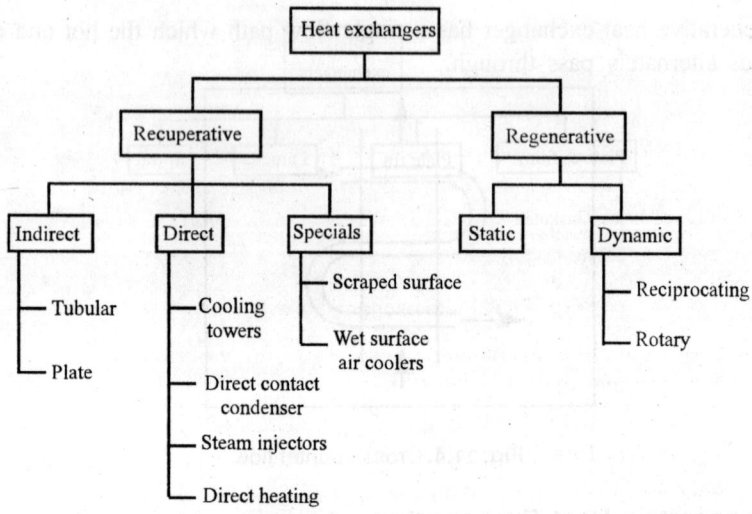

Fig. 11.5. Heat exchanger classification.

Indirect heat exchangers

In this type, the streams are separated by a wall, usually metal. Examples of these are tubular exchangers (Fig. 11.6), and plate exchangers (Fig. 11.7).

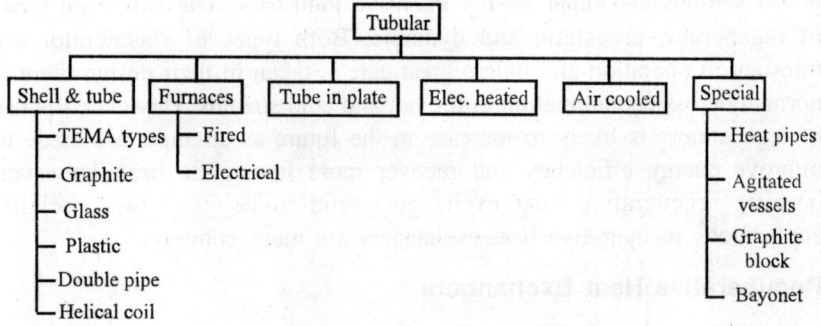

Fig. 11.6. Tubular exchanger classification.

Tubular heat exchangers are very popular due to the flexibility the designer has to allow for a wide range of pressures and temperatures. Tubular heat exchangers can be sub-divided into a number of categories, of which the shell and tube exchanger is the most common.

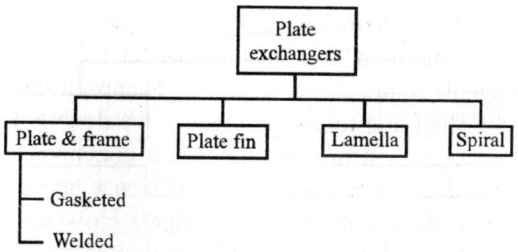

Fig. 11.7. Plate exchanger classification.

A shell and tube exchanger consists of a number of tubes mounted inside a cylindrical shell. Fig. 11.8 illustrates a typical unit that may be found in a petrochemical plant. Two fluids can exchange heat, one fluid flows over the outside of the tubes while the second fluid flows through the tubes. The fluids can be single or two phase and can flow in a parallel or a cross/counter flow arrangement. The shell and tube exchanger consists of four major parts :

1. *Front end:* This is where the fluid enters the tubeside of the exchanger.

2. *Rear end:* This is where the tubeside fluid leave the exchanger or where it is returned to the front header in exchangers with multiple tubeside passes.

3. *Tube bundle:* This comprises of the tubes, tube sheets, baffles and tie rods etc. to hold the bundle together.

4. *Shell:* This contains the tube bundle.

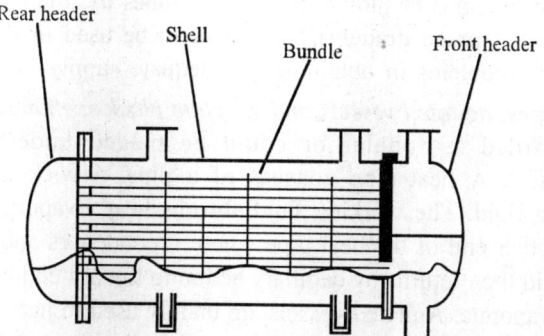

Fig. 11.8. Shell and tube exchanger.

The popularity of shell and tube exchangers has resulted in a standard being developed for their designation and use. This is the *Tubular Exchanger Manufacturers Association* (TEMA) standard. In general shell and tube exchangers are made of metal but for specialist applications (e.g.:

involving strong acids of pharmaceuticals) other materials such as graphite, plastic and glass may be used. It is also normal for the tubes to be straight but in some cryogenic applications helical or Hampson coils are used. A simple form of the shell and tube exchanger is the double pipe exchanger. This exchanger consists of a one or more tubes contained within a larger pipe. In its most complex form there is little difference between a multi-tube double pipe and a shell and tube exchanger. However, double pipe exchangers tend to be modular in construction and so several units can be bolted together to achieve the required duty. Other types of tubular exchanger include :

1. *Furnaces:* The process fluid passes through the furnace in straight or helically wound tubes and the heating is either by burners or electric heaters.

2. *Tubes in plate:* These are mainly found in heat recovery and air conditioning applications. The tubes are normally mounted in some form of duct and the plates act as supports and provide extra surface area in the form of fins.

3. *Electrically heated:* In this case the fluid normally flows over the outside of electrically heated tubes.

4. *Air cooled heat exchangers:* These are consist of bundle of tubes, a fan system and supporting structure. The tubes can have various type of fins in order to provide additional surface area on the air side. Air is either sucked up through the tubes by a fan mounted above the bundle (induced draught) or blown through the tubes by a fan mounted under the bundle (forced draught). They tend to be used in locations where there are problems in obtaining an adequate supply of cooling water.

5. *Heat pipes, agitated vessels and graphite block exchangers:* These can be regarded as tubular or could be placed under recuperative "Specials". A heat pipe consists of a pipe, a wick material and a working fluid. The working fluid absorbs heat, evaporates and passes to the other end of the heat pipe where it condenses and releases heat. The fluid then returns by capillary action to the hot end of the heat pipe to re-evaporate. Agitated vessels are mainly used to heat viscous fluids. They consist of a vessel with tubes on the inside and a agitator such as a propeller or a helical ribbon impeller. The tubes carry the hot fluid and the agitator is introduced to ensure uniform heating of the cold fluid. Carbon block exchangers are normally used when corrosive fluids need to be heated or cooled. They consist of solid blocks of carbon which have holes drilled in them for the fluids to pass through.

The blocks are then bolted together with headers to form the heat exchanger.

Plate heat exchangers separate the fluid exchanging heat by the means of plates. These normally have enhanced surfaces such as fins or embossing and are either bolted together, brazed or welded. Plate heat exchangers are mainly found in the cryogenic and food processing industries. However, because of their high surface area to volume ratio, low inventory of fluids and their ability to handle more than two streams, they are also starting to be used in the chemical industry.

Plate and frame heat exchangers consist of two rectangular end members which hold together a number of embossed rectangular plates with holes on the corner for the fluids to pass through. Each of the plates is separated by a gasket which seals the plates and arranges the flow of fluids between the plates (Fig. 11.9). This type of exchanger is widely used in the food industry because it can easily be taken apart to clean. If leakage to the environment is a concern it is possible to weld two plate together to ensure that the fluid flowing between the welded plates can not leak. However, as there are still some gaskets present it is still possible for leakage to occur. Brazed plate heat exchangers avoid the possibility of leakage by brazing all the plates together and then welding on the inlet and outlet ports.

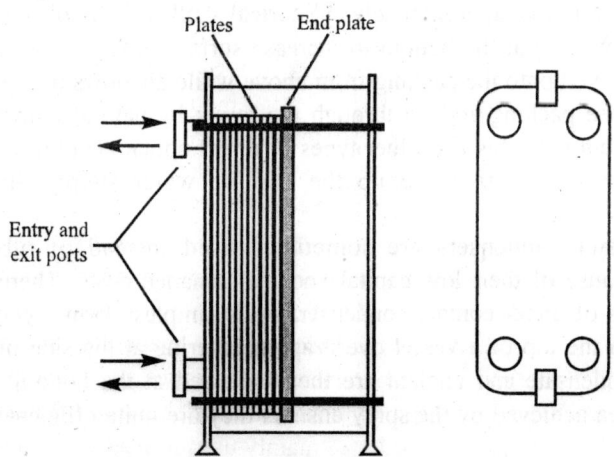

Fig. 11.9. Plate and frame exchanger.

Plate fin exchangers consist of fins or spacers sandwiched between parallel plates. The fins can be arranged so as to allow any combination of cross flow or parallel flow between adjacent plates. It is also possible to pass up to 12 fluid streams through a single exchanger by careful arrangement of headers. They are normally made of aluminium or stainless steel and brazed together. Their main use is in gas liquefaction due to their ability to operate with close temperature approaches.

Lamella heat exchangers are similar in some respects to a shell and tube. Rectangular tubes with rounded corners are stacked close together to form a bundle which is placed inside a shell. One fluid passes through the tubes while the fluid flows in parallel through the gaps between the tubes. They tend to be used in the pulp and paper industry where larger flow passages are required.

Spiral plate exchangers are formed by winding two flat parallel plates together to form a coil. The ends are then sealed with gaskets or are welded. They are mainly used with viscous, heavily fouling fluids or fluids containing particles or fibres.

Direct Contact

This category of heat exchanger does not use a heat transfer surface, because of this, it is often cheaper than indirect heat exchangers. However, to use a direct contact heat exchanger with two fluids they must be immiscible or if a single fluid is to be used it must undergo a phase change.

The most easily recognisable form of direct contact heat exchanger is the natural draught cooling tower found at many power stations. These units comprise of a large approximately cylindrical shell (usually over 100m in height) and packing at the bottom to increase surface area. The water to be cooled is sprayed onto the packing from above while air flows in through the bottom of the packing and up through the tower by natural buoyancy. The main problem with this and other types of direct contact cooling tower is the continuous need to make up the cooling water supply due to evaporation.

Direct contact condensers are sometimes used instead of tubular condensers because of their low capital and maintenance costs. There are many variations of direct contact condenser. In its simplest form a coolant is sprayed from the top of a vessel over vapour entering at the side of the vessel. The condensate and coolant are then collected at the bottom. The high surface area achieved by the spray ensures they are quite efficient heat exchangers.

Steam injection is used for heating fluids in tanks or in pipelines. The steam promotes heat transfer by the turbulence created by injection and transfers heat by condensing. Normally no attempt is made to collect the condensate.

Direct heating is mainly used in dryers where a wet solid is dried by passing it through a hot air stream. Another form of direct heating is submerged combustion. This was developed mainly for the concentration and crystallisation of corrosive solutions. The fluid is evaporated by the flame and exhaust gases being aimed down into the fluid which is held in some form of tank.

Specials

The wet surface air cooler is similar in some respects to an air cooled heat exchanger. However, in this type of unit water is sprayed over the tubes and a fan sucks air and the water down over the tube bundle. The whole system is enclosed and the warm damp air is normally vented to atmosphere.

Scraped surface exchangers consist of jacketed vessel which the fluid passes through and a rotating scraper which continuously removes deposit from the inside walls of the vessel. These units are used in the food and pharmaceutical industry in process where deposits form on the heated walls of the jacketed vessel.

Static regenerators

Static regenerators or fixed bed regenerators have no moving parts except for valves. In this case the hot gas passes through the matrix for a fixed time period at the end of which a reversal occurs, the hot gas is shut off and the cold gas passes through the matrix. The main problem with this type of unit is that both the hot and cold flow are intermittent. To overcome this and have continuous operation at least two static regenerators are required or a rotary regenerator could be used.

Rotary regenerator

In a rotary regenerator cylindrical shaped packing rotates about the axis of a cylinder between a pair of gas seals. Hot and cold gas flows simultaneously through ducting on either side of the gas seals and through the rotating packing.

Thermal analysis

The thermal analysis of any heat exchanger involves the solution of the basic transfer equation.

$$d\dot{Q} = \alpha(T_h - T_c)dA \qquad \qquad ...(11.1)$$

This equation calculates the amount of heat $d\dot{Q}$ transferred through the area dA, where T_h and T_c are the local temperatures of the hot and cold fluids, α is the local heat transfer coefficient and dA is the local incremental area on which α is based. For a flat wall

$$\alpha_w = \delta_w/\lambda_w \qquad \qquad ...(11.2)$$

where δ_w is the wall thickness and λ_w its thermal conductivity.

For single phase flow past the wall α for each of the streams is a function of Re and Pr. When condensing or boiling is taking place α may also be a function of the temperature difference. Once the heat transfer coefficient for each stream and the wall are known the overall heat transfer coefficient U is then given by

$$1/U = 1/\alpha_h + r_w + 1/\alpha_c \qquad ...(11.3)$$

where the wall resistance r_w is given by $1/\alpha_w$. The total rate of heat transfer between the hot and cold fluids is then given by

$$\dot{Q}_T = UA_T\,(T_h - T_c) \qquad ...(11.4)$$

This equation is for constant temperatures and heat transfer coefficients. In most heat exchangers this is not the case and so a different form of the equation is used

$$\dot{Q}_T = UA_T\varDelta T_M \qquad ...(11.5)$$

where $\dot{Q}_T$ is the total heat load, U is the mean overall heat transfer coefficient and $\varDelta T_M$ the mean temperature difference.

Calculation of U and $\varDelta T_M$ requires information on the exchanger type, the geometry (e.g. the size of the passages in a plate or the diameter of a tube), flow orientation, pure counter current flow or cross-flow, etc. The total duty $\dot{Q}_T$ can then be calculated using an assumed value of A_T and compared with the required duty. Changes to the assumed geometry can then be made and U, $\varDelta T_M$ and $\dot{Q}_T$ recalculated to eventually iterate to a solution where $\dot{Q}_T$ is equal to the required duty. However, in performing the thermal analysis a check should also be made at each iteration that the allowable pressure drop is not exceeded. Computer programs such as TASC from HTFS (Heat Transfer and Fluid Flow Service) perform these calculations automatically and optimise the design.

Mechanical considerations

All heat exchanger types have to undergo some form of mechanical design. Any exchanger that operates at above atmospheric pressure should be designed according to the locally specified pressure vessel design code such as ASME VIII (American Society of Mechanical Engineers) or BS 5500 (British Standard). These codes specify the requirements for a pressure vessel, but they do not deal with any specific features of a particular heat exchanger type. In some cases specialist standards exist for certain types of heat exchanger.

Advantages and disadvantages in utilising direct contactors

The exchange of heat between two fluid streams can, in general, be accomplished using either direct contact or surface-type heat exchangers. There are, however, several limitations to the use of direct contactors. First, if two fluid streams are placed in direct contact, they will mix, unless the streams are immiscible.

Thus, stream contamination will occur depending on the degree of miscibility. The two streams must also be at the same pressure in a direct contactor, which could lead to additional costs. The advantages in utilising a direct contactor include the lack of surfaces to corrode or foul, or otherwise degrade the heat transfer performance. Other advantages include the potentially superior heat transfer for a given volume of heat exchanger due to the larger heat transfer surface area achievable and the ability to transfer heat at much lower temperature differences between the two streams.

Still another advantage is the much lower pressure drop associated with direct contactors as compared to their tubular counterparts. A final advantage is the much lower capital cost as direct contact heat exchangers can be constructed out of little more than a pressure vessel, inlet nozzles for the fluid streams, and exit ports. Of course, it is sometimes advantageous to provide internals.

Varieties of direct contact heat exchangers

A typical direct contactor provides heat transfer between two fluid streams. The processes include the simple heating or cooling of one fluid by the other; cooling with the vapourisation of the coolant; cooling of a gas-vapour mixture with partial condensation; cooling of a vapour or vapour mixture with total condensation; and cooling of a liquid with partial or complete solidification. Most of the direct-contact applications can be accomplished with the following devices: (i) spray columns; (ii) baffle tray columns; (iii) sieve tray or bubble tray columns; (iv) packed columns; (v) pipeline contactors; and (vi) mechanically agitated contactors.

Figs. 11.10–11.15 illustrate the general configurations of (i) through (vi), respectively. Except for the turbulent pipe contactor, all of the devices are counter-current devices and depend upto the relative buoyancy of the dispersed phase through a continuous phase.

While the figures illustrate a less-dense dispersed phase being introduced at the bottom of the column, it is possible for the dispersed phase to be denser and introduced to the top, with the configuration internals appropriately revised.

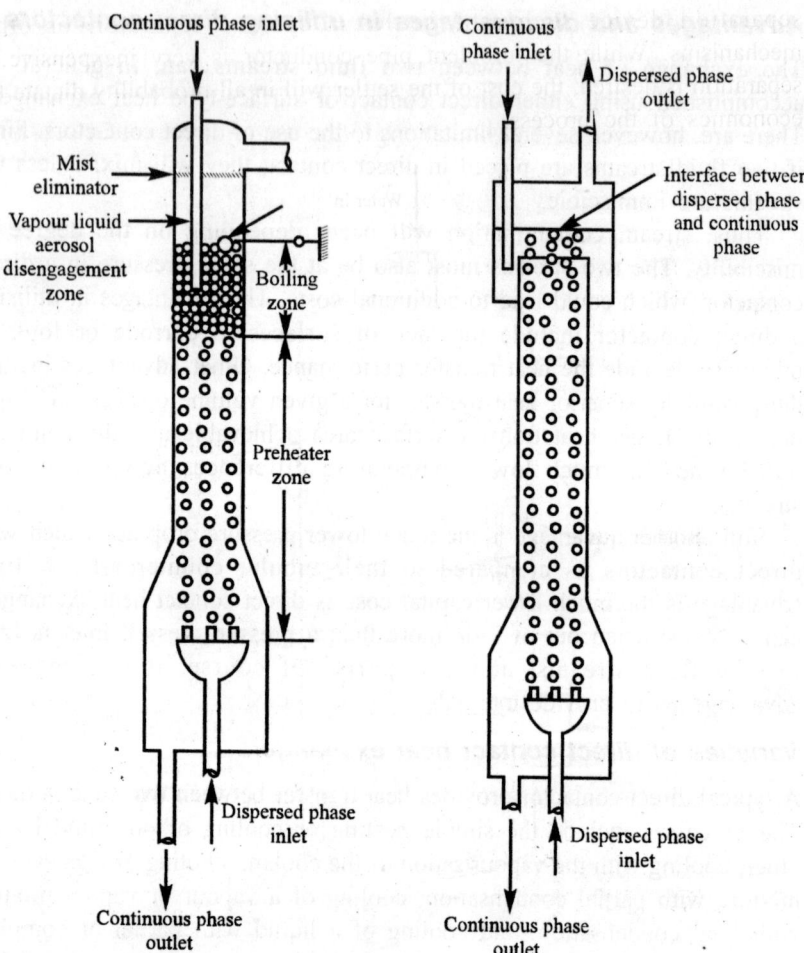

Fig. 11.10. Schematic diagrams of spray columns for evaporation and for sensible heating of the dispersed lighter phase.

The turbulent pipe contactor is a parallel-flow device and has the limits of efficiency of all such systems, whether they be direct contact or surface-type heat exchangers. That is, the maximum temperature achieved by the cool stream is that of the mixing cup temperature.

The size of the turbulent pipe contactor is dictated by the relative mass flow rate and the nature of the turbulence. Turbulence promoters can be installed to enhance the turbulence and, thereby reduce the length of contactor necessary to essentially obtain the mixing cup temperature. If separation of the stream is desired, the contactor must be followed by a separation device such as a settler, a cyclone separator, or other mechanisms. While the turbulent pipe conductor is very inexpensive, if separation is desired, the cost of the settler will in all probability dictate the economics of the process.

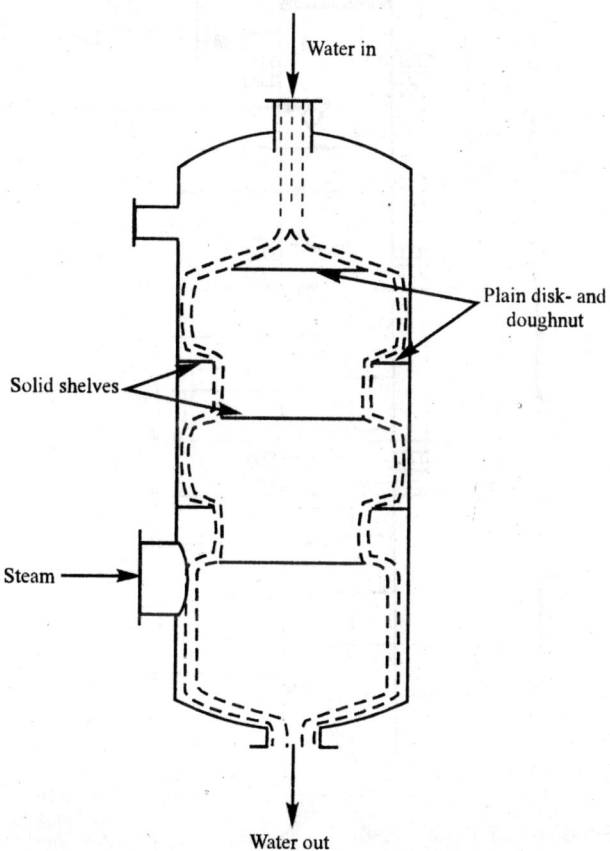

Water in

Plain disk- and doughnut

Solid shelves

Steam

Water out

Fig. 11.11. Schematic diagram of a disk and doughnut baffle tray column for use as a steam condenser.

The remaining apparatus all have the heat transfer take place between a continuous phase and a clearly defined disperse phase in the form of drops, bubbles, jets, sheets, or thin supported films in the case of packed beds. Heat exchangers with mechanical agitators (Fig. 11.15), while often superior as heat or mass transfer equipment are more difficult to design as the dispersed phase may have a wide range of drop or bubble sizes. Thus, empirical data from the manufacturer to establish performance is necessary. Further, problems may result in seals at the penetration point of the drive shafts. Special designs may therefore be necessary.

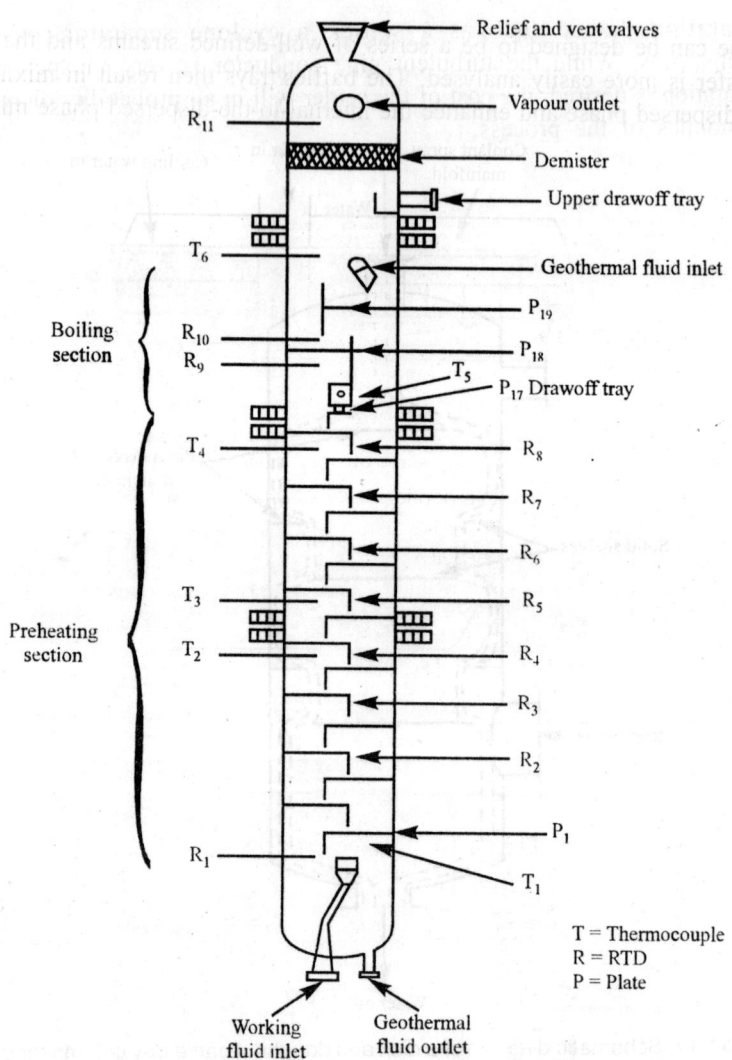

Fig. 11.12. Schematic diagram of a sieve tray column used for extracting heat from geothermal brine.

Baffle tray columns may have similar problems in defining the nature of the curtain of the dispersed phase. Depending on flow rates and baffle design, the dispersed phase may be a sheet, a series of rivlets or defined streams, which can break up into drops (Fig. 11.16). The effective thermal diffusivity to molecular diffusivity as a function of drop peclet number is shown in Fig. 11.17 and tray in a sieve tray column is shown in Fig. 11.18. If the baffles are, in fact, trays with serrated or notched rims, the dispersed phase can be designed to be a series of well-defined streams and the heat transfer is more easily analysed. The baffles/trays then result in mixing of the dispersed phase and enhance the internal-to-the-dispersed phase mixing.

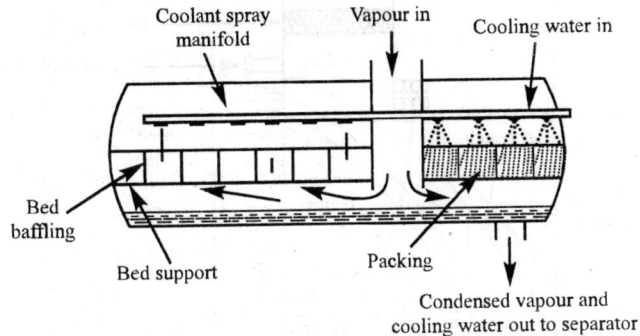

Fig. 11.13. A possible configuration of a packed bed condenser.

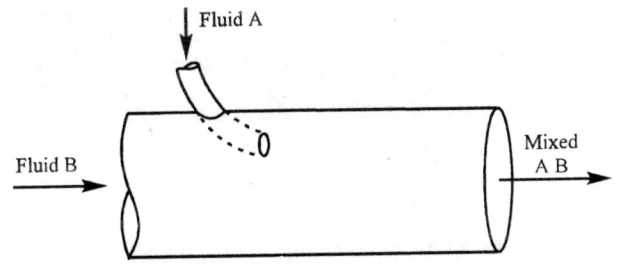

Fig. 11.14. Turbulent pipe contactor.

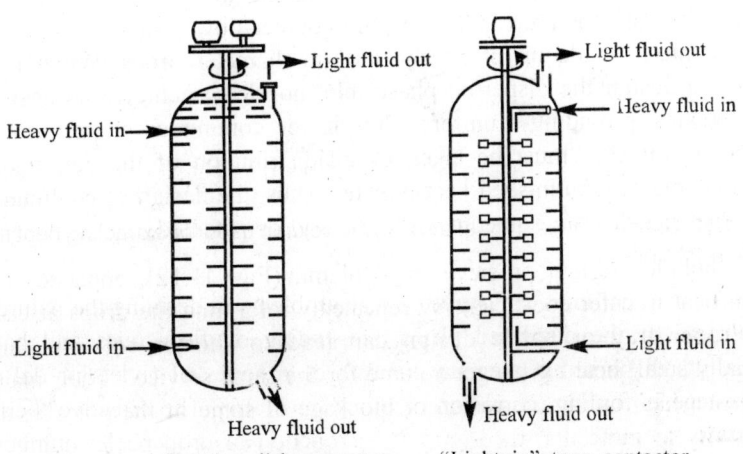

Rotating disk contactor   "Lightnin" type contactor

Fig. 11.15. Typical mechanically agitated towers.

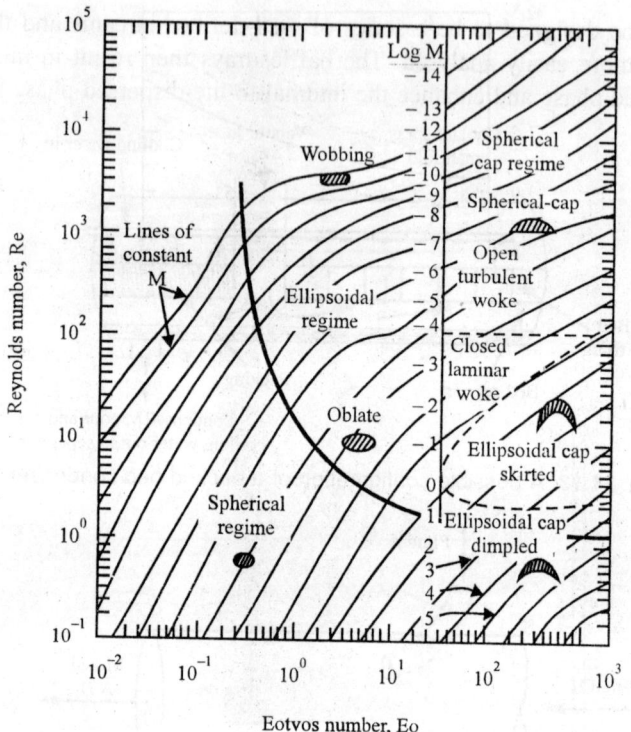

Fig. 11.16. Drop characterisation map.

The spray column shown in Fig. 11.10 is an open column whose only internals are the inlet nozzles for the dispersed and continuous phase. Ideally, such columns are capable of pure counter-flow operation, with the dispersed phase made up of nearly uniform diameter drops. While it is possible to design the dispersed phase inlet nozzle to achieve the desired characteristics, providing a uniform flow in the continuous phase is more difficult. Great care must be taken or maldistribution of the continuous phase may lead to diminished heat transfer. Thus, the design of continuous phase inlet nozzles are sometimes proprietary, or patented.

The bubble column or sieve tray column (Fig. 11.12), enhances the internal heat transfer coefficient by repeatedly reforming the drops at each tray. Proper tray or baffle design can lead to shorter columns, and potentially small heat exchanger volume for the same service. Their major disadvantage is fouling, corrosion or blockage of some of the holes in the sieve tray.

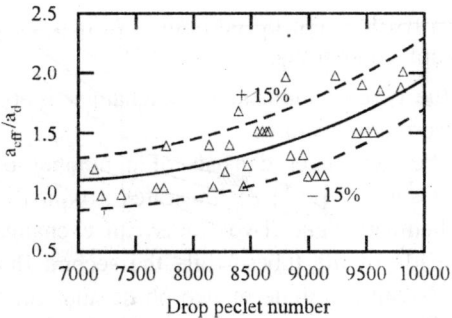

Fig. 11.17. Effective thermal diffusivity to molecular diffusivity as a function of drop peclet number.

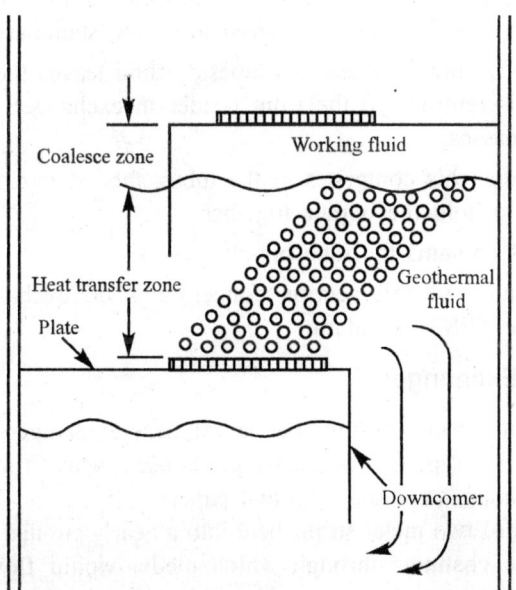

Fig. 11.18. Schematic diagram of a tray in a sieve tray column.

Shell and Tube Heat Exchangers

Shell and tube heat exchangers are one of the most popular types of exchanger due to the flexibility the designer has to allow for a wide range of pressures and temperatures. There are two main categories of shell and tube exchanger :

1. Those that are used in the petrochemical industry which tend to be covered by standards from TEMA, Tubular Exchanger Manufacturers Association.

2. Those that are used in the power industry such as feedwater heaters and power plant condensers.

Regardless of the type of industry the exchanger is to be used in there are a number of common features.

A shell and tube exchanger consists of a number of tubes mounted inside a cylindrical shell (Fig. 11.8) illustrates a typical unit that may be found in a petrochemical plant. Two fluids can exchange heat, one fluid flows over the outside of the tubes while the second fluid flows through the tubes. The fluids can be single or two phase and can flow in a parallel or a cross/counter flow arrangement. The shell and tube exchanger consists of four major parts :

1. *Front header:* this is where the fluid enters the tubeside of the exchanger. It is sometimes referred to as the stationary header.
2. *Rear header:* this is where the tubeside fluid leaves the exchanger or where it is returned to the front header in exchangers with multiple tubeside passes.
3. *Tube bundle:* this comprises of the tubes, tube sheets, baffles and tie rods etc. to hold the bundle together.
4. *Shell:* this contains the tube bundle.

The remainder of this section concentrates on exchangers that are covered by the TEMA standard.

Spiral Heat Exchangers

It is a popular misconception that spiral heat exchangers are a recent development. This type of heat exchanger is used today in many industries including chemical, steel and pulp and paper.

It consisted of two metal strips, bent into a nearly circular shape to form two concentric channels through which media would flow in opposite directions. Channel spacing was achieved by a steel bar along the length.

Today, practical maximum capacity of a standard spiral heat exchanger is 400–600m^2. Currently the spiral is manufactured in a winding process using a D-shaped mandrel with the two strips being welded to a central plate and distance studs have replaced the steel bars. Alternatively tubular centres are becoming more common. Usually, alternate edges of the passages are closed and covers fitted to both sides of the spiral assembly.

Heat transfer relationships: Turbulent flow (Re > approx 500) in spiral channels

Heat transfer data for spiral heat exchangers is empirically correlated using a conventional Dittus-Boelter type relationship for turbulent flow. A channel curvature component is added in order to take into account the somewhat improved heat transfer generated by secondary flow effects. The equation takes the following form :

$$\text{Nu } Pr^{-1/3}(\eta/\eta_w)^{-0.17} = 0.023(1 + 3.54.d_H/D)Re^{0.8} \quad ...(11.6)$$

where Nu is Nusselt number, Pr is Prandtl number, η is bulk fluid viscosity, η_w is fluid viscosity at the wall, d_H is channel hydraulic diameter, D is diameter of the spiral and Re is Reynolds number.

The term d_H/D represents the local channel curvature, which, for a channel of constant spacing, will vary from a maximum at the centre of the body to a minimum at the periphery.

Non-turbulent flow (Re < approx 500) in spiral channels

Test data and, to a certain extent, results from installed units indicate the existence of two regions for spiral flow :

Pure laminar flow :

$$\text{Nu}(\mu/\mu_w) - 0.17 = 7.6 \quad ...(11.7)$$

Transition-laminar flow :

$$\text{Nu } Pr^{-1/3}(\mu/\mu_w)^{-0.17} = 0.33 \ (d_H/D)^{1/6}Re^{0.475} \quad ...(11.8)$$

For non-turbulent flow therefore, the higher of the two Nu values obtained from the above equations is the one that applies.

Different Spiral Types

Type I

The main feature of this exchanger type is that there is a single passage for each fluid. In actual operation, the cold fluid enters at the periphery and flows towards the centre where it exits via the cover. The hot fluid goes in the opposite directions, giving counter-current flow. The single channel makes the unit well suited to handling fouling liquids with the original design being known as the Type I. A further unit is used for final cooling of the treated sludge.

Type II

The Type II spiral can handle the growing demand for vapourising and condensing capabilities within the process industries. Although it operates on the same basic principle as the Type I, it differs most significantly in terms of channel geometry. It has only one medium flowing spirally. The other flows crosswise, parallel to the axis of the spiral element. The spiral channel is closed on both sides with the crossflow fluid flowing through the spiral annulus.

Type II spirals are used in duties involving large volumes of vapour, vapour/gas or vapour/liquid mixtures. The channel geometry makes it possible to combine high liquid velocity in the spiral passage with very low pressure drop on the vapour/mixture side. They are also occasionally used in liquid/liquid applications where one side has to cope with a much larger volume of liquid than the other, such as some fermenter cooling cases.

Type III

When the Type II is used as a condenser it achieves very little sub-cooling of the vapours or condensate. For applications where this is a necessary part of the process, a different type of spiral had to be developed—the Type III.

The unit is constructed with (normally) alternately welded channels. The lower face of the body is fitted with a cover, while the upper face is fitted with a distribution cone such that the outer turns are closed and the inner turns are open to the cross-flow of the fluid entering the unit. The periphery of the unit is provided with an upper connection for the removal of residual gas/vapour, and a lower connection for the condensate. The cooling medium side is in spiral flow throughout.

The function of the unit is that of condensing a vapour or vapour mixture with or without non-condensable gas in which it is required to cool the residual vapour/gas mixture to as low a temperature as possible and thus obtain maximum possible condensation. A secondary feature is that the condensate is effectively subcooled, the outer turns being in counter-current flow to the coolant. That the flow is in the spiral mode in the outer turns results in higher heat and mass transfer coefficients than would be obtained with the vapour in crossflow only. The SHE type III is best suited for vapour mixtures at moderate pressure containing small to moderate amounts of non-condensable gas. Operation at very low absolute pressure ("high vacuum") is seldom feasible due to the resulting excessive pressure drop in the outer turns.

Type G

The process industry uses columns and reactors extensively and Type G spiral was developed to meet the need for a custom built unit which would be vertically mounted onto a column or reactor. The advantage of this arrangement is that it eliminates the need for a separate condenser and, more importantly, all of the large vapour pipework and reflux drum associated with it.

In this model, vapour enters through an open centre tube and then rises. In the upper shell extension, its direction of flow is reversed and it condenses downwards in crossflow in the spiral element. Meanwhile, coolant is pumped through a peripheral connection, flows through the spiral channel towards the centre and, finally, exits via a pipe in the upper shell extension. For minimal sub-cooling, the condensate is allowed to enter the lower shell extension. However, when subcooling is required, a baffle plate fitted to the lower face of the spiral element forces the condensate to flow in the lower parts of the channels in countercurrent to the coolant.

Overall Heat Transfer Coefficient

The operating principles of most heat exchangers can be explained in terms of the simple concentric tube apparatus pictured in Fig. 11.19. If possible, the hot fluid should flow through the inside pipe and the fluid to be heated (cold fluid) should flow in the opposite directions in the annular space between the two pipes.

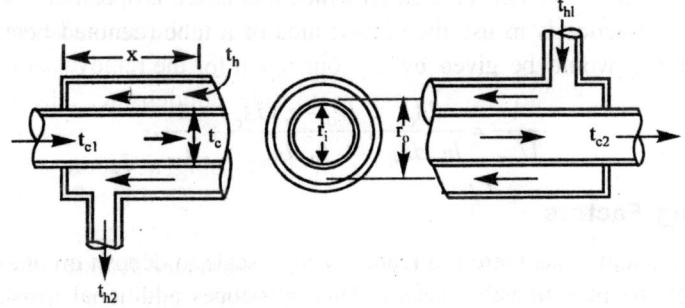

Fig. 11.19. Concentric tube apparatus illustrating a counterflow heat exchanger.

With this arrangement the outer surface of the equipment will be at the lowest possible temperature, and heat loss to the surroundings will be at a minimum. At any distance x from the left end of the outer pipe the temperature will be a maximum at the centre of the inner tube and will decrease in a radial direction to a minimum at the inner surface of the outer tube (assuming negligible heat loss). The heat transfer from the hot fluid to a short length of the inner pipe at any such point would be given in terms of the local heat transfer coefficient h_{hx} and a local temperature difference Δt_h. For this local temperature difference it is convenient to use the fluid bulk temperature minus the inner wall temperature; thus $dq_h = h_{hx}dA_h$ $(t_h - t_{wh})$. Similarly on the cold fluid side at the location x, $dq_c = h_{cx}dA_c$ $(t_{wc} - t_c)$ where t_{wc} denotes the temperature of the outer surface of the inner pipe, and t_c the bulk temperature of the cold fluid. If both flow rates are steady, conservation of energy requires that $dq_h = dq_c = dq$. Since this is also the heat rate through the wall, $dq = k_w dA_w$ $(t_{wh} - t_{wc})/(r_o - r_i)$, where k_w is the thermal conductivity of the wall, A_w the mean area, and $(r_o - r_i)$ the wall thickness. Solving for the individual temperature differences and then adding these differences will yield

$$t_h - t_c = dq \left(\frac{1}{h_{hx}dA_h} + \frac{r_o - r_i}{k_w dA_w} + \frac{1}{h_{cx}dA_c} \right) \qquad ...(11.9)$$

In this way the intermediate wall temperatures are eliminated and the heat rate can be expressed by a simple equation having the form of Newton's law of cooling

$$dq = U_x dA(t_h - t_c) \qquad ...(11.10)$$

That is, we introduce a local overall heat transfer coefficient U_x which, when multiplied by the appropriate heat transfer area dA, is equivalent to the reciprocal of the quantity in parentheses in equation (11.19).

There is no physical interpretation to be associated with U_x; it has a definite value only after the area on which it is based is specified. The most common practice is to use the outside area of a tube (denoted here by A_c) so that U_{ox} would be given by

$$\frac{1}{U_{ox}} = \frac{dA_c}{h_{hx}dA_h} + \frac{(r_o - r_i)dA_c}{k_w dA_w} + \frac{1}{h_{cx}} \qquad ...(11.11)$$

Fouling Factors

In most installations there is a tendency for a scale to deposit on one or both sides of the heat transfer surface. This introduces additional resistance in the heat flow path resulting in a temperature distribution as shown in Fig. 11.20.

Because the scale thickness and thermal conductivity are difficult to assess, it is more convenient to specify the effect on the heat transfer in terms of an equivalent heat transfer coefficient h_s or resistance $r_s = 1/h_s$ based on the area on which the deposit has formed.

Thus, the hot fluid scale resistance would be given by $R_{sh} = 1/h_{sh}dA_h$, and the cold fluid resistance by $R_{sc} = 1/h_{ac}dA_c$, and the local over-all coefficient by

$$\frac{1}{U_{ox}} = dA_c \left[\frac{1}{h_{hx}dA_h} + R_{sh} + \frac{r_o - r_i}{k_w dA_w} + R_{sc} + \frac{1}{h_{cx}dA_c} \right] \qquad ...(11.12)$$

Table 11.1 shows values of the scale resistance which have been found in representative installations. These are commonly referred to as fouling factors.

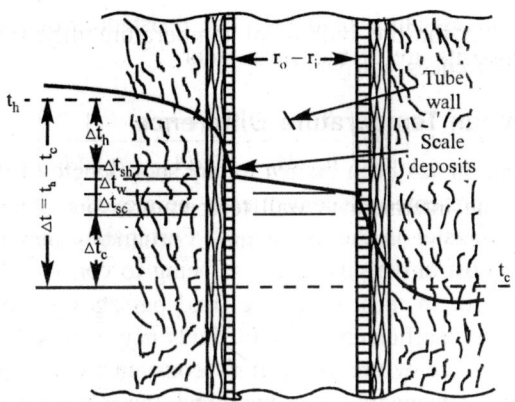

Fig. 11.20. Temperature variation through a heat exchanger wall with scale deposits on both sides.

Table 11.1. Heat exchanger fouling factors, ($r = 1/h$ in °Ft^2 hr/Btu).

Water

Temperature of heating medium	up to 240°F		240–400°F	
Temperature of water	125°F or less		above 125°F	
Water velocity, ft/sec	3 and less	over 3	3 and less	over 3
Distilled	0.0005	0.0005	0.0005	0.0005
Sea water	0.0005	0.0005	0.001	0.001
City or well water	0.001	0.001	0.002	0.002
Treated boiler feed water	0.001	0.0005	0.001	0.001
Mississippi river water	0.003	0.002	0.004	0.003
Liquid gasoline, organic vapours				0.0005
Refrigerating liquids, cooling brine, oil-bearing steam				0.001
Refrigerating vapours, distillate bottoms above 20° API, Air				0.002
Fuel oil, salty crude oil, residual bottoms less than 20° API				0.005
Diesel exhaust gas, coke-oven gas, cracking unit residuum				0.010

The relative magnitude of the resistance added by scale deposits may be seen by comparison with the resistances of the hot and cold fluid heat transfer coefficients. The steam condensed on a copper pipe and the coefficient was 2000 Btu/hr ft^2.F. This is equivalent to a thermal resistance of 1/2000 = 0.0005 per unit area. From Table 11.1 we see that the scale deposit of even the cleanest fluids adds the equivalent of this resistance. The coefficient for the water flowing inside the pipe was 260 Btu/hr ft^2 F, which is equivalent to a resistance of 0.004. The resistances of river water and fuel oil scale deposits are of the same order.

Logarithmic-Mean Temperature Difference

So far our discussion has been limited to any short length of the exchanger. Can the concepts presented be conveniently employed in the calculation of the total energy transfer in the exchanger? Fortunately they can, the most important and almost universal procedure being to consider U constant all along the exchanger. In most cases the fluid passages are of such length that, although local coefficients for both fluids may vary considerably near the entrance regions, they are very nearly constant for the greater portion of the passage. In exchangers where this condition is not too well satisfied, use of an average value of U has been found to give acceptable results. Hence, for an average overall coefficient U for the entire exchanger based on outside area of the inner pipe, A_o,

$$\frac{1}{U} = \left[\frac{1}{h_i A_i / A_o} + \frac{r_{si}}{A_i / A_o} + \frac{r_o - r_i}{k_w A_m / A_o} + r_o + \frac{1}{h_o} \right] \qquad ...(11.13)$$

where the subscripts i and o are used to denote the two fluid sides.

Let us consider a counterflow exchanger in which the flow rate and specific heat of each fluid are constant and heat losses are negligible. Since the fluid stream bulk temperatures will change in the manner shown in Fig. 11.21, the total heat transfer q must be determined by integration. Considering a differential length dx for which the heat exchange area is dA, we may write

$$\frac{dq}{\Delta t} = U dA \qquad ...(11.14)$$

Although we are taking U as constant, both dq and Δt are functions of the fluid temperatures. If the cold fluid flow direction is taken as positive (Fig. 11.21).

$$dq = (Wc_p)_c dt_c = C_c dt_c \qquad ...(11.15a)$$

and

$$= (Wc_p)_h dt_h = C_h dt_h \qquad ...(11.15b)$$

The product of the fluid flow rate and its specific heat Wc_p, is called the fluid capacity rate [Btu/hr F] and is denoted by C. To relate dq and Δt, note that

$$d(\Delta t) = d(t_h - t_c) = dt_h - dt_c \qquad ...(11.16)$$

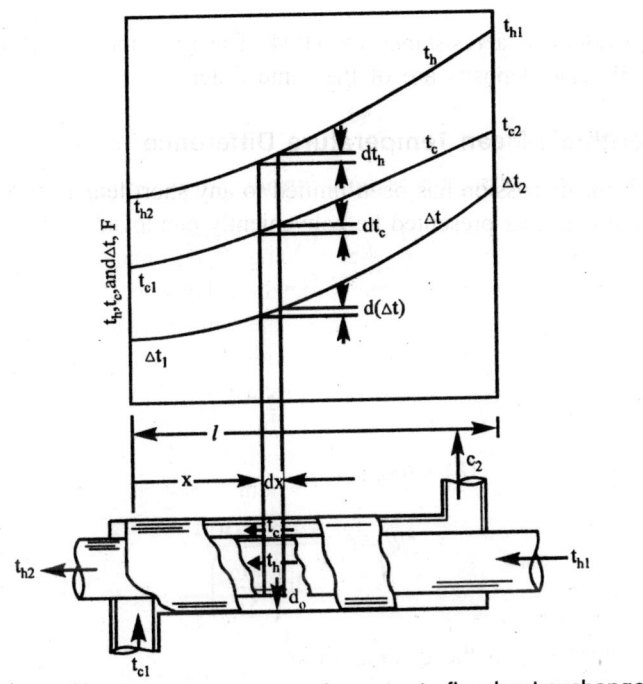

Fig. 11.21. Variation of temperatures in a counterflow heat exchanger.

Solving equation (11.15) for dt_h and dt_c and substituting in equation (11.16), we obtain

$$d(\Delta t) = dq \left[\frac{1}{C_h} - \frac{1}{C_c} \right]$$

or

$$dq = d(\Delta t) \Big/ \left[\frac{1}{C_h} - \frac{1}{C_c} \right] \qquad \text{...(11.17)}$$

Replacing dq in equation (11.14) with this expression gives

$$\frac{1}{\left[\dfrac{1}{C_h} - \dfrac{1}{C_c} \right]} \frac{d(\Delta t)}{\Delta t} = UdA \qquad \text{...(11.18)}$$

Integrating from 0 to l (or 0 to A) yields

$$\frac{1}{\left[\dfrac{1}{C_h} - \dfrac{1}{C_c}\right]} \ln \frac{\Delta t_2}{\Delta t_1} = UA \qquad \text{...(11.19)}$$

The total heat exchange q is now introduced by noting

$$1/C_h = (t_{h1} - t_{h2})/q$$
$$1/C_c = (t_{c2} - t_{c1})/q$$

and

$$\frac{1}{C_h} - \frac{1}{C_c} = \frac{1}{q}[t_{h1} - t_{h2} - t_{c2} + t_{c1}] = \frac{1}{q}[\Delta t_2 - \Delta t_1]$$

so that equation (11.18) becomes

$$q = UA \frac{(\Delta t_2 - \Delta t_1)}{\ln\left(\dfrac{\Delta t_2}{\Delta t_1}\right)} \qquad \text{...(11.20)}$$

Comparing this with the desired form

$$q = UA\Delta t_m$$

we see that

$$\Delta t_m = \Delta t_{lm} = \frac{(\Delta t_2 - \Delta t_1)}{\ln\left(\dfrac{\Delta t_2}{\Delta t_1}\right)} \qquad \text{...(11.21)}$$

which is defined as the logarithmic-mean over-all temperature difference. It is strictly correct when U is constant, the specific heats are constant, and the flow is steady.

Parallelflow, Counterflow and Crossflow Exchangers

A second method of operating the concentric tube exchanger of Fig. 11.21 would be to have both fluids flowing in the same direction. It would then be called a parallelflow exchanger and the temperature would vary as indicated in Fig. 11.22. Following the same procedure as above gives the logarithmic mean temperature difference

$$\Delta t_{lmp} = \frac{(t_{h2} - t_{c2}) - (t_{h1} - t_{c1})}{\ln\left(\dfrac{t_{h2} - t_{c2}}{t_{h1} - t_{c1}}\right)} = \frac{\Delta t_2 - \Delta t_1}{\ln\left(\dfrac{\Delta t_2}{\Delta t_1}\right)} \qquad \text{...(11.22)}$$

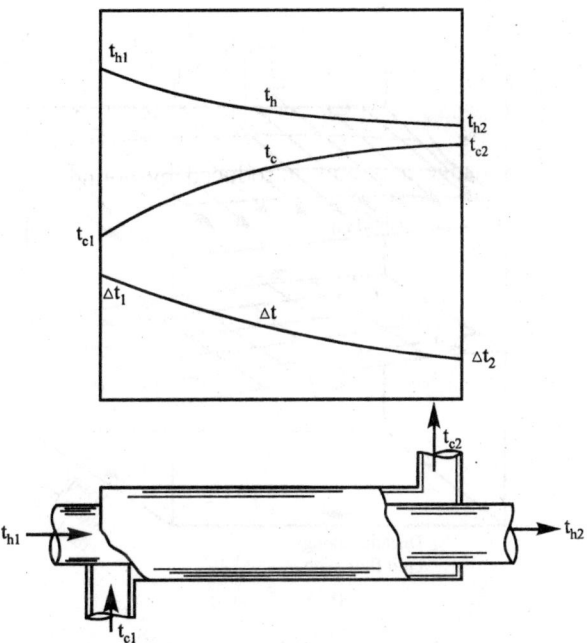

Fig. 11.22. Temperature variation in a parallelflow heat exchanger.

It should be observed that the expression for the logarithmic-mean temperature difference for counterflow or parallelflow exchangers is the same and involves only the temperature differences at the entrance and exit of the exchanger. Note, however, that for a given set of entrance and exit temperatures, the counterflow temperature difference is greater and therefore less transfer area is required.

The third general arrangement of a heat exchange system is to have the two fluids flow at right angles to each other on either side of the separating surface. One fluid flowing through a bank of parallel tubes with the second fluid flowing normally through the tube bank is an example of such a system; a second example, is the flat-plate type exchanger shown in Fig. 11.23a. Because there is no opportunity for mixing of the fluids, temperatures in the flat-plate exchanger vary with in the direction of flow and across any section normal to the flow (Fig. 11.23b).

The mathematical analysis of the mean temperature difference for cross-flow exchangers is rather complex. Solutions involving the use of series have been obtained by Nusselt and Smith. The results have been presented in tabular form for practical calculations.

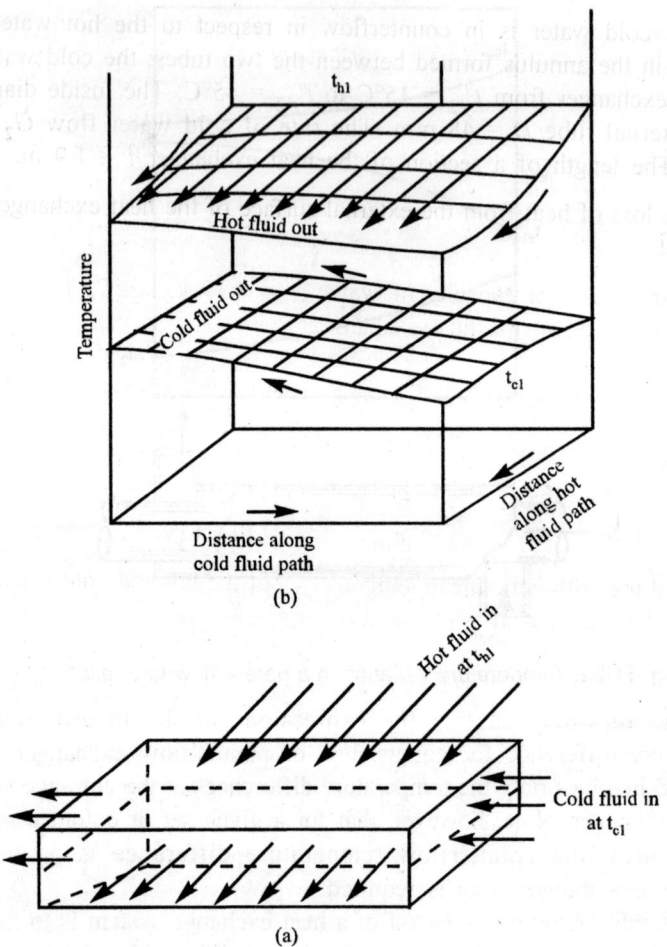

Fig. 11.23. Temperature variation in a flat plate crossflow exchanger with neither fluid mixed.

SOLVED EXAMPLES

Example 11.1. Determine the heating surface and the number of sections of a water-to-water heat exchanger of the "tube-in-tube" type. The heating (hot) water flows in the internal steel tube [λ_w = 45 W/m·°C)] with diameters $d_2/d_1 = 35/32$ mm and its inlet temperature $t'_{f_1} = 95$°C. The rate of hot water flow $G_1 = 2130$ kg/h.

The cold water is in counterflow in respect to the hot water and it moves in the annulus formed between the two tubes; the cold water heats in the exchanger from $t'_{f2} = 15°C$ to $t''_{f2} = 45°C$. The inside diameter of the external tube $D = 48$ mm. The rate of cold water flow $G_2 = 3200$ kg/h. The length of a section of the heat exchanger $l = 1.9$ m.

The loss of heat from the external surface of the heat exchanger can be ignored.

Solution. The heat capacity of water $c_p \approx 4.19$ kJ/kg.°C. The amount of heat transferred in the heat exchanger

$$Q = G_2 c_{p2}(t''_{f2} - t'_{f2}) = \frac{3200}{3600} 4.19(45 - 15) = 111 \text{ kW}.$$

The outlet temperature of the heating water

$$t''_{f1} = t'_{f1} - \frac{Q}{G_1 c_{p1}} = 95 - \frac{111 \times 3600}{2130 \times 4.19} = 50°C.$$

Find the arithmetic mean temperatures of the hot and cold fluids and the physical properties of water at these temperatures :

$$t_{f1} = 0.5 (t'_{f1} + t''_{f1}) = 0.5 (95 + 50) = 72.5°C;$$

at that temperature

$$\rho_{f1} = 976 \text{ kg/m}^3; \quad \nu_{f1} = 0.403 \times 10^{-6} \text{ m}^2/\text{s};$$
$$\lambda_{f1} = 0.670 \text{ W/m.°C}; \quad Pr_{f1} = 2.47.$$

$$t_{f2} = 0.5 (t'_{f2} + t''_{f2}) = 0.5 (15 + 45) = 30°C;$$

at that temperature

$$\rho_{f2} = 996 \text{ kg/m}^3; \quad \nu_{f2} = 0.805 \times 10^{-6} \text{ m}^2/\text{s};$$
$$\lambda_{f2} = 0.618 \text{ W/m.°C}; \quad Pr_{f2} = 5.42.$$

The velocities of the hot and cold fluids

$$w_1 = \frac{4G_1}{\rho_{f1} \pi d_1^2 3600} = \frac{4 \times 2130}{976 \times 3.14(3.2 \times 10^{-2})^2 3600} = 0.755 \text{ m / s.}$$

$$w_2 = \frac{4G_2}{\rho_{f2} \pi(D^2 - d_2^2)3600} = \frac{4 \times 3200}{996 \times 3.14(4.8^2 - 3.5^2) \, 10^{-4} \times 3600} =$$

$$= 1.06 \text{ m/s.}$$

The Reynolds number for the hot fluid

$$\mathrm{Re}_{f1} = \frac{w_1 d_1}{\nu_{f1}} = \frac{0.755 \times 3.2 \times 10^{-2}}{0.403 \times 10^{-6}} = 6 \times 10^4.$$

The hot water is in turbulent flow so that the Nusselt number and the heat-transfer coefficient are calculated by formula (5.7).

The Nusselt number

$$\mathrm{Nu}_{f1} = 0.021 \mathrm{Re}_{f1}^{0.8} \, \mathrm{Pr}_{f1}^{0.43} \left(\frac{\mathrm{Pr}_{f1}}{\mathrm{Pr}_{w1}} \right)^{0.25}.$$

Since the wall temperature is not known, then for a first approximation we assume a wall temperature

$$t_{w1} \approx 0.5 \, (t_{f1} + t_{f2}) = 0.5 \, (72.5 + 30) = 51.2°C.$$

At that temperature the Prandtl number $\mathrm{Pr}_{w1} = 3.5$; then

$$\mathrm{Nu}_{f1} = 0.021 \, (6 \times 10^4)^{0.8} (2.47)^{0.43} \left(\frac{2.47}{3.5} \right)^{0.25} = 188.$$

The coefficient of heat transfer from the hot water to the wall of a tube

$$\alpha_1 = \mathrm{Nu}_{f1} \frac{\lambda_{f1}}{d_1} = 188 \frac{0.670}{3.2 \times 10^{-2}} = 3940 \ W/m^2.°C.$$

The Reynolds number for the cold (heated) water

$$\mathrm{Re}_{f2} = \frac{w_2 d_e}{\nu_{f2}} = \frac{1.06 \times 1.3 \times 10^{-2}}{0.805 \times 10^{-6}} = 1.71 \times 10^4,$$

where the equivalent diameter for the annular channel

$$d_e = D - d_2 = 48 - 35 = 13 \ mm.$$

The cold (heated) water is in turbulent flow so that the Nusselt number and the heat-transfer coefficient are calculated by equation (5.12) derived for heat transfer in fluids in turbulent flow in annular channels :

$$\mathrm{Nu}_{f2} = 0.017 \mathrm{Re}_{f2}^{0.8} \, \mathrm{Pr}_{f2}^{0.4} \left(\frac{\mathrm{Pr}_{f2}}{\mathrm{Pr}_{w2}} \right)^{0.25} \left(\frac{D}{d_2} \right)^{0.18}$$

Assuming as a first approximation that $t_{w2} \approx t_{w1}$ and, consequently, that $Pr_{w2} \approx Pr_{w1} \approx 3.5$, we obtain :

$$\mathrm{Nu}_{f2} = 0.017(1.71 \times 10^4)^{0.8}(5.42)^{0.4}\left(\frac{5.42}{3.5}\right)^{0.25}\left(\frac{48}{35}\right)^{0.18} = 95.$$

The coefficient of heat transfer from the tube to the cold water

$$\alpha_2 = \mathrm{Nu}_{f2}\frac{\lambda_{f2}}{d_e} = 95\frac{0.618}{1.3 \times 10^{-2}} = 4500 \ \mathrm{W/m^2.°C.}$$

The overall heat-transfer coefficient

$$k = \frac{1}{\frac{1}{\alpha_1} + \frac{\delta_w}{\lambda_w} + \frac{1}{\alpha_2}} = \frac{1}{\frac{1}{3940} + \frac{1.5\times10^{-3}}{45} + \frac{1}{4500}} = 1970 \ \mathrm{W/m^2.°C.}$$

Since in the case considered $\dfrac{t'_{f1}-t''_{f2}}{t''_{f1}-t'_{f2}} = \dfrac{50}{35} < 1.5$, then calculations can be based with sufficient accuracy on the arithmetic mean temperature difference

$$\Delta t_a = t_{f1} - t_{f2} = 72.5 - 30 = 42.5°C.$$

The rate of heat flow from the hot to the cold water

$$q = k\Delta t_a = 1970 \times 42.5 = 8.37 \times 10^4 \ \mathrm{W/m^2.}$$

The heating surface

$$F = \frac{Q}{q} = \frac{111}{83.7} = 1.33 \ \mathrm{m^2.}$$

The number of sections

$$n = \frac{F}{\pi d_1 l} = \frac{1.33}{\pi \times 3.2 \times 10^{-2} \times 1.9} \approx 7$$

The surface temperatures of a tube

$$t_{w1} = t_{f1} - \frac{q}{\alpha_1} = 72.5 - \frac{83700}{3940} = 51.3°C;$$

$$t_{w2} = t_{f2} - \frac{q}{\alpha_2} = 30 + \frac{83700}{4500} = 48.6°C.$$

At these temperatures $Pr_{w1} = 3.47$ and $Pr_{w2} = 3.65$ and the corrections, allowing for a change in the physical properties of the fluid over the cross-section of flow are as follows :

$$\left(\frac{Pr_{f1}}{Pr_{w1}}\right)^{0.25} = \left(\frac{2.47}{3.47}\right)^{0.25} = 0.915 \text{ (assumed equal to 0.92);}$$

$$\left(\frac{Pr_{f2}}{Pr_{w2}}\right)^{-0.25} = \left(\frac{5.42}{3.65}\right)^{0.25} = 1.10 \text{ (assumed equal to 1.12).}$$

The coincidence is sufficiently accurate; it may be assumed that $F = 1.33 \text{ m}^2$ and the number of sections $n = 7$.

Example 11.2. In a tubular two-pass air heater of a steam boiler air in the amount $G_2 = 21.5$ kg/s must be heated from an inlet temperature $t_{f2}' = 30°C$ to an outlet temperature $t_{f2}'' = 260°C$.

Determine the required heating surface, the height of tubes in one pass, l_1, and the number of tubes arranged across and along the flow of air.

The flue gas (13% CO_2, 11% H_2O), in the amount of $G_1 = 19.6$ kg/s, moves in the steel tubes ($\lambda_w = 46.5$ W/m.°C) having diameters $d_2/d_1 = 53/50$ mm with a mean velocity $w_1 = 14$ m/s. At the air heater inlet flue gas temperature $t_{f1}' = 380°C$.

Air flows across the bank of tubes with a mean velocity in the narrow cross-section of the bank $w_2 = 8$ m/s. The tubes are in a staggered arrangement in the bank with pitches $s_1 = s_2 = 1.3 \, d_2$.

Solution. The arithmetic mean temperature of air

$$t_{f2} = 0.5 \, (t_{f2}' + t_{f2}'') = 0.5 \, (30 + 260) = 145°C.$$

At that temperature the physical properties of air are as follows : $\rho_{f2} = 0.844$ kg/m^3, $c_{pf2} = 1.01$ kJ/kg.°C, $\lambda_{f2} = 3.52 \times 10^{-2}$ W/m.°C, $v_{f2} = 28.3 \times 10^{-6}$ m^2/s, $Pr_{f2} = 0.684$.

The amount of heat transferred from the flue gas to the air $Q = G_2 c_{pf2}$ $(t_{f2}'' - t_{f2}') = 21.5 \times 1.01 \, (260 - 30) = 5000$ kW.

Determine the temperature of the gas at the air heater outlet.

As a first approximation the mean temperature of the flue gas in the air heater is assumed to be $t_{f1} = 300°C$, and at that temperature $c_{pf1} \approx 1.12$ kJ/kg.°C

and

$$t''_{f1} = t'_{f1} - \frac{Q}{G_1 c_{pf1}} = 380 - \frac{5000}{19.6 \times 1.12} = 152°C,$$

then

$$t_{f1} = 0.5 \, (t'_{f1} + t''_{f1}) = 0.5 \, (380 + 152) = 266°C.$$

At that temperature $c_{pf1} = 1.11$ kJ/kg.°C and as a result of a second approximation we find that

$$t''_{f1} = 150°C \text{ and } t_{f1} = 265°C.$$

At the temperature $t_{f1} = 265°C$ the physical properties of flue gas of the given composition are as follows :

$$\rho_{f1} = 0.622 \text{ kg/m}^3; \ c_{pf1} = 1.11 \text{ kJ/kg.°C};$$

$$\lambda_{f1} = 0.0454 \text{ W/m°C}; \ v_{f1} = 41.2 \times 10^{-6} \text{ m}^2/\text{s}; \ Pr_{f1} = 0.66.$$

The Reynolds number for the flue gas

$$Re_{f1} = \frac{w_1 d_1}{v_{f1}} = \frac{14 \times 0.05}{41.2 \times 10^{-6}} = 17000.$$

The Nusselt number and the coefficient of heat transfer from the flue gas to the walls of tubes are calculated by the following equation :

$$Nu_{f1} = 0.021 \, Re_{f1}^{0.8} \, Pr_{f1}^{0.43} = 0.021 \, (1.7 \times 10^4)^{0.8} \, (0.66)^{0.43} = 43.5;$$

$$\alpha_1 = Nu_{f1} \frac{\lambda_{f1}}{d_1} = 43.5 \frac{0.0454}{0.05} = 39.5 \text{ W / m}^2.°C.$$

The Reynolds number for air

$$Re_{f2} = \frac{w_2 d_2}{v_{f2}} = \frac{8 \times 0.053}{28.3 \times 10^{-6}} = 15000.$$

For cross flow the Nusselt number and the coefficient of heat transfer from the walls of tubes to the air are calculated by the following equation :

$$\mathrm{Nu}_{f2} = 0.41 \mathrm{Re}_{f2}^{0.6} \mathrm{Pr}_{f2}^{0.33} \varepsilon_s =$$

$$= 0.41(1.5 \times 10^4)^{0.6}(0.684)^{0.33} = 115,$$

where, for tubes in a staggered arrangement, and $s_1/s_2 < 2$ $\varepsilon_s = (s_1/s_2)^{1.6}$ and since $s_1 = s_2$ then $\varepsilon_s = 1$;

$$\alpha_2 = \mathrm{Nu}_{f2} \frac{\lambda_{f2}}{d_2} = 115\frac{0.0352}{0.053} = 76.2 \ \mathrm{W/m^2.°C}.$$

The overall heat-transfer coefficient

$$k = \frac{1}{\frac{1}{\alpha} + \frac{\delta_w}{\lambda_w} + \frac{1}{\alpha_2}} = \frac{1}{\frac{1}{39.5} + \frac{1.5 \times 10^{-3}}{46.5} + \frac{1}{76.2}} = 26 \ \mathrm{W/m^2.°C}.$$

Since

$$\frac{t'_{f1} - t''_{f2}}{t''_{f1} - t'_{f2}} = \frac{380 - 150}{260 - 30} = 1,$$

then the mean temperature difference (or temperature drop)

$$\Delta t_{\log. \ count} \approx \Delta t_a = t_{f1} - t_{f2} = 265 - 145 = 120°C.$$

For the flue gas and air flows considered from the graph we find :

$$P = \frac{t''_{f2} - t'_{f2}}{t'_{f1} - t'_{f2}} = \frac{260 - 30}{380 - 30} = 0.658$$

and

$$R = \frac{t'_{f1} - t''_{f1}}{t''_{f2} - t'_{f2}} = \frac{\cdot 380 - 150}{260 - 30} = 1.0;$$

$$\varepsilon = 0.88;$$

consequently,

$$\Delta t = \varepsilon \Delta t_{count} = 0.88 \times 120 = 105.5°C$$

The heating surface of the air heater

$$F = \frac{Q}{k\Delta t} = \frac{5 \times 10^6}{26 \times 105.5} = 1830 \ \mathrm{m^2}.$$

The total number of tubes

$$n = \frac{4G_1}{\rho_{f1}\pi d_1^2 w_1} = \frac{4 \times 19.6}{0.622 \times 3.14(5 \times 10^{-2})^2 14} = 1080.$$

The height of tubes in one pass

$$l_1 = \frac{F}{2\pi d_1 n} = \frac{1830}{2 \times 3.14 \times 0.05 \times 1080} = 5.4 \text{ m}$$

The clear passage for air flow

$$f_1 = \frac{G_2}{\rho_{f_2} w_2} = \frac{21.5}{0.844 \times 8} = 3.2 \text{ m}^2$$

The number of tubes arranged across the flow

$$n_1 = \frac{f}{l_1 (s_1 - d_2)} = \frac{3.2}{5.4(1.3 \times 0.053 - 0.053)} \approx 38.$$

The number of tubes exposed to longitudinal flow

$$n_2 = \frac{n}{n_1} = \frac{1080}{38} \approx 29.$$

Example 11.3. Calculate and determine the main dimensions of a vertical four-pass tubular steam-to-water heat exchanger designed to heat water in the amount of $G_1 = 30$ t/h from an inlet temperature $t''_{f1} = 20°C$ to an outlet temperature $t''_{f1} = 95°C$.

The heated water flows in brass tubes ($\lambda = 104.5$ W/m.°C) having diameters $d_2/d_1 = 14/12$ mm with a velocity $w = 1.5$ m/s. The hot fluid is dry saturated steam at a pressure $p = 127.5$ kPa, condensing on the outer surface of the water-carrying tubes.

The loss of heat to surroundings is to be assumed equal to 2 per cent of the amount of heat added.

Solution. The amount of heat transferred from steam to water

$$Q = G_1 c_{p1}(t''_{f1} - t'_{f1}) = \frac{3 \times 10^4}{3600} 4.187(95 - 20) = 2620 \text{ kW}.$$

Find the rate of heat flow G_2. At the pressure $p = 127.5$ kPa $t_s = 106.6°C$, $i'' = 2685$ kJ/kg, $i' = 447$ kJ/kg and

$$G_2 = \frac{Q}{0.98(i'' - i')} = \frac{2620 \times 10^3}{0.98 (2685 - 447)} = 1.2 \text{ kg / s}.$$

To calculate the coefficient of heat transfer to the outer surface of the tube due to steam condensation, we must know the temperature of the tube outer surface t_{w2} and the height of tube H. Since these quantities are not know, calculations should be carried out by the method of successive approximations. Determine the logarithmic mean temperature difference

$$\Delta t_{log} = \frac{t''_{f1} - t'_{f1}}{\ln \frac{t_s - t'_{f1}}{t_s - t''_{f1}}} = \frac{95 - 20}{2.3 \log \frac{106.6 - 20}{106.6 - 95}} = 37.4°\text{C}.$$

for a first approximation assume

$$t_{w_2} \approx t_s - \frac{\Delta t_{log}}{2} = 106.6 - \frac{37.4}{2} \approx 88°\text{C}.$$

In addition, assume a height of tubes $H = 2$ m. The reduced length of a tube

$$Z = \Delta t_2 H A.$$

At a saturation temperature $t_s = 106.6°\text{C}$, we find $A = 57.6$ 1/m.°C and $B = 6.71 \times 10^{-3}$ m/W. Then, $Z = (t_s - t_{w2}) HA = (106.6 - 88) 2 \times 57.6 = 2140 < 2300$.

The condensate is in laminar flow along the entire tube and the calculation can be done from the following equation :

$$\text{Re} = 3.8 Z^{0.78} = 3.8 \ (2140)^{0.78} = 1520;$$

$$\alpha_2 = \frac{\text{Re}}{\Delta t_2 HB} = \frac{1520}{18.6 \times 2 \times 6.71 \times 10^{-3}} = 6080 \ \text{W/m}^2.°\text{C}.$$

Determine the coefficient of heat transfer from the condensing steam to the heated water.

The arithmetic mean temperature of water

$$t_{f1} = 0.5 \ (t'_{f1} + t''_{f1}) = 0.5 \ (20 + 95) = 57.5°\text{C};$$

and at that temperature

$$v_{f1} = 0.498 \times 10^{-6} \ \text{m}^2/\text{s}; \quad \lambda_{f1} = 0.665 \ \text{W/m.°C};$$

$$\rho_{f1} = 984 \ \text{kg/m}^3; \quad \text{and} \ \text{Pr}_{f1} = 3.12;$$

$$\text{Re}_{f1} = \frac{wd_1}{v_1} = \frac{1.5 \times 1.2 \times 10^{-2}}{0.498 \times 10^{-6}} = 3.62 \times 10^4.$$

The water is in turbulent flow; calculation formula (5.7) is used.

The drop of temperature across the wall of a tube is set approximately at 1°C, then, $t_{w1} \approx t_{w2} - 1 = 87°C$ and $Pr_{w1} \approx 2.03$.

$$Nu_{f1} = 0.021 Re_{f1}^{0.8} Pr_{f1}^{0.43} (Pr_{f1} / Pr_{w1})^{0.25}$$

$$= 0.021(3.62 \times 10^4)^{0.8}(3.12)^{0.43}\left(\frac{3.12}{2.03}\right)^{0.25} = 165;$$

$$\alpha_1 = Nu_{f1}\frac{\lambda_{f1}}{d_1} = 165\frac{0.655}{1.2 \times 10^{-2}} = 9000 \text{ W} / m^2.°C.$$

The overall heat-transfer coefficient

$$k = \frac{1}{\frac{1}{\alpha_1} + \frac{\delta}{\lambda} + \frac{1}{\alpha_2}} = \frac{1}{\frac{1}{9000} + \frac{1 \times 10^{-3}}{104.5} + \frac{1}{6080}} = 3630 \text{ W} / m^2.°C.$$

The mean rate of heat flow

$$q = k\Delta t_{log} = 3630 \times 37.4 \times 10^{-3} = 135.8 \text{ kW/m}^2.$$

For a first approximation the heating surface

$$F = \frac{Q}{q} = \frac{2620}{135.8} = 19.3 \text{ m}^2.$$

The number of tubes in one pass

$$m = \frac{4G_1}{\rho_{f1}w\pi d_1^2} = \frac{4 \times 8.34}{984 \times 1.5 \times 3.14(1.2 \times 10^{-2})^2} = 50.$$

There are four passes in the heat exchanger and the total number of tubes $n = 4 \times 50 = 200$.

For a first approximation the height of tubes

$$H = \frac{F}{\pi d_{av}n} = \frac{19.3}{3.14 \times 1.3 \times 10^{-2} \times 200} = 2.37 \text{ m}.$$

The surface temperature of tubes

$$t_{w2} = t_s - \frac{q}{\alpha_2} = 106.6 - \frac{135.8 \times 10^3}{6080} = 84.3° C;$$

$$t_{w1} = t_{w2} - \frac{q}{\lambda}\delta = 84.3 - \frac{135.8 \times 10^3}{104.5} 1 \times 10^{-3} \approx 83°C.$$

Since the obtained numerical values of H, t_{w2} and t_{w1} do not coincide with these assumed for the first approximation, the calculation is repeated, assuming $H = 2.4$ m, $t_{w2} = 84.3°C$ and $t_{w1} = 83°C$. The second calculation gives : $\alpha_1 = 8950$ W/m².°C, $\alpha_2 = 6030$ W/m².°C, $k = 3490$ W/m².°C, $q = 130$ kW/m², $F = 20$ m²; for the second approximation the height of tubes $H = 2.45$ m.

For the second approximation the tube surface temperatures $t_{w2} = 85°C$ and $t_{w1} = 83.8°C$. The values obtained coincide with sufficient accuracy with these previously assumed, so we assume finally that $F = 20$ m² and $H = 2.5$ m.

CHAPTER 12

Mass Transfer

INTRODUCTION

In mass transfer, mass is being transferred from one phase to another distinct phase; the basic mechanism is the same whether the phases are gas, solid or liquid. This includes distillation, absorption, liquid-liquid extraction, membrane separation, adsorption and leaching.

The transmission of heat between a solid phase and a fluid is limited to a boundary layer or thin film adjacent to the surface. Likewise, the transfer of material in processes such as humidification is also dependent upon the movement of material in thin films. The transfer of mass in this case takes place by two mechanisms, actual movement of the bulk fluid (convection), and molecular or eddy movement. This latter mechanism is known as diffusion. These procedures may operate simultaneously. Mass transfer through the film is mainly by molecular or eddy diffusion; through the main body of the fluid convection is more important. The mechanisms of heat conduction and diffusion are quite similar under rather special conditions.

The rate of mass transfer per unit area depends upon the product of the diffusivity and the concentration gradient. However, the rate of transfer of heat per unit area by conduction under steadystate conditions has been shown to be equal to the product of the thermal conductivity and the rate of change of temperature with respect to distance. If the relation is multiplied by a unity factor of $c_p \rho / c_p \rho$, then the relation may be altered, so that the rate of heat transfer is equal to the product of the thermal diffusivity and the rate of change of the quantity of $\rho c_p t$ with respect to direction. This latter quantity is the heat content per unit of material, which might be termed a thermal concentration. The similarity between the rates of heat transfer and mass transfer is that both are proportional to concentration gradients. Also, the proportionality constants for thermal diffusivity and mass diffusion are analogous. Factors, such as agitation of the fluid, increase both the specific rate of diffusion and heat transfer since the thickness of the boundary layer is thereby decreased, and turbulence is increased.

There are a number of unit operations which are related since they involve transfer of material as well as transfer of heat through fluid films. These operations are humidification or dehumidification, drying, rectification, gas absorption, extraction, and crystallisation. In the unit operations of humidification or dehumidification water vapour diffuses in a gas film.

SIMILARITY OF MASS, HEAT, AND MOMENTUM TRANSFER PROCESSES

We have noted that the various unit operations could be classified into three fundamental transfer (or "transport") processes : momentum transfer, heat transfer, and mass transfer. The fundamental process of momentum transfer occurs in such unit operations as fluid flow, mixing, sedimentation, and filtration. Heat transfer occurs in conductive and convective transfer of heat, evaporation, distillation, and drying.

The third fundamental transfer process, mass transfer, occurs in distillation, absorption, drying, liquid-liquid extraction, adsorption, and membrane processes. When mass is being transferred from one distinct phase to another or through a single phase, the basic mechanisms are the same whether the phase is a gas, liquid, or solid. This was also shown in heat transfer, where the transfer of heat by conduction followed Fourier's law in a gas, solid, or liquid.

General Molecular Transport Equation

All three of the molecular transport processes of momentum, heat, and mass are characterised by

$$\text{rate of a transfer process} = \frac{\text{driving force}}{\text{resistance}} \qquad ...(12.1)$$

This can be written as follows for molecular diffusion of the property momentum, heat, and mass :

$$\psi_z = -\delta \frac{d\Gamma}{dz} \qquad ...(12.2)$$

Molecular diffusion equations for momentum heat, and mass transfer

Newton's equation for momentum transfer for constant density can be written as follows :

$$\tau_{zx} = -\frac{\mu}{\rho}\frac{d(v_x \rho)}{dz} \qquad ...(12.3)$$

where τ_{zx} is momentum transferred/s.m^2, μ/ρ is kinematic viscosity in m^2/s, z is distance in m, and $v_x \rho$ is momentum/m^3, where the momentum has units of kg.m/s.

Fourier's law for heat conduction can be written as follows for constant ρ and c_p :

$$\frac{q_z}{A} = -\alpha \frac{d(\rho c_p T)}{dz} \qquad ...(12.4)$$

where q_z/A is heat flux in W/m^2, α is the thermal diffusivity in m^2/s, and $\rho c_p T$ is J/m^3.

The equation for molecular diffusion of mass is Fick's law. It is written as follows for constant total concentration in a fluid :

$$J_{Az}^* = -D_{AB}\frac{dc_A}{dz} \qquad ...(12.5)$$

where J_{Az}^* is the molar flux of component A in the z direction due to molecular diffusion in kg mol A/s.m^2, D_{AB} the molecular diffusivity of the molecule A in B in m^2/s, c_A the concentration of A in kg mol/m^3, and z the distance of diffusion in m. In cgs units J_{Az}^* is g mol A/s.cm^2, D_{AB} is cm^2/s, and c_A is g mol A/cm^3. In English units, J_{Az}^* is lb mol/h. ft^2, D_{AB} is ft^2/h, and c_A is lb mol/ft^3.

The similarity of Eqs. (12.3), (12.4), and (12.5) for momentum, heat, and mass transfer should be obvious. All the fluxes on the left-hand side of the three equations have as units transfer of a quantity of momentum, heat, or mass per unit time per unit area. The transport properties μ/ρ, α, D_{AB} all have units of m^2/s, and the concentrations are represented as momentum/m^3, J/m^3, or kg mol/m^3.

Turbulent diffusion equations for momentum, heat, and mass transfer

The equations can be written by discussing the similarities among momentum, heat, and mass transfer in turbulent transfer. For turbulent momentum transfer and constant density.

$$\tau_{zx} = -\left(\frac{\mu}{\rho} + \varepsilon_t\right)\frac{d(v_x \rho)}{dz} \qquad ...(12.6)$$

For turbulent heat transfer for constant ρ and c_p,

$$\frac{q_z}{A} = -(\alpha + \alpha_t)\frac{d(\rho c p)}{dz} \qquad ...(12.7)$$

For turbulent mass transfer for constant c,

$$J_{Az}^* = -(D_{AB} + \varepsilon_M)\frac{dc_A}{dz} \qquad ...(12.8)$$

In these equations ε_t is the turbulent or eddy momentum diffusivity in m^2/s, α_t the turbulent or eddy thermal diffusivity in m^2/s, and ε_M the turbulent or eddy mass diffusivity in m^2/s. Again, these equations are quite similar to each other. Many of the theoretical equations and empirical correlations for turbulent transport to various geometries are also quite similar.

Examples of Mass-Transfer Processes

Mass transfer is important in many areas of science and engineering. Mass transfer occurs when a component in a mixture migrates in the same phase or from phase to phase because of a difference in concentration between two points. Many familiar phenomena involve mass transfer. Liquid in an open pail of water evaporates into still air because of the difference in concentration of water vapour at the water surface and the surrounding air. There is a "driving force" from the surface to the air. A piece of sugar added to a cup of coffee eventually dissolves by itself and diffuses to the surrounding solution. When newly cut and moist green timber is exposed to the atmosphere, the wood will dry partially when water in the timber diffuses through the wood, to the surface, and then to the atmosphere. In a fermentation process nutrients and oxygen dissolved in the solution diffuse to the micro-organisms. In a catalytic reaction the reactants diffuse from the surrounding medium to the catalyst surface, where reaction occurs.

Many purification processes involve mass transfer. In uranium processing, a uranium salt in solution is extracted by an organic solvent. Distillation to separate alcohol from water involves mass transfer. Removal of SO_2 from flue gas is done by absorption in a basic liquid solution.

We can treat mass transfer in a manner somewhat similar to that used in heat transfer with Fourier's law of conduction. However, an important difference is that in molecular mass transfer one or more of the components of the medium is moving. In heat transfer by conduction the medium is usually stationary and only energy in the form of heat is being transported. This introduces some differences between heat and mass transfer that will be discussed in this chapter.

Fick's Law for Molecular Diffusion

Molecular diffusion or molecular transport can be defined as the transfer or movement of individual molecules through a fluid by means of the random, individual movements of the molecules. We can imagine the molecules travelling only in straight lines and changing direction by bouncing off other molecules after collisions. Since the molecules travel in a random path, molecular diffusion is often called a random-walk process.

In Fig. 12.1 the molecular diffusion process is shown schematically. A random path that molecule A might take in diffusing through B molecules from point (1) to (2) is shown. If there are a greater number of A molecules near point (1) than at (2), then, since molecules diffuse randomly in both directions, more A molecules will diffuse from (1) to (2) than from (2) to (1). The net diffusion of A is from high- to low-concentration regions.

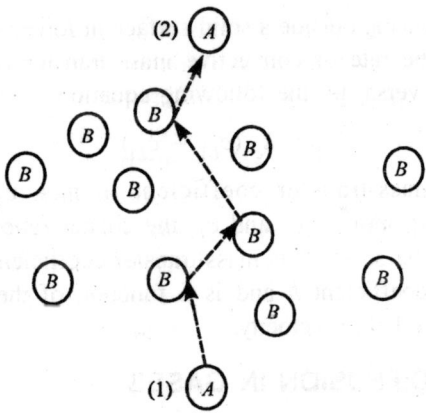

Fig. 12.1. Schematic diagram of molecular diffusion process.

As another example, a drop of blue liquid dye is added to a cup of water. The dye molecules will diffuse slowly by molecular diffusion to all parts of the water. To increase this rate of mixing of the dye, the liquid can be mechanically agitated by a spoon and convective mass transfer will occur. The two modes of heat transfer, conduction and convective heat transfer, are analogous to molecular diffusion and convective mass transfer.

First, we will consider the diffusion of molecules when the whole bulk fluid is not moving but is stationary. Diffusion of the molecules is due to a concentration gradient. The general Fick's law equation can be written as follows for a binary mixture of A and B :

$$J^*_{Az} = -cD_{AB} \frac{dx_A}{dz} \qquad ...(12.9)$$

where c is total concentration of A and B in kg mol $A + B/m^3$, and x_A is the mole fraction of A in the mixture of A and B. If c is constant, then since $c_A = cx_A$,

$$c \, dx_A = d(cx_A) = dc_A \qquad \qquad ...(12.10)$$

Substituting into Eq. (12.9) we obtain Eq. (12.5) for constant total concentration.

$$J_{Az}^* = -D_{AB} \frac{dc_A}{dz} \qquad \qquad ...(12.11)$$

This equation is the more commonly used one in many molecular diffusion processes. If c varies some, an average value is often used with Eq. (12.5).

Convective Mass-Transfer Coefficient

When a fluid is flowing outside a solid surface in forced convection motion, we can express the rate of convective mass transfer from the surface to the fluid, or vice versa, by the following equation :

$$N_A = k_c \, (c_{L1} - c_{Li}) \qquad \qquad ...(12.12)$$

where k_c is a mass-transfer coefficient in m/s, c_{L1} the bulk fluid concentration in kg mol A/m^3, and c_{Li} the concentration in the fluid next to the surface of the solid. This mass-transfer coefficient is very similar to the heat-transfer coefficient h and is a function of the system geometry, fluid properties, and flow velocity.

MOLECULAR DIFFUSION IN GASES

Equimolar Counterdiffusion in Gases

In Fig. 12.2 a diagram is given of two gases A and B at constant total pressure P in two large chambers connected by a tube where molecular diffusion at steadystate is occurring. Stirring in each chamber keeps the concentrations in each chamber uniform. The partial pressure $p_{A1} > p_{A2}$ and $p_{B2} > p_{B1}$. Molecules of A diffuse to the right and B to the left. Since the total pressure P is constant throughout, the net moles of A diffusing to the right must equal the net moles of B to the left. If this is not so, the total pressure would not remain constant. This means that

$$J_{Az}^* = -J_{Bz}^* \qquad \qquad ...(12.13)$$

The subscript z is often dropped when the direction is obvious. Writing Fick's law for B for constant c,

$$J_B^* = -D_{BA}\frac{dc_B}{dz} \qquad ...(12.14)$$

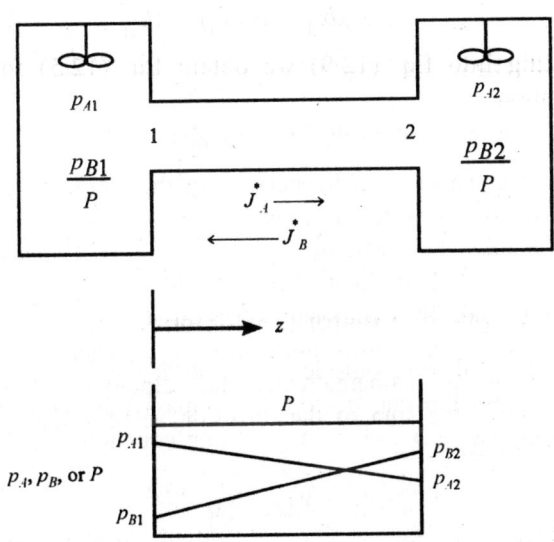

Fig. 12.2. Equimolar counterdiffusion of gases A and B.

Now since $P = p_A + p_B =$ constant, then

$$c = c_A + c_B \qquad ...(12.15)$$

Differentiating both sides,

$$dc_A = -dc_B \qquad ...(12.16)$$

Equating Eqs. (12.5) to (12.14),

$$J_A^* = -D_{AB}\frac{dc_A}{dz} = -J_B^* = -(-)D_{BA}\frac{dc_B}{dz} \qquad ...(12.17)$$

Substituting Eq. (12.14) into (12.17) and canceling like terms,

$$D_{AB} = D_{BA} \qquad ...(12.18)$$

This shows that for a binary gas mixture of A and B the diffusivity coefficient D_{AB} for A diffusing in B is the same as D_{BA} for B diffusing into A.

General Case for Diffusion of Gases A and B Plus Convection

Up to now we have considered Fick's law for diffusion in a stationary fluid; i.e., there has been no net movement or convective flow of the entire phase of the binary mixture A and B. The diffusion flux J_A^* occurred because of the concentration gradient. The rate at which moles of A passed a fixed point to the right, which will be taken as a positive flux, is J_A^* kg mol A/s.m^2. This flux can be converted to a velocity of diffusion of A to the right by

$$J_A^*(\text{kg mol } A / \text{s.m}^2) = v_{Ad} c_A \left(\frac{\text{m}}{\text{s}} \frac{\text{kg mol } A}{\text{m}^3} \right) \qquad ...(12.19)$$

where v_{Ad} is the diffusion velocity A in m/s.

Now let us consider what happens when the whole fluid is moving in bulk or convective flow to the right. The molar average velocity of the whole fluid relative to a stationary point is v_M m/s. Component A is still diffusing to the right, but now its diffusion velocity v_{Ad} is measured relative to the moving fluid. To a stationary observer A is moving faster than the bulk of the phase, since its diffusion velocity v_{Ad} is added to that of the bulk phase v_M. Expressed mathematically, the velocity of A relative to the stationary point is the sum of the diffusion velocity and the average or convective velocity.

$$v_A = v_{Ad} + v_M \qquad ...(12.20)$$

where v_A is the velocity of A relative to a stationary point. Expressed pictorially,

$$\xrightarrow{\hspace{2cm}} v_A$$

$$\xrightarrow{\hspace{1cm}} \xrightarrow{\hspace{1cm}}$$
$$\quad v_{Ad} \qquad v_M$$

Multiplying Eq. (12.20) by c_A,

$$c_A v_A = c_A v_{Ad} + c_A v_M \qquad ...(12.21)$$

Each of the three terms represents a flux. The first term, $c_A v_A$, can be represented by the flux N_A kg mol A/s.m^2. This is the total flux of A relative to the stationary point. The second term is J_A^*, the diffusion flux relative to the moving fluid. The third term is the convective flux of A relative to the stationary point. Hence, Eq. (12.21) becomes

$$N_A = J_A^* + c_A v_M \qquad ...(12.22)$$

Let N be the total convective flux of the whole stream relative to the stationary point. Then,

$$N = c v_M = N_A + N_B \qquad ...(12.23)$$

Or, solving for v_M,

$$v_M = \frac{N_A + N_B}{c} \qquad ...(12.24)$$

Substituting Eq. (12.24) into (12.22),

$$N_A = J_A^* + \frac{c_A}{c}(N_A + N_B) \qquad ...(12.25)$$

Since J_A^* is Fick's law, Eq. (12.9),

$$N_A = -cD_{AB}\frac{dx_A}{dz} + \frac{c_A}{c}(N_A + N_B) \qquad ...(12.26)$$

Equation (12.26) is the final general equation for diffusion plus convection to use when the flux N_A is used, which is relative to a stationary point. A similar equation can be written for N_B,

$$N_B = -cD_{BA}\frac{dx_B}{dz} + \frac{c_B}{c}(N_A + N_B) \qquad ...(12.27)$$

To solve Eq. (12.26) or (12.27), the relation between the flux N_A and N_B must be known. Equations (12.26) and (12.27) hold for diffusion in a gas, liquid, or solid.

For equimolar counterdiffusion, $N_A = -N_B$ and the convective term in Eq. (12.26) becomes zero. Then, $N_A = J_A^* = -N_B = -J_B^*$.

Special Case for A Diffusing through Stagnant, Non-diffusing B

The case of diffusion of A through stagnant or non-diffusing B at steady-state often occurs. In this case one boundary at the end of the diffusion path is impermeable to component B, so it cannot pass through.

One example shown in Fig. 12.3(a) is in evaporation of a pure liquid such as benzene (A) at the bottom of a narrow tube, where a large amount of inert or non-diffusing air (B) is passed over the top. The benzene vapour (A) diffuses through the air (B) in the tube.

The boundary at the liquid surface at point 1 is impermeable to air, since air is insoluble in benzene liquid. Hence, air (B) cannot diffuse into or away from the surface. At point 2 the partial pressure $p_{A2} = 0$, since a large volume of air is passing by.

Another example shown in Fig. 12.3(b) occurs in the absorption of NH_3 (A) vapour which is in air (B) by water. The water surface is impermeable to the air, since air is only very slightly soluble in water. Thus, since B cannot diffuse, $N_B = 0$.

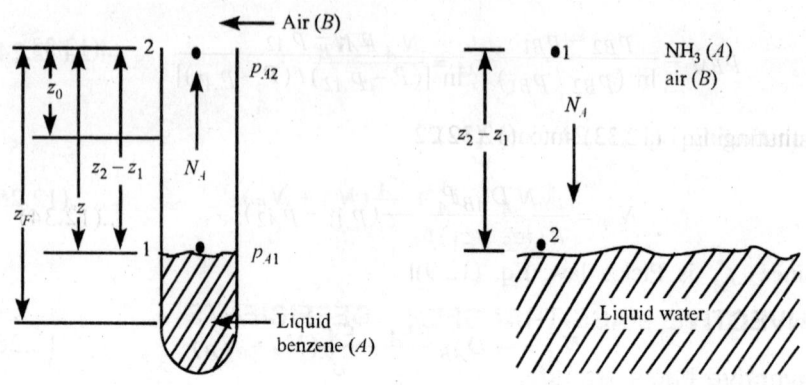

Fig. 12.3. Diffusion of A through stagnant, non-diffusing B: (a) benzene evaporating into air; and (b) ammonia in air being absorbed into water.

To derive the case for A diffusing in stagnant, non-diffusing B, $N_B = 0$ is substituted into the general Eq. (12.26).

$$N_A = -cD_{AB}\frac{dx_A}{dz} + \frac{c_A}{c}(N_A + 0) \qquad ...(12.28)$$

Keeping the total pressure P constant, substituting $c = P/RT$, $p_A = x_A P$, and $c_A/c = p_A/P$ into Eq. (12.28),

$$N_A = -\frac{D_{AB}}{RT}\frac{dp_A}{dz} + \frac{p_A}{P}N_A \qquad ...(12.29)$$

Rearranging and integrating,

$$N_A\left(1 - \frac{p_A}{P}\right) = -\frac{D_{AB}}{RT}\frac{dp_A}{dz} \qquad ...(12.30)$$

$$N_A\int_{z_1}^{z_2} dz = -\frac{D_{AB}}{RT}\int_{p_{A1}}^{p_{A2}}\frac{dp_A}{1 - p_A/P} \qquad ...(12.31)$$

$$N_A = \frac{D_{AB}P}{RT(z_2 - z_1)}\ln\frac{P - p_{A2}}{P - p_{A1}} \qquad ...(12.32)$$

Equation (12.32) is the final equation to be used to calculate the flux of A. However, it is often written in another form as follows. A log mean value of the inert B is defined as follows. Since $P = p_{A1} + p_B = p_{A2} + p_{B2}$, $p_{B1} = P - p_{A1}$, and $p_{B2} = P - p_{A2}$,

$$p_{BM} = \frac{p_{B2} - p_{B1}}{\ln (p_{B2} / p_{B1})} = \frac{p_{A1} - p_{A2}}{\ln [(P - p_{A2}) / (P - p_{A1})]} \qquad ...(12.33)$$

Substituting Eq. (12.33) into (12.32),

$$N_A = \frac{D_{AB} P}{RT(z_2 - z_1) p_{BM}} (p_{A1} - p_{A2}) \qquad ...(12.34)$$

CONVECTIVE MASS-TRANSFER COEFFICIENTS

Convective Mass Transfer

In the previous sections we have emphasised molecular diffusion in stagnant fluids or fluids in familiar flow. In many cases the rate of diffusion is slow, and more rapid transfer is desired.

To have a fluid in convective flow usually requires the fluid to be flowing by another immiscible fluid or by a solid surface. An example is a fluid flowing in a pipe, where part of the pipe wall is made by a slightly dissolving solid material such as benzoic acid. The benzoic acid dissolves and is transported perpendicular to the main stream from the wall. When a fluid is in turbulent flow and is flowing past a surface, the actual velocity of small particles of fluid cannot be described clearly as in laminar flow. In laminar flow the fluid flows in streamlines and its behaviour can usually be described mathematically. However, in turbulent motion there are no streamlines, but there are large eddies or "chunks" of fluid moving rapidly in seemingly random fashion.

When a solute A is dissolving from a solid surface there is a high concentration of this solute in the fluid at the surface, and its concentration, in general, decreases as the distance from the wall increases. However, minute samples of fluid adjacent to each other do not always have concentrations close to each other. This occurs because eddies having solute in them move rapidly from one part of the fluid to another, transferring relatively large amounts of solute. This turbulent diffusion or eddy transfer is quite fast in comparison to molecular transfer.

Three regions of mass transfer can be visualised. In the first, which is adjacent to the surface, a thin viscous sub-layer film is present. Most of the mass transfer occurs by molecular diffusion, since few or no eddies are present. A large concentration drop occurs across this film as a result of the slow diffusion rate.

The transition or buffer region is adjacent to the first region. Some eddies are present and the mass transfer is the sum of turbulent and molecular diffusion. There is a gradual transition in this region from the transfer by mainly molecular diffusion at the one end to mainly turbulent at the other end.

In the turbulent region adjacent to the buffer region, most of the transfer is by turbulent diffusion, with a small amount of molecular diffusion. The concentration decrease is very small here since the eddies tend to keep the fluid concentration uniform.

A typical plot for the mass transfer of a dissolving solid from a surface to a turbulent fluid in a conduit is given in Fig. 12.4. The concentration drop from c_{A1} adjacent to the surface is very abrupt close to the surface and then levels off. This curve is very similar to the shapes found for heat and momentum transfer. The average or mixed concentration $\bar{c}_A$ is shown and is slightly greater than the minimum c_{A2}.

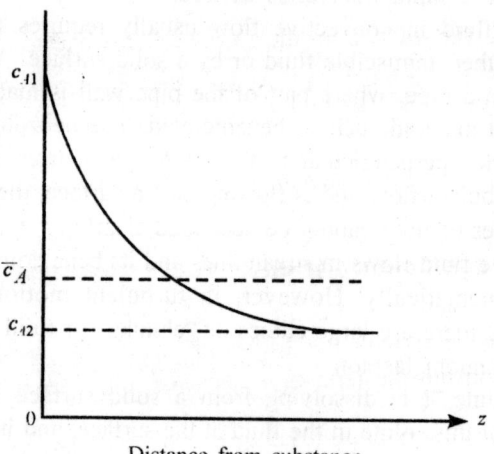

Distance from substance

Fig. 12.4. Concentration profile in turbulent transfer from a surface to a fluid.

Types of Mass-Transfer Coefficients

Definition of mass-transfer coefficient

Since our understanding of turbulent flow is incomplete, we attempt to write the equations for turbulent diffusion in a manner similar to that for molecular diffusion. For turbulent mass transfer for constant c, Eq. (12.8) is written as

$$J_A^* = -(D_{AB} + \varepsilon_M)\frac{dc_A}{dz} \qquad ...(12.35)$$

where D_{AB} is the molecular diffusivity in m²/s and ε_M is the mass eddy diffusivity in m²/s. The value of ε_M is a variable and is near zero at the interface or surface and increases as the distance from the wall increases. We then use an average value $\bar{\varepsilon}_M$ since the variation of ε_M is not generally known. Integrating Eq. (12.35) between points 1 and 2,

$$J_{A1}^* = \frac{D_{AB} + \bar{\varepsilon}_M}{z_2 - z_1}(c_{A1} - c_{A2}) \qquad ...(12.36)$$

The flux J_{A1}^* is based on the surface area A_1 since the cross-sectional area may vary. The value of $z_2 - z_1$, the distance of the path, is often not known. Hence, Eq. (12.36) is simplified and is written using a convective mass-transfer coefficient k'_c.

$$J_{A1}^* = k'c(c_{A1} - c_{A2}) \qquad ...(12.37)$$

where J_{A1}^* is the flux of A from the surface A_1 relative to the whole bulk phase, k'_c is $(D_{AB} + \bar{\varepsilon}_M)/(z_2 - z_1)$ an experimental mass-transfer coefficient in kg mol/s.m². (kg mol/m³) or simplified as m/s, and c_{A2} is the concentration at point 2 in kg mol A/m³ or more usually the average bulk concentration $\bar{c}_{A2}$. This defining of a convective mass-transfer coefficient k'_c is quite similar to the convective heat-transfer coefficient h.

Mass-transfer coefficient for equimolar counterdiffusion

Generally, we are interested in N_A, the flux of A relative to stationary coordinates. We can start with the following, which is similar to that for molecular diffusion but the term ε_M is added.

$$N_A = -c(D_{AB} + \varepsilon_M)\frac{dx_A}{dz} + x_A(N_A + N_B) \qquad ...(12.38)$$

For the case of equimolar counterdiffusion, where $N_A = -N_B$, and integrating at steadystate, calling $k'_c = (D_{AB} + \bar{\varepsilon}_M)/(z_2 - z_1)$,

$$N_A = k'_c(c_{A1} - c_{A2}) \qquad ...(12.39)$$

Equation (12.39) is the defining equation for the mass-transfer coefficient. Often, however, we define the concentration in terms of mole fraction if a liquid or gas or in terms of partial pressure if a gas. Hence, we can define the mass-transfer coefficient in several ways. If y_A is mole fraction in a gas phase and x_A in a liquid phase, then Eq. (12.39) can be written as follows for equimolar counterdiffusion :

Gases : $N_A = k'_c(c_{A1} - c_{A2}) = k'_G (p_{A1} - p_{A2})$

$$= k'_y(y_{A1} - y_{A2}) \qquad ...(12.40)$$

Liquids : $N_A = k'_c(c_{A1} - c_{A2}) = k'L(c_{A1} - c_{A2})$

$$= k'_x(x_{A1} - x_{A2}) \qquad ...(12.41)$$

All of these mass-transfer coefficients can be related to each other. For example, using Eq. (12.40) and substituting $y_{A1} = c_{A1}/c$ and $y_{A2} = c_{A2}/c$ into the equation,

$$N_A = k'_c(c_{A1} - c_{A2}) = k'_y(y_{A1} - y_{A2})$$

$$= k'_y \left(\frac{c_{A1}}{c} - \frac{c_{A2}}{c} \right) = \frac{K'_y}{c}(c_{A1} - c_{A2}) \qquad ...(12.42)$$

Hence,

$$k'_c = \frac{k'_y}{c} \qquad ...(12.43)$$

These relations among mass-transfer coefficients, and the various flux equations, are given in Table 12.1.

Mass-transfer coefficient for A diffusing through stagnant, non-diffusing B

For A diffusing through stagnant, non-diffusing B where $N_B = 0$, Eq. (12.38) gives for steadystate

$$N_A = \frac{K'_c}{x_{BM}}(c_{A1} - c_{A2}) = k_c(c_{A1} - c_{A2}) \qquad ...(12.44)$$

where the x_{BM} and its counterpart y_{BM} are similar to Eq. (12.33) and k_c is the mass-transfer coefficient for A diffusing through stagnant B. Also,

$$x_{BM} = \frac{x_{B2} - x_{B1}}{\ln (x_{B2}/x_{B1})} \qquad y_{BM} = \frac{y_{B2} - y_{B1}}{\ln (y_{B2}/y_{B1})} \qquad ...(12.45)$$

Rewriting Eq. (12.44) using other units,

(Gases) : $N_A = k_c(c_{A1} - c_{A2}) = k_G(p_{A1} - p_{A2})$

$$= k_y(y_{A1} - y_{A2}) \qquad ...(12.46)$$

(Liquids) : $N_A = k_c(c_{A1} - c_{A2}) = k_L(c_{A1} - c_{A2})$

$$= k_x(x_{A1} - x_{A2}) \qquad ...(12.47)$$

Again all the mass-transfer coefficients can be related to each other and are given in Table 12.1. For example, setting Eq. (12.44) equal to (12.47),

$$N_A = \frac{k_c'}{x_{BM}}(c_{A1} - c_{A2}) = k_x(x_{A1} - x_{A2}) = k_x\left(\frac{c_{A1}}{c} - \frac{c_{A2}}{c}\right) \quad ...(12.48)$$

Hence,

$$\frac{k_c'}{x_{BM}} = \frac{k_x}{c} \quad ...(12.49)$$

Table 12.1. Flux equations and mass-transfer coefficients.

Flux equations for equimolar counterdiffusion

Gases : $N_A = k_c'(c_{A1} - c_{A2}) = k_G'(p_{A1} - p_{A2}) = k_y'(y_{A1} - y_{A2})$

Liquids : $N_A = k_c'(c_{A1} - c_{A2}) = k_L'(c_{A1} - c_{A2}) = k_x'(x_{A1} - x_{A2})$

Flux equations for A diffusing through stagnant, non-diffusing B

Gases : $N_A = k_c(c_{A1} - c_{A2}) = k_G(p_{A1} - p_{A2}) = k_y(y_{A1} - y_{A2})$

Liquids : $N_A = k_c(c_{A1} - c_{A2}) = k_L(c_{A1} - c_{A2}) = k_x(x_{A1} - x_{A2})$

Conversions between mass-transfer coefficients

Gases : $k_c' = k_c'\dfrac{P}{RT} = k_c\dfrac{p_{BM}}{RT}$

$= k_G'P = k_G p_{BM} = k_y y_{BM} = k_y' = k_c' y_{BM} c = k_G y_{BM} P$

Liquids : $k_c' c = k_L' c = k_L c \, x_{BM} c = k_L' \rho / M = k_x' = k_x x_{BM}$

(where ρ is density of liquid and M is molecular weight)

Units of mass-transfer coefficients

	SI Units	Cgs Units	English Units
k_c, k_L, k_c', k_L'	$\dfrac{\text{m}}{\text{s}}$	$\dfrac{\text{cm}}{\text{s}}$	$\dfrac{\text{ft}}{\text{h}}$
k_x, k_y, k_x', k_y'	$\dfrac{\text{kg mol}}{\text{s.m}^2 \cdot \text{mol frac}}$	$\dfrac{\text{g mol}}{\text{s.m}^2 \cdot \text{mol frac}}$	$\dfrac{\text{lb mol}}{\text{h.ft}^2 \cdot \text{mol frac}}$
k_G, k_G'	$\dfrac{\text{kg mol}}{\text{s.m}^2 \cdot \text{Pa}}$ $\dfrac{\text{kg mol}}{\text{s.m}^2 \cdot \text{atm}}$	$\dfrac{\text{g mol}}{\text{s.cm}^2 \cdot \text{atm}}$	$\dfrac{\text{lb mol}}{\text{h.ft}^2 \cdot \text{atm}}$

Methods to Determine Mass-Transfer Coefficients

Many different experimental methods have been employed to experimentally obtain mass-transfer coefficients. In determining the mass-transfer coefficient to a sphere, Steele and Geankoplis used a solid sphere of benzoic acid held rigidly by a rear support in a pipe. Before the run the sphere was weighed. After flow of the fluid for a timed interval, the sphere was removed, dried, and weighed again to give the amount of mass transferred, which was small compared to the weight of the sphere. From the mass transferred and the area of the sphere, the flux N_A was calculated. Then the driving force $(c_{AS} - 0)$ was used to calculate k_L, where c_{AS} is the solubility and the water contained no benzoic acid.

Another method used is to flow gases over various geometries wet with evaporating liquids. For mass transfer from a flat plate, a porous blotter wet with the liquid serves as the plate.

MASS-TRANSFER COEFFICIENTS FOR VARIOUS GEOMETRIES

Dimensionless Numbers Used to Correlate Data

The experimental data for mass-transfer coefficients obtained using various kinds of fluids, different velocities, and different geometries are correlated using dimensionless numbers similar to those of heat and momentum transfer.

The most important dimensionless number is the Reynolds number N_{Re}, which indicates degree of turbulence.

$$N_{Re} = \frac{Lv\rho}{\mu} \qquad ...(12.50)$$

where L is diameter D_p for a sphere, diameter D for a pipe, or length L for a flat plate. The velocity v is the mass average velocity if in a pipe. In a packed bed the superficial velocity v' in the empty cross-section is often used or sometimes $v = v'/\varepsilon$ is used, where v is interstitial velocity and ε void fraction of bed.

The Schmidt number is

$$N_{Sc} = \frac{\mu}{\rho D_{AB}} \qquad ...(12.51)$$

The viscosity μ and density ρ used are the actual flowing mixture of solute A and fluid B. If the mixture is dilute, properties of the pure fluid B can be used. The Prandtl number $c_p\mu/k$ for heat transfer is analogous to the Schmidt number for mass transfer. The Schmidt number is the ratio of the shear component for diffusivity μ/ρ to the diffusivity for mass transfer D_{AB}, and it physically relates the relative thickness of the hydrodynamic layer and mass-transfer boundary layer.

The Sherwood number, which is dimensionless, is

$$N_{Sh} = k_c' \frac{L}{D_{AB}} = k_c y B_M \frac{L}{D_{AB}} = \frac{K'_x}{c} \frac{L}{D_{AB}} = \dots \qquad \dots(12.52)$$

Other substitutions from Table 12.1 can be made for k'_c in Eq. (12.52).

The Stanton number occurs often and is

$$N_{St} = \frac{K_c}{v} = \frac{K'_y}{G_M} = \frac{K'_G P}{G_M} = \dots \qquad \dots(12.53)$$

Again, substitution for k'_c can be made. $G_M = v\rho/M_{av} = vc$.

Often the mass-transfer coefficient is correlated as a dimensionless J_D factor which is related to k'_c and N_{Sh} as follows :

$$J_D = \frac{k_c'}{v}(N_{Sc})^{2/3} = \frac{k'_G P}{G_M}(N_{Sc})^{2/3} = \dots = N_{Sh}/(N_{Re}N_{Sc}^{1/3}) \qquad \dots(12.54)$$

For heat transfer a dimensionless J_H factor is as follows :

$$J_H = \frac{h}{c_p G}(N_{Pr})^{2/3} \qquad \dots(12.55)$$

Analogies among Mass, Heat, and Momentum Transfer

In molecular transport of momentum, heat, or mass there are many similarities. The molecular diffusion equations of Newton for momentum, Fourier for heat, and Fick for mass are very similar and we can say that we have analogies among these three molecular transport processes. There are also similarities in turbulent transport, where the flux equations were written using the turbulent eddy momentum diffusivity ε_t, the turbulent eddy thermal diffusivity α_t, and the turbulent eddy mass diffusivity ε_M. However, these similarities are not as well defined mathematically or physically and are more difficult to relate to each other.

A great deal of effort has been devoted in the literature to developing analogies among these three transport processes for turbulent transfer so as to allow prediction of one from any of the others.

Reynolds analogy

Reynolds was the first to note similarities in transport processes and relate turbulent momentum and heat transfer. Since then, mass transfer has also been related to momentum and heat transfer. We derive this analogy from Eqs. (12.6) to (12.8) for turbulent transport. For fluid flow in a pipe for heat transfer from the fluid to the wall, Eq. (12.7) becomes as follows, --

where z is distance from the wall :

$$\frac{q}{A} = -\rho c_p (x + x_i) \frac{dT}{dz} \qquad \qquad ...(12.56)$$

For momentum transfer, Eq. (12.6) becomes

$$\tau = -\rho \left(\frac{\mu}{\rho} + \varepsilon_t \right) \frac{dv}{dz} \qquad \qquad ...(12.57)$$

Next we assume α and μ/ρ are negligible and that $\alpha_t - \varepsilon_t$. Then dividing Eq. (12.56) by (12.57)

$$\left(\frac{\tau}{q/A} \right) c_p dT = dv \qquad \qquad ...(12.58)$$

If we assume that heat flux q/A in a turbulent system is analogous to momentum flux τ, the ratio $\tau/(q/A)$ must be constant for all radial positions. We now integrate between conditions at the wall where $T = T_t$ and $v = 0$ to some point in the fluid where T is the same as the bulk T and assume that the velocity at this point is the same as v_{av}, the bulk velocity. Also, q/A is understood to be the flux at the wall, as is the shear at the wall, written as τ_s. Hence,

$$\frac{\tau_s}{q/A} c_p (T - T_i) = v_{av} - 0 \qquad \qquad ...(12.59)$$

Also, substituting $q/A = h(T - T_i)$ and $\tau_s = f v_{av}^2 \, \rho/2$ in Eq. (12.59),

$$\frac{f}{2} = \frac{h}{c_p v_{av} \rho} = \frac{h}{c_p G} \qquad \qquad ...(12.60)$$

In a similar manner using Eq. (12.8) for J_A^* and $J_A^* = k'_c (c_A - c_{Ai})$, we can relate this to Eq. (12.58) for momentum transfer. Then, the complete Reynolds analogy is

$$\frac{f}{2} = \frac{h}{c_p G} = \frac{k'_c}{v_{av}} \qquad \qquad ...(12.61)$$

Experimental data for gas streams agree approximately with Eq. (12.61) if the Schmidt and Prandtl numbers are near 1.0 and only skin friction is present in flow past a flat plate or inside a pipe. When liquids are present and/or form drag is present, the analogy is not valid.

Other analogies

The Reynolds analogy assumes that the turbulent diffusivities ε_r, α_r, and ε_M are all equal and that the molecular diffusivities μ/ρ, α, and D_{AB} are negligible compared to the turbulent diffusivities. When the Prandtl number $(\mu/\rho)/\alpha$ is 1.0, then $\mu/\rho = \alpha$; also, for $N_{Sc} = 1.0$, $\mu/\rho = D_{AB}$. Then, $(\mu/\rho + \varepsilon_t) = (\alpha + \alpha_t) = (D_{AB} + \varepsilon_M)$ and the Reynolds analogy can be obtained with the molecular terms present. However, the analogy breaks down when the viscous sublayer becomes important since the eddy diffusivities diminish to zero and the molecular diffusivities become important.

Prandtl modified the Reynolds analogy by writing the regular molecular diffusion equation for the viscous sublayer and a Reynolds-analogy equation for the turbulent core region. Then since these processes are in series, these equations were combined to produce an overall equation. The results also are poor for fluids where the Prandtl and Schmidt numbers differ from 1.0.

Von Kármán further modified the Prandtl analogy by considering the buffer region in addition to the viscous sublayer and the turbulent core. Again, an equation is written for molecular diffusion in the viscous sublayer using only the molecular diffusivity and a Reynolds analogy equation for the turbulent core. Both the molecular and eddy diffusivity are used in an equation for the buffer layer, where the velocity in this layer is used to obtain an equation for the eddy diffusivity. These three equations are then combined to give the von Kármán analogy.

Chilton and Colburn J-factor analogy

The most successful and most widely used analogy is the Chilton and Colburn J-factor analogy. This analogy is based on experimental data for gases and liquids in both the laminar and turbulent flow regions and is written as follows :

$$\frac{f}{2} = J_H = \frac{h}{c_p G}(N_{Pr})^{2/3} = J_D = \frac{k_c'}{v_{av}}(N_{Sc})^{2/3} \qquad ...(12.62)$$

Although this is an equation based on experimental data for both laminar and turbulent flow, it can be shown to satisfy the exact solution derived from laminar flow over a flat plate.

Equation (12.62) has been shown to be quite useful in correlating momentum, heat, and mass transfer data. It permits the prediction of an unknown transfer coefficient when one of the other coefficients is known. In momentum transfer the friction factor is obtained for the total drag or friction loss, which includes form drag or momentum losses due to blunt objects and also skin friction. For flow past a flat plate or in a pipe where no form drag is present, $f/2 = J_H = J_D$. When form drag is present, such as in flow in packed beds or past other blunt objects, $f/2$ is greater than J_H or J_D and $J_H \cong J_D$.

Derivation of Mass-Transfer Coefficients in Laminar Flow

When a fluid is flowing in laminar flow and mass transfer by molecular diffusion is occurring, the equations are very similar to those for heat transfer by conduction in laminar flow. The phenomena of heat and mass transfer are not always completely analogous since in mass transfer several components may be diffusing. Also, the flux of mass perpendicular to the direction of the flow must be small so as not to distort the laminar velocity profile.

In theory it is not necessary to have experimental mass-transfer coefficients for laminar flow, since the equations for momentum transfer and for diffusion can be solved. However, in many actual cases it is difficult to describe mathematically the laminar flow for geometries, such as flow past a cylinder or in a packed bed. Hence, experimental mass-transfer coefficients are often obtained and correlated. A simplified theoretical derivation will be given for two cases in laminar flow.

Mass transfer in laminar flow in a tube

We consider the case of mass transfer from a tube wall to a fluid inside in laminar flow, where, for example, the wall is made of solid benzoic acid which is dissolving in water. This is similar to heat transfer from a wall to the flowing fluid where natural convection is negligible. For fully developed flow, the parabolic velocity derived can be written as :

$$v_x = v_{max}\left[1 - \left(\frac{r}{R}\right)^2\right] = 2v_{av}\left[1 - \left(\frac{r}{R}\right)^2\right] \qquad ...(12.63)$$

where v_x is the velocity in the x direction at the distance r from the centre. For steady-state diffusion in a cylinder, a mass balance can be made on a differential element where the rate in by convection plus diffusion equals the rate out radially by diffusion to give

$$v_x \frac{\partial c_A}{\partial x} = D_{AB}\left(\frac{1}{r}\frac{\partial c_A}{\partial r^2} + \frac{\partial^2 c_A}{\partial r^2} + \frac{\partial^2 c_A}{\partial x^2}\right) \qquad ...(12.64)$$

Then, $\partial^2 c_A/\partial x^2 = 0$ if the diffusion in the x direction is negligible compared to that by convection. Combining Eqs. (12.63) and (12.64), the final solution is a complex series similar to the Graetz solution for heat transfer and a parabolic velocity profile.

If it is assumed that the velocity is flat as in rodlike flow, the solution is more easily obtained. A third solution, called the approximate Leveque solution, has been obtained, where there is a linear velocity profile near the wall and the solute diffuses only a short distance from the wall into the fluid. This is similar to the parabolic velocity profile solution at high flow rates.

Diffusion in a laminar falling film

The equation for the velocity profile in a falling film shown in Fig. 12.5(a). We will consider mass transfer of solute A into a laminar falling film, which is important in wetted-wall columns, in developing theories to explain mass transfer in stagnant pockets of fluids, and in turbulent mass transfer. The solute A in the gas is absorbed at the interface and then diffuses a distance into the liquid so that it has not penetrated the whole distance $x = \delta$ at the wall. At steadystate the inlet concentration $c_A = 0$. At a point z distance from the inlet the concentration profile of c_A is shown in Fig. 12.5(a).

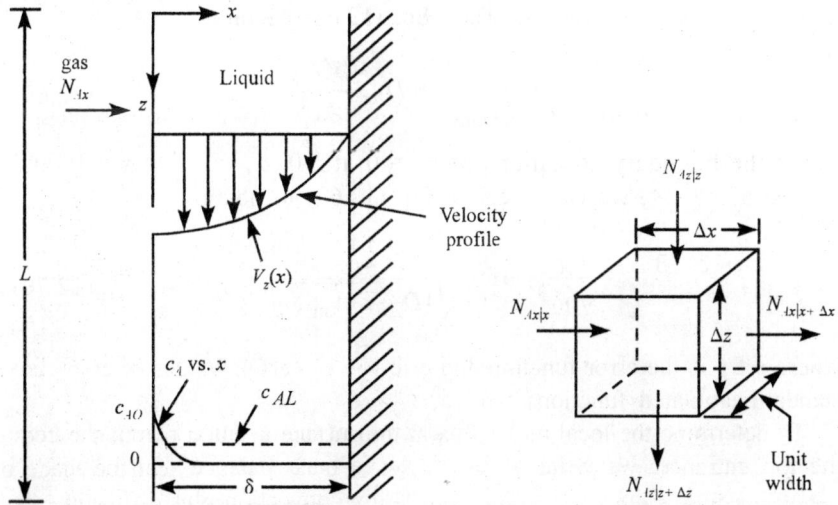

Fig. 12.5. Diffusion of solute A in a laminar falling film: (a) velocity profile and concentration profile, (b) small element for mass balance.

A mass balance will be made on the element shown in Fig. 12.5(a). For steadystate, rate of input = rate of output.

$$N_{Ax}|_x (1 \, \Delta z) + N_{Az}|_z (1 \, \Delta x) = N_{Ax}|_{x+\Delta x} (1 \, \Delta z) = N_{Az}|_{z+\Delta z} (1 \, \Delta x) \quad ...(12.65)$$

For a dilute solution the diffusion equation for A in the x direction is

$$N_{Ax} = -D_{AB} \frac{\partial c_A}{\partial x} + \text{zer} \, \text{} \text{ convection} \quad ...(12.66)$$

For the z direction the diffusion is negligible.

$$N_{Az} = 0 + c_A v_z \quad ...(12.67)$$

Dividing Eq. (12.65) by $\Delta x \, \Delta z$, letting Δx and Δz approach zero, and substituting Eqs. (12.67) and (12.68) into the result, we obtain

$$v_z \frac{\partial c_A}{\partial z} = D_{AB} \frac{\partial^2 c_A}{\partial x^2} \quad ...(12.68)$$

Since the velocity profile is parabolic and is $v_z = v_{z \, max} [1 - (x/\delta)^2]$. Also, $v_{z \, max} = (3/2) v_{z \, av}$. If the solute has penetrated only a short distance into the fluid, i.e., short contact times of t seconds equals z/v_{max}, then the A that has diffused has been carried along at the velocity $v_{x \, max}$ or v_{max} if the subscript z is dropped. Then Eq. (12.68) becomes

$$\frac{\partial c_A}{\partial (z / v_{max})} = D_{AB} \frac{\partial^2 c_A}{\partial x^2} \quad ...(12.69)$$

Using the boundary conditions of $c_A = 0$ at z 0, $c_A = c_{A0}$ at $x = 0$, and $c_A = 0$ at $x = \infty$, we can integrate Eq. (12.69) to obtain

$$\frac{c_A}{c_{A0}} = \text{erfc} \left(\frac{x}{\sqrt{4 D_{AB} z / v_{max}}} \right) \quad ...(12.70)$$

where erf y is the error function and erfc $y = 1 - $ erf y. Values of erf y are standard tabulated functions.

To determine the local molar flux at the surface $x = 0$ at position z from the top entrance, we write

$$N_{Ax}(z)|_{x=0} = -D_{AB} \frac{\partial c_A}{\partial x}\bigg|_{x=0} = c_{A0} \sqrt{\frac{D_{AB} v_{max}}{\pi z}} \quad ...(12.71)$$

The total moles of A transferred per second to the liquid over the entire length $z = 0$ to $z = L$, where the vertical surface is unit width is

$$N_A(L.1) = (1)\int_0^L (N_{Ax}|_{x=0})\, dz$$

$$= (1)\int_0^L c_{A0}\left(\frac{D_{AB}v_{max}}{\pi}\right)^{1/2}\frac{1}{z^{1/2}}\, dz$$

$$= (L.1)c_{A0}\sqrt{\frac{4D_{AB}v_{max}}{\pi L}} \qquad\qquad ...(12.72)$$

The term L/v_{max} is t_L, time of exposure of the liquid to the solute A in the gas. This means the rate of mass transfer is proportional to $D_{AB}^{0.5}$ and $1/t_L^{0.5}$. This is the basis for the penetration theory in turbulent mass transfer where pockets of liquid are exposed to unsteady-state diffusion (penetration) for short contact times.

Mass Transfer for Flow Inside Pipes

Mass transfer for laminar flow inside pipes

When a liquid or a gas is flowing inside a pipe and the Reynolds number $Dv\rho/\mu$ is below 2100, laminar flow occurs. Experimental data obtained for mass transfer from the walls for gases (G2, L1) are plotted in Fig. 12.6 for values of $W/D_{AB}\rho L$ less than about 70. The ordinate is $(c_A - c_{A0})/(c_{Ai} - c_{A0})$, where c_A is the exit concentration, c_{A0} inlet concentration, and c_{Ai} concentration at the interface between the wall and the gas. The dimensionless abscissa is $W/D_{AB}\rho L$ or $N_{Re}N_{Sc}(D/L)(\pi/4)$, where W is flow in kg/s and L is length of mass-transfer section in m. Since the experimental data follow the rodlike plot, that line should be used. The velocity profile is assumed fully developed to parabolic form at the entrance.

For liquids that have small values of D_{AB}, data follow the parabolic flow line, which is as follows for $W/D_{AB}\rho L$ over 400.

$$\frac{c_A - c_{A0}}{c_{Ai} - c_{A0}} = 5.5\left(\frac{W}{D_{AB}\rho L}\right)^{-2/3} \qquad\qquad ...(12.73)$$

Mass transfer for turbulent flow inside pipes

For turbulent flow for $Dv\rho/\mu$ above 2100 for gases or liquids flowing inside a pipe,

$$N_{Sh} = k_c'\frac{D}{D_{AB}} = \frac{k_c p_{BM}}{P}\frac{D}{D_{AB}} = 0.023\left(\frac{Dv\rho}{\mu}\right)^{0.83}\left(\frac{\mu}{\rho D_{AB}}\right)^{0.33} \qquad ...(12.74)$$

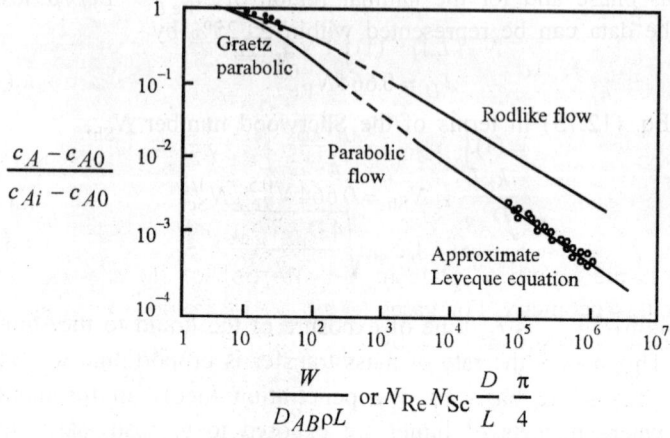

Fig. 12.6. Data for diffusion in a fluid in streamline flow inside a pipe: filled circles; vapourisation data of Gilliland and Sherwood; open circles, dissolving-solids data of Linton and Sherwood.

The equation holds for N_{Sc} of 0.6 to 3000 (G2, L1). Note that the N_{Sc} for gases is in the range 0.5–3 and for liquids is above 100 in general. Equation (12.74) for mass transfer for heat transfer inside a pipe are similar to each other.

Mass transfer for flow inside wetted-wall towers

When a gas is flowing inside the core of a wetted-wall tower the same correlations that are used for mass transfer of a gas in laminar or turbulent flow in a pipe are applicable. This means that Eqs. (12.73) and (12.74) can be used to predict mass transfer for the gas. For the mass transfer in the liquid film flowing down the wetted-wall tower, Eqs. (12.71) and (12.72) can be used for Reynolds numbers of $4\Gamma/\mu$ up to about 1200, and the theoretically predicted values should be multiplied by about 1.5 because of ripples and other factors These equations hold for short contact times or Reynolds numbers above about 100.

Mass Transfer for Flow Outside Solid Surfaces

Mass transfer in flow parallel to flat plates

The mass transfer and vapourisation of liquids from a plate or flat surface to a flowing stream is of interest in the drying of inorganic and biological materials, in evaporation of solvents from paints, for plates in wind tunnels, and in flow channels in chemical process equipment.

When the fluid flows past a plate in a free stream in an open space the boundary layer is not fully developed. For gases or evaporation of liquids in the gas phase and for the laminar region of $N_{Re.L} = Lv\rho/\mu$ less than 15000, the data can be represented within $\pm$ 25% by

$$J_D = 0.664 N_{Re.L}^{-0.5} \qquad \qquad ...(12.75)$$

Writing Eq. (12.75) in terms of the Sherwood number N_{Sh},

$$\frac{k_c'L}{D_{AB}} = N_{Sh} = 0.664 N_{Re.L}^{0.5} N_{Sc}^{1/3} \qquad \qquad ...(12.76)$$

where L is the length of plate in the direction of flow. Also, $J_D = J_M$ $= f/2$ for this geometry. For gases and $N_{Re.L}$ of 15000–1300000, the data are represented within $\pm$ 30% by $J_D = J_H = f/2$ as

$$J_D = 0.036 N_{Re.L}^{-0.2} \qquad \qquad ...(12.77)$$

Experimental data for liquids are correlated within about $\pm$ 40% by the following for a $N_{Re.L}$ of 600–50000 (L2) :

$$J_D = 0.99 N_{Re.L}^{-0.5} \qquad \qquad ...(12.78)$$

Mass transfer for flow past single spheres

For flow past single spheres and for very low $N_{Re} = D_p v \rho/\mu$, where v is the average velocity in the empty test section before the sphere, the Sherwood number, which is $k_c' Dp/D_{AB}$, should approach a value of 2.0. Where D_p is the sphere diameter,

$$N_A = \frac{2D_{AB}}{D_p}(c_{A1} - c_{A2}) = k_c(c_{A1} - c_{A2}) \qquad \qquad ...(12.79)$$

The mass-transfer coefficient k_c, which is k_c' for a dilute solution, is then

$$k_c' = \frac{2D_{AB}}{D_p} \qquad \qquad ...(12.80)$$

Rearranging,

$$\frac{k_c' D_p}{D_{AB}} = N_{Sh} = 2.0 \qquad \qquad ...(12.81)$$

Of course, natural convection effects could increase k_c'.

For gases for a Schmidt number range of 0.6–2.7 and a Reynolds number range of 1–48000, a modified equation can be used.

$$N_{Sh} = 2 + 0.552 N_{Re}^{0.53} N_{Sc}^{1/3} \qquad ...(12.82)$$

This equation also holds for heat transfer where the Prandtl number replaces the Schmidt number and the Nusselt number hD_p/k replaces the Sherwood number.

For liquids and a Reynolds number range of 2 to about 2000, the following can be used.

$$N_{Sh} = 2 + 0.95 N_{Re}^{0.50} N_{Sc}^{1/3} \qquad ...(12.83)$$

For liquids and a Reynolds number of 2000–17000, the following can be used.

$$N_{Sh} = 0.347 N_{Re}^{0.62} N_{Sc}^{1/3} \qquad ...(12.84)$$

Mass transfer to packed beds

Mass transfer to and from packed beds occurs often in processing operations, including drying operations, adsorption or desorption of gases or liquids by solid particles such as charcoal, and mass transfer of gases and liquids to catalyst particles. Using a packed bed a large amount of mass-transfer area can be contained in a relatively small volume.

The void fraction in a bed is ε, m^3 volume void space divided by the m^3 total volume of void space plus solid. The values range from 0.3 to 0.5 in general. Because of flow channeling, non-uniform packing, etc., accurate experimental data are difficult to obtain and data from different investigations can deviate considerably.

For a Reynolds number range of 10–10000 for gases in a packed bed of spheres, the recommended correlation with an average deviation of about ± 20% and a maximum of about ± 50% is

$$J_D = J_H = \frac{0.4548}{\varepsilon} N_{Re}^{-0.4069} \qquad ...(12.85)$$

It has been shown that J_D and J_H are approximately equal. The Reynolds number is defined as $N_{Re} = D_p v' \rho/\mu$, where D_p is diameter of the spheres and v' is the superficial mass average velocity in the empty tube without packing. For Eqs. (12.85), (12.88) and Eqs. (12.54), (12.55), v' is used.

For mass transfer of liquids in packed beds, the correlations of Wilson and Geankoplis should be used. For a Reynolds number $D_p v' \rho/\mu$ range of

0.0016–55 and a Schmidt number range of 165–70600, the equation to use is

$$J_D = \frac{1.09}{\varepsilon} N_{Re}^{-2/3} \qquad \qquad ...(12.86)$$

For liquids and a Reynolds number range of 55–1500 and a Schmidt number range of 165–10690,

$$J_D = \frac{0.250}{\varepsilon} N_{Re}^{-0.31} \qquad \qquad ...(12.87)$$

Or, as an alternate, Eq. (12.85) can be used for liquids for a Reynolds number range of 10–1500.

For fluidised beds of spheres, Eq. (12.85) can be used for gases and liquids and a Reynolds number range of 10–4000. For liquids in a fluidised bed and a Reynolds number range of 1–10,

$$\varepsilon J_D = 1.1068 N_{Re}^{-0.72} \qquad \qquad ...(12.88)$$

If packed beds of solids other than spheres are used, approximate correction factors can be used with Eqs. (12.85)–(12.86) for spheres. This is done, for example, for a given non-spherical particle as follows. The particle diameter to use in the equations to predict J_D is the diameter of a sphere with the same surface area as the given solid particle. The flux to these particles in the bed is then calculated using the area of the given particles. An alternative approximate procedure to use is given elsewhere.

Calculation method for packed beds

To calculate the total flux in a packed bed, J_D is first obtained and then k_c in m/s from the J_D. Then knowing the total volume V_b m³ of the bed (void plus solids), the total external surface area A m² of the solids for mass transfer is calculated using Eqs. (12.89) and (12.90).

$$a = \frac{6(1-\varepsilon)}{D_p} \qquad \qquad ...(12.89)$$

where a is the m² surface area/m³ total volume of bed when the solids are spheres.

$$A = aV_b \qquad \qquad ...(12.90)$$

To calculate the mass-transfer rate the log mean driving force at the inlet and outlet of the bed should be used.

$$N_A A = A k_c \frac{(c_{Ai} - c_{A1}) - (c_{Ai} - c_{A2})}{\ln \frac{c_{Ai} - c_{A1}}{c_{Ai} - c_{A2}}} \qquad \qquad ...(12.91)$$

where the final term is the log mean driving force: c_{Ai} is the concentration at the surface of the solid, in kg mol/m^3; c_{A1} is the inlet bulk fluid concentration; and c_{A2} is the outlet. The material-balance equation on the bulk stream is

$$N_A A = V(c_{A2} - c_{A1}) \qquad ...(12.92)$$

where V is volumetric flow rate of fluid entering in m^3/s. Equations (12.91) and (12.92) must both satisfied. The use of these two equations is similar to the use of the log mean temperature difference and heat balance in heat exchangers. These two equations can also be used for a fluid flowing in a pipe or past a flat plate, where A is the pipe wall area or plate area.

Mass transfer for flow past single cylinders

Experimental data have been obtained for mass transfer from single cylinders when the flow is perpendicular to the cylinder. The cylinders are long and mass transfer to the ends of the cylinder is not considered. For the Schmidt number range of 0.6 to 2.6 for gases and 1000 to 3000 for liquids and a Reynolds number range of 50 to 50000, data of many references have been plotted and the correlation to use is as follows :

$$J_D = 0.600(N_{Re})^{-0.487} \qquad ...(12.93)$$

The data scatter considerably by up to ± 30%. This correlation can also be used for heat transfer with $J_D = J_H$.

Liquid metals mass transfer

In recent years several correlations for mass-transfer coefficients to liquid metals have appeared in the literature. It has been found that with moderate safety factors, the correlations for non-liquid metals mass transfer may be used for liquid metals mass transfer. Care must be taken to ensure that the solid surface is wetted. Also, if the solid is an alloy, there may exist a resistance to diffusion in the solid phase.

MASS TRANSFER TO SUSPENSIONS OF SMALL PARTICLES

Mass transfer from or to small suspended particles in an agitated solution occurs in a number of process applications. In liquid-phase hydrogenation, hydrogen diffuses from gas bubbles, through an organic liquid, and then to small suspended catalyst particles. I. fermentations, oxygen diffuses from small gas bubbles, through the aqueous medium, and then to small suspended micro-organisms.

For a liquid-solid dispersion, increased agitation over and above that necessary to freely suspend very small particles has very little effect on the mass-transfer coefficient k_L to the particle. When the particles in a mixing vessel are just completely suspended, turbulence forces balance those due to gravity, and the mass-transfer rates are the same as for particles freely moving under gravity. With very small particles of say a few μm or so, which is the size of many micro-organisms in fermentations and some catalyst particles, their size is smaller than eddies, which are about 100 μm or so in size. Hence, increased agitation will have little effect on mass transfer except at very high agitation.

For a gas-liquid-solid dispersion, such as in fermentation, the same principles hold. However, increased agitation increases the number of gas bubbles and hence the interfacial area. The mass-transfer coefficients from the gas bubble to the liquid and from the liquid to the solid area relatively unaffected.

Equations for Mass Transfer to Small Particles

Mass transfer to small particles < 0.6 mm

Equations to predict mass transfer to small particles in suspension have been developed which cover three size ranges of particles. The equations for particles < 0.6 mm (600 μm) is discussed first.

The following equation has been shown to hold to predict mass-transfer coefficients from small gas bubbles such as oxygen or air to the liquid phase or from the liquid phase to the surface of small catalyst particles, micro-organisms, other solids, or liquid drops.

$$k_c' = \frac{2D_{AB}}{D_p} + 0.31 N_{Sc}^{-2/3} \left(\frac{\Delta\rho\mu_c g}{\rho_c^2} \right)^{1/3} \qquad ...(12.94)$$

where D_{AB} is the diffusivity of the solute A in solution in m²/s, D_p is the diameter of the gas bubble or the solid particle in m, μ_c is the viscosity of the solution in kg/m.s, g = 9.80665 m/s², $\Delta\rho = (\rho_c - \rho_p)$ or $(\rho_p - \rho_c)$, ρ_c is the density of the continuous phase in kg/m³, and ρ_p is the density of the gas or solid particle. The value of $\Delta\rho$ is always positive.

The first term on the right in Eq. (12.94) is the molecular diffusion term, and the second term is due to free fall or rise of the sphere by gravitational forces. This equation has been experimentally checked for dispersions of low-density solids in agitated dispersions and for small gas bubbles in agitated systems.

Mass transfer to large gas bubbles > 2.5 mm

For large gas bubbles or liquid drops >2.5 mm, the mass-transfer coefficient can be predicted by

$$k'_L = 0.42 N_{Sc}^{-0.5} \left(\frac{\Delta \rho \mu_c g}{\rho_c^2} \right)^{1/3} \qquad ...(12.95)$$

Large gas bubbles are produced when pure liquids are aerated in mixing vessels and sieve-plate columns. In this case of mass-transfer coefficient k'_L or k_L is independent of the bubble size and is constant for a given set of physical properties.

For the same physical properties the large bubble (Eq. 12.95) gives values of k_L about three to four times larger than (12.94) for small particles. Again, Eq. (12.95) shows that the k_L is essentially independent of agitation intensity in an agitated vessel and gas velocity in a sieve-tray tower.

Mass transfer to particles in transition region

In mass transfer in the transition region between small and large bubbles in the size range 0.6 to 2.5 mm, the mass-transfer coefficient can be approximated by assuming that it increases linearly with bubble diameter.

Mass transfer to particles in highly turbulent mixers

In the preceding three regions, the density difference between phases is sufficiently large to cause the force of gravity to primarily determine the mass-transfer coefficient. This also includes solids just completely suspended in mixing vessels.

When agitation power is increased beyond that needed for suspension of solid or liquid particles and the turbulence forces become larger than the gravitational forces, Eq. (12.94) is not followed and Eq. 12.96 should be used where small increases in k'_L are observed.

$$k'_L N_{Sc}^{2/3} = 0.13 \left(\frac{(P/V)\mu_c}{\rho_c^2} \right)^{1/4} \qquad ...(12.96)$$

where P/V is power input per unit volume. The data deviate substantially by up to 60% from this correlation. In the case of gas-liquid dispersions it is quite impractical to exceed gravitational forces by agitation systems.

The experimental data are complicated by the fact that very small particles are easily suspended and if their size is of the order of the smallest eddies, the mass-transfer coefficient will remain constant until a large increase in power input is added above that required for suspension.

MASS TRANSFER BETWEEN PHASES

Introduction to Interphase Mass Transfer

Mass transfer from a fluid phase to another phase which is primarily a solid phase. The solute is usually transferred from the fluid phase by convecting mass transfer and through the solid by diffusion. Here we shall be concerned with mass transfer of solute from one fluid phase by convection and then through a second fluid phase by convection e.g., the solute may diffuse through a gas phase and then diffuse through and be absorbed in an adjacent and immiscible liquid phase. This occurs in the case of absorption of ammonia from air by water. The two phases are in direct contact with each other, such as in a packed, tray, or spray-type tower, and the interfacial area between the phase is usually not well defined. In two phase mass transfer, a concentration gradient will exist in each phase, causing mass transfer to occur. At the interphase between the two fluid phases, equilibrium exists in most cases.

Equilibrium Relations

Even when mass transfer is occurring equilibrium relations are important to determine concentration profiles for reducing rates of mass transfer. Here, these equilibrium relations and equilibrium distillation coefficients between two phases are used in discussion of mass transfer between phases.

CONCENTRATION PROFILES IN INTERPHASE MASS TRANSFER

In the majority of mass-transfer systems, two phases, which are essentially immiscible in each other, are present and also an interface between these two phases. Assuming the solute A is diffusing from the bulk gas phase G to the liquid phase L, it must pass through phase G, through the interface, and then into phase L in series. A concentration gradient must exist to cause this mass transfer through the resistances in each phase, as shown in Fig. 12.7. The average or bulk concentration of A in the gas phase in mole fraction units is y_{AG}, where $y_{AG} = p_A/P$, and x_{AL} in the bulk liquid phase in mole fraction units.

The concentration in the bulk gas phase y_{AG} decreases to y_{Ai} at the interface. The liquid concentration starts at x_{Ai} at the at the interface and falls to x_{AL}. At the interface, since there would be no resistance to transfer across this interface, y_{Ai} and x_{Ai} are in equilibrium and are related by the equilibrium distribution relation

$$y_{Ai} = f(x_{Ai}) \qquad \qquad \ldots(12.99)$$

where y_{Ai} is a function of x_{Ai}. They are related by an equilibrium plot such as Fig. 12.7. If the system follows Henry's law, $y_A P$ or p_A and x_A are related at the interface.

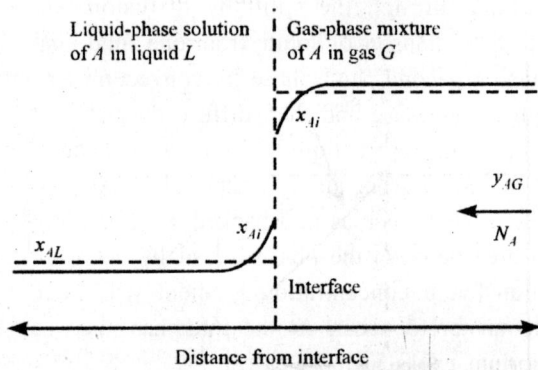

Fig. 12.7. Concentration profile of solute A diffusing through two phases.

Experimentally, the resistance at the interface has been shown to be negligible for most casts of mass transfer where chemical reactions do not occur, such as absorption of common gases from air to water and extraction of organic solutes from one phase to another. However, there are some exceptions. Certain surface-active compounds may concentrate at the interface and cause an "interfacial resistance" that slows down the diffusion of solute molecules. Theories to predict when interfacial resistance may occur are still obscure and unreliable.

MASS TRANSFER USING FILM MASS-TRANSFER COEFFICIENTS AND INTERFACE CONCENTRATIONS

Equimolar Counterdiffusion

For equimolar counterdiffusion the concentrations of Fig. 12.7 can be plotted on an xy diagram in Fig. 12.8. Point P represents the bulk phase compositions x_{AG} and x_{AL} of the two phases and point M the interface concentrations y_{Ai} and x_{Ai}. For A diffusing from the gas to liquid and B in equimolar counterdiffusion from liquid to gas,

$$N_A = k_y'(y_{AG} - y_{Ai}) = k_x'(x_{Ai} - x_{AL}) \qquad \ldots(12.100)$$

where k_y' is the gas-phase mass-transfer coefficient in kg mol/s.m². mol frac (g mol/s. cm².mol frac, lb mol/h.ft². mol frac) and k_x the liquid-phase mass-transfer coefficient in kg mol/s.m². mol frac (g mol/s.cm².mol frac, lb mol frac, lb mol/h.ft².mol frac). Rearranging Eq. (12.100),

$$-\frac{k'_x}{k'_y} = \frac{y_{AG} - y_{Ai}}{x_{AL} - x_{Ai}} \qquad ...(12.101)$$

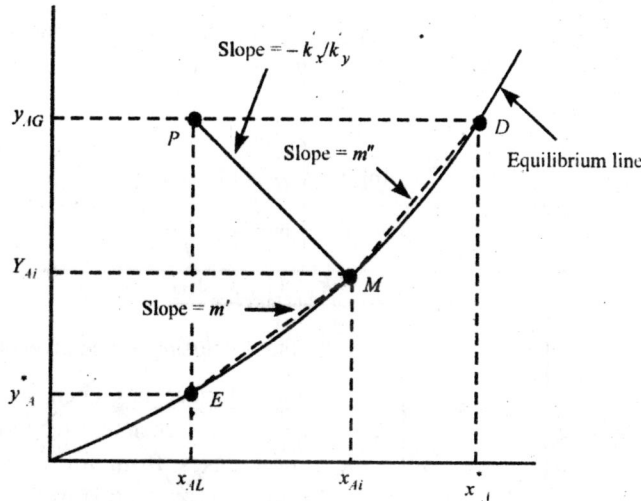

Fig. 12.8. Concentration driving forces and interface concentrations in interphase mass transfer (equimolar counterdiffusion).

The driving force in the gas phase is $(y_{AG}-y_{Ai})$ and in the liquid phase it is $(x_{Ai} - x_{AL})$. The slope of the line PM is $- k_x / k_y$. This means if the two film coefficients k'_x and k'_y are known the interface compositions can be determined by drawing line PM with a slope $-k'_x/k'_y$ intersecting the equilibrium line.

The bulk-phase concentrations y_{AG} and x_{AL} can be determined by simply sampling the mixed bulk gas phase and sampling the mixed bulk liquid phase. The interface concentrations are determined by Eq. (12.101).

Diffusion of A through Stagnant or Non-diffusing B

For the common case of A diffusing through a stagnant gas phase and then through a stagnant liquid phase, the concentrations are shown in Fig. 12.9, where P again represents bulk-phase compositions and M interface compositions. The equations for A diffusing through a stagnant gas and then through a stagnant liquid are

$$N_A = k_y\,(y_{AG} - y_{Ai}) = k_x'(x_{Ai} - x_{AL}) \qquad ...(12.102)$$

Now,

$$k_y = \frac{k_y'}{(1-y_A)_{iM}} \qquad k_x = \frac{k_x'}{(1-x_A)_{iM}} \qquad ...(12.103)$$

where

$$(1-y_A)_{iM} = \frac{(1-y_{Ai})-(1-y_{AG})}{\ln[(1-y_{Ai})/(1-y_{AG})]} \qquad ...(12.104)$$

$$(1-x_A)_{iM} = \frac{(1-x_{AL})-(1-x_{Ai})}{\ln[(1-x_{AL})/(1-x_{Ai})]} \qquad ...(12.105)$$

Then,

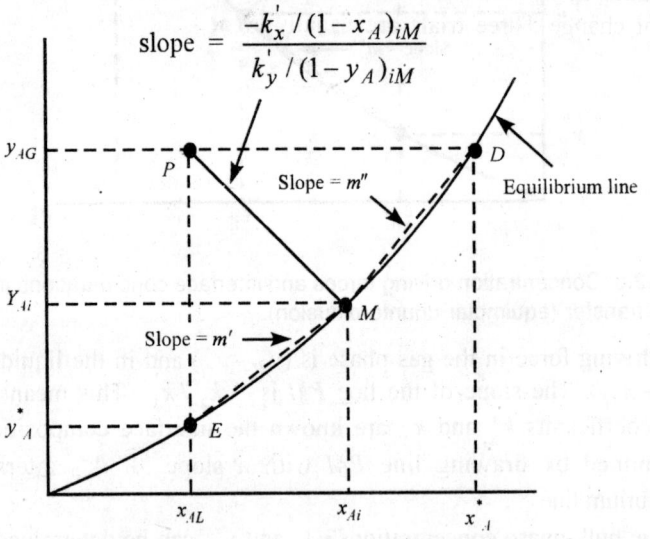

Fig. 12.9. Concentration driving forces and interface concentrations in interphase mass transfer (*A* diffusing through stagnant *B*).

$$N_A = \frac{k_y'}{(1-y_A)_{iM}}(y_{AG} - y_{Ai}) = \frac{k_x'}{(1-x_A)_{iM}}(x_{Ai} - x_{AL}) \qquad ...(12.106)$$

Note that $(1 - y_A)_{iM}$ is the same as y_{BM} of Eq. (12.45) but is written for the interface, and $(1 - xA)_{iM}$ is the same as x_{BM} of Eq. (12.45). Using Eq. (12.106) and rearranging,

$$\frac{-k'_x / (1 - x_A)_{iM}}{k'_y / (1 - y_A)_{iM}} = \frac{y_{AG} - y_{Ai}}{x_{AL} - x_{Ai}} \qquad ...(12.107)$$

The slope of the line PM in Fig. 12.9 to obtain the interface compositions is given by the left-hand side of Eq. (12.107). This differs from the slope of Eq. (12.101) for equimolar counterdiffusion by the terms $(1 - y_A)_{iM}$ and $(1 - x_A)_{iM}$. When A is diffusing through stagnant B and the solutions are dilute, $(1 - y_A)_{iM}$ and $(1 - x_A)_{iM}$ are close to 1.

A trial-and-error method is needed to use Eq. (12.107) to get the slope, since the left-hand side contains y_{Ai} and x_{Ai} that are being sought. For the first trial $(1 - y_A)_{iM}$ and $(1 - x_A)_{iM}$ are assumed to be 1.0 and Eq. (10.107) is used to get the slope and y_{A1} and x_{Ai} values. Then for the second trial, these values of y_{Ai} and x_{Ai} are used to calculate a new slope to get new values of y_{Ai} and x_{Ai}. This is repeated until the interface compositions do not change Three trials are usually sufficient.

Appendices

Appendices

APPENDIX I

Notations

SI units are given first (followed by English and/or cgs units).

a	particle radius, m (ft)
a	area, m^2 (ft^2); also m^2 area/m^3 volume bed or packing (ft^2/ft^3)
a_e	acceleration from centrifugal force, m/s^2 (ft/s^2)
a_v	specific surface area of particle, m^{-1} (ft^{-1})
A	cross-sectional area, m^2 (ft^2, cm^2); also area, m^2 (ft^2)
A	absorption factor $= L/mV$, dimensionless, also filter area. m^2 (ft^2)
A_m	membrane area, cm^2, m^2 (ft^2)
A_w	solvent permeability constant, kg/s·m^2 atm
A_s	solute permeability constant, m/s (ft/h)
b	length, m (ft, cm)
B	flow rate of dry solid, kg/h (lb$_m$/h); also filtration constant, s/m^3 (s/ft^3)
B	physical property of membrane, atm^{-1}
c	concentration, kg/m^3, kg mol/m^3 (lb$_m$/ft^3, g mol/cm^3)
c	concentration of absorbate in fluid, kg/m^3 (lb$_m$/ft^3)
c_A	concentration of A, kg mol/m^3 (lb mol/ft^3, g mol/cm^3)
c_b	break-point concentration, kg/m^3 (lb$_m$/ft^3)
c_g	concentration of solute in gel, kg solute/m^3 (lb$_m$/ft^3, g/cm^3)
c_s	concentration of solute at surface of membrane, kg solute/m^3 (lb$_m$/ft^3, g/cm^3)
c_S	concentration of A in solid, kg mol A/m^3 (lb$_m$/ft^3, g mol/cm^3)
c_s	concentration of solids in slurry, kg/m^3 (lb$_m$/ft^3)
c_S	humid heat, kJ/kg dry air.K (btu/lb$_m$ dry air.°F)
c_x	concentration of solids in slurry, mass fraction

455

c_p heat capacity at constant pressure, J/kg.K, kJ/kg.K, kJ/kg mol.K (btu/lb$_m$.°F, cal/g.°C)

$\bar{c}_w$ mean solvent concentration in membrane, kg/m^3 (lb$_m$/ft^3)

c'_A deviation of concentration from mean concentration $\bar{c}_A$, kg mol/m^3

c_P concentration of P, kg P/m^3

c_v heat capacity at constant volume, J/kg.K

C filtration constant, N/m^2 (lb$_f$ft^2); also, number of components

C fluid heat capacity, W/K (btu/h.°F)

C height of bottom of agitator above tank bottom, m (ft)

C_p pitot tube coefficient, dimensionless

C_D drag coefficient, dimensionless

C_v, C_0 Venturi coefficient, orifice coefficient, dimensionless

D molecular diffusivity, m^2/s (ft^2/h, cm^2/s); also diameter, m (ft)

D decimal reduction time, min; also distillate flow rate, kg/h, kg mol/h (lb$_m$/h)

D_{AB} molecular diffusivity, m^2/s (ft^2/h, cm^2/s)

D_p particle diameter, m (ft)

D_{KA} Knudsen diffusivity m^2/s (ft^2/h, cm^2/s)

D_{NA} transition-region diffusivity, m^2/s (ft^2/h, cm^2/s)

$D_{A\,eff}$ effective diffusivity, m^2/s (ft^2/h, cm^2/s)

$D_{p,m}$ effective mean diameter for mixture, m (ft)

D_a diameter of agitator, m (ft)

D_t diameter of tank, m (ft)

D_{pc} critical diameter, m (ft)

D_{AP} diffusivity of A in protein solution, m^2/s

E activation energy, J/kg mol (cal/g mol)

E energy for size reduction, kW.h/ton; also tray efficiency, dimensionless

E total energy, J/kg (ft.lb$_f$/lb$_m$)

E fraction unaccomplished change, dimensionless

E emitted radiation energy, W/m^2 (btu/h.ft^2)

E axial dispersion coefficient, m^2/s (ft^2/h)

$E_{B\lambda}$ monochromatic emissive power, W/m^3 (btu/h.ft^3)

f fraction of feed vapourised; also cycle fraction, dimensionless

f Fanning friction factor, dimensionless

f_t mixing factor, dimensionless

F	number of degrees of freedom
F	frictional loss, J/kg (ft.lb$_f$/lb$_m$)
F	flow rate, kg/h, kg mol/h (lb$_m$/h)
F	force, N (lb$_f$, dyn)
F_T	correction factor for temperature difference, dimensionless
F_0	process time at 121.1°C (250°F), min
$\mathscr{F}_{12}$	geometric view factor for gray surfaces, dimensionless
F_{12}	geometric view factor, dimensionless
g	standard acceleration of gravity
g_c	gravitational conversion factor
G	mass velocity $= v\rho$, kg/s.m^2, kg/h.m^2 (lb$_m$/h.ft^2)
G	growth constant, mm/h
G'	mass velocity $= v'\rho$, kg/s.m^2, kg/h.m^2(lb$_m$/h.ft^2)
G	irradiation on a body, W/m^2 (btu/h.ft^2)
h	constant spacing in x for Simpson's rule
h	head, J/kg (ft.lb$_f$/lb$_m$); also height of fluid, m (ft)
h	heat-transfer coefficient, W/m^2.K (btu/h.ft^2.°F)
h	enthalpy of liquid, J/kg, kJ/kg, kJ/kg mol (btu/lb$_m$)
h_{fg}	latent heat of vapourisation, J/kg, kJ/kg (btu/lb$_m$)
h_c	contact resistance coefficient, W/m^2.K (btu/h.ft^2.°F)
H	distance, m (ft)
H	Henry's law constant, atm/mol frac
H	head, J/kg (ft.lb$_f$/lb$_m$); also height of fluid, m (ft)
H	enthalpy, J/kg, kJ/kg, kJ/kg mol (btu/lb$_m$); also enthalpy kJ/kg dry air (btu/lb$_m$ dry air)
H	humidity, kg water vapour/kg dry air (lb water vapour/lb dry air)
H	enthalpy of vapour, J/kg, kJ/kg, kJ/kg mol (btu/lb$_m$)
H	equilibrium relation, kg mol/m^3.atm
H	bed length, m (ft); also effective length of spindle, m (ft)
H'	enthalpy, kJ/kg dry solid (btu/lb$_m$); also enthalpy, kJ/kg dry air (btu/lb$_m$ dry air)
H_B	length of bed used up to break point, m (ft)
H_G, H_L	height of transfer unit, m (ft)
H_{OG}	
H_{OL}	
H_P, H_R	percentage humidity, percentage relative humidity, respectively

H_T total bed length, m (ft)

H_{UNB} length of unused bed, m (ft)

i unit vector along x axis

I intensity of turbulence, dimensionless; also current, amp

I_B radiation intensity of black body, W/m^2.sr (btu/h.ft^2.sr)

I_λ intensity of radiation, W/m^2 (btu/h.ft^2)

j unit vector along y axis

J width of baffle, m (ft)

j_A mass flux of A relative to mass average velocity, kg/s.m^2

j_A^* mass flux of A relative to molar average velocity, kg/s.m^2

J_A molar flux of A relative to mass average velocity, kg mol/s.m^2

j_A^* molar flux vector of A relative to molar average velocity, kg mol/s.m^2 (lb mol/h.ft^2, g mol/s.cm^2)

J_D, J_H mass-transfer and heat-transfer factors, dimensionless

k unit vector along z axis

k, k' reaction velocity constant, h^{-1}, min^{-1}, or s^{-1}

k thermal conductivity, W/m.K (btu/h.ft.°F)

k reaction velocity constant, h^{-1}, min^{-1}, or s^{-1}

k'_1 first-order heterogeneous reaction velocity constant m/s^2 (cm/s)

k'_c mass-transfer coefficient, kg mol/s.m^2.conc diff (lb mol/h.ft^2.

k_c conc diff, g mol/s.cm^2.conc diff); kg mol/s.m^2.Pa,

k_G, k_x kg mol/s.m^2.atm (lb mol/h.ft^2. atm).

k_c mass-transfer coefficient, m/s (ft/h, cm/s)

$k_G a$ volumetric mass-transfer coefficient, kg mol/s.m^3.conc diff (lb

$k_x a$ mol/h.ft^3. conc diff, g mol/s.cm^3. conc diff); kg mol/s.m^3.Pa

$k_y a$

k_s surface reaction coefficient, kg mol/s.m^2.mol frac

k_w, k'_w wall constants, dimensionless

K'_y overall mass-transfer coefficient, kg mol/s.m^2.mol frac (lb

K'_x mol/h.ft^2. mol frac, g mol/s. cm^2. mol frac)

K consistency index, N.s^n /m^2 (lb$_f$. s^n/ft^2)

K', K_s equilibrium distribution coefficient, dimensionless

K' consistency index, N.s^n/m^2 (lb$_f$. s$^{n'}$/ft^2)

K equilibrium distribution coefficient, dimensionless

K_p filtration constant, s/m^6 (s/ft^6)

K_V filtration constant, N/m^5 (lb$_f$/ft^5)

K_c, K_{ex} contraction, expansion, fitting loss coefficient, dimensionless

K_f

L	length, m (ft); also amount, kg, kg mol(lb_m); also kg/h.m^2
L	liquid flow rate, kg/h, kg mol/h (lb_m/h)
L	Prandtl mixing length, m (ft)
L	mean beam length, m (ft)
L_S	dry solid weight, kg dry solid (lb_m dry solid)
L'	flow rate, kg inert/h, kg mol inert/h (lb_m inert/h)
m, m' m''	slope of equilibrium line, dimensionless
m	flow rate, kg/s, kg/h (lb_m/s)
m	dimensionless ratio $= k/hx_1$; also, position m
m	ratio, kg wet cake/kg dry cake (lb wet cake/lb dry cake)
M	molecular weight, kg/kg mol (lb_m/lb mol)
M	total mass, kg (lb_m); also parameter $= (\Delta x)^2/\alpha \Delta t$, dimensionless
M	modulus $= (\Delta x)^2/D_{AB} \Delta t$, dimensionless
M	flow rate, kg/h, kg mol/h(lb_m/h)
M	amount of adsorbent, kg(lb_m)
n	exponent, dimensionless; also flow behaviour index, dimensionless
n'	slope of line for power-law fluid, dimensionless
n	position parameter; also dimensionless ratio $= x/x_1$
n	total amount, kg mol (lb mol, g mol)
n_A	flux of A relative to stationary coordinates, kg/s.m^2
N	rpm or rps; also number of radiation shields parameter $= h \Delta x/k$ dimensionless; also number of viable organisms, dimensionless
N	total flux relative to stationary coordinates, kg mol/s.m^2 (lb mol/h.ft^2, g mol/s.cm^2)
N	modulus $= k_c \Delta x/D_{AB}$, dimensionless; also number of stages
N	concentration of solid B, kg solid B/kg solution (lb solid B/lb solution)
N	number of equal temperature subdivisions, dimensionless
N_s	solute flux kg/s.m^2 (lb_m/h.ft^2)
N_w	solvent flux kg solvent/s.m^2 (lb_m/h.ft^2)
N_A	molar flux vector of A relative to stationary coordinates, kg mol/s.m^2 (lb mol/h. ft^2, g mol/s.cm^2)
$\overline{N}_A$	mass transfer of A relative to stationary coordinates, kg mol/s (lb mol/h, g mol/s)
N_G, N_L	number of transfer units, dimensionless
$N_{OG,}$ N_{OL}	

N_{Bi}	Biot number = hx_1/k, dimensionless
N_{Pe}	Peclet number = $N_{Re}N_{Pr}$, dimensionless
N_{Nu}, δ	Nusselt number = $h\delta/k$, dimensionless
N_{Fr}	Froude number = v^2/gL, dimensionless
N_{Kn}	Knudsen number = $\lambda/2\bar{r}$, dimensionless
N_{Nu}	Nusselt number = hL/k, dimensionless
N_{Nu}, x	Nusselt number = $h_x x/k$, dimensionless
N_{Pr}	Prandtl number = $c_p\mu/k$, dimensionless
N_{Re}	Reynolds number = $Dv\rho/\mu$, dimensionless
N'_{Re}	Reynolds number = $D^2_a N\rho/\mu$, dimensionless
N_{Re}, $_L$	Reynolds number = $Lv_\infty\rho/\mu$, dimensionless
N_{Re}, $_x$	Reynolds number = $xv_\infty\rho/\mu$, dimensionless
N_{Eu}	Euler number = $p/\rho v^2$, dimensionless
N_{Sc}	Schmidt number = $\mu/\rho D_{AB}$, dimensionless
N_{Sh}	Sherwood number = $k'_c D/D_{AB}$, dimensionless
N_Q	flow number, dimensionless
N_{St}	Stanton number = k'_c/v, dimensionless
NTU	number of transfer units, dimensionless
O	flow rate, kg/h, kg mol/h (lb_m/h)
p	pressure, N/m^2, Pa (lb_f/ft^2, atm, psia, mm Hg)
p_A	partial pressure of A, N/m^2, Pa (lb_f/ft^2, atm, psia, mm Hg)
p_h	total pressure in high-pressure side (feed), cm Hg. pa (atm)
p_l	total pressure in low-pressure side (permeate) cm Hg, Pa (atm)
p_M	permeability in solid, m/s (ft/h)
P	total pressure, N/m^2, Pa(lb_f/ft^2, atm, psia, mm H)
P	power, W (ft.lb_f/s, hp)
P	flow rate, kg/h, kg/min; also number of phases at equilibrium
P	momentum vector, kg.m/s (lb_m.ft/s)
P_A	vapour pressure of pure A, N/m^2, Pa (lb_f ft^2, atm, psia, mm Hg)
P'_A	permeability of A, cm^3 (STP). cm/s.cm^2. cm Hg
P_w	solvent membrane permeability, kg solvent/s.m.atm (lb_m/h.ft.atm)
P_m	permeability, kg mol/s.m.atm (lb mol/h.ft.atm)
P_M	permeability, m^3 (STP)/(s.m^2 C.S. atm/m)
P'_M	permeability, cm^3 (STP) (s.cm^2. C.S. atm.cm)
P''_M	permeability, cm^3 (STP)/(s.cm^2 C.S. cm Hg/mm)
q	heat-transfer rate, W (btu/h); also net energy added to system, W (btu/h); also J (btu)

q	flow rate, m^3/s (ft^3/s)
q	heat flux vector, W/m^2 $(btu/h.ft^2)$
q	rate of heat generation, W/m^3 $(btu/h.ft^3)$
q'	flow rate in Darcy's law, cm^3/s
q	kg adsorbate/kg adsorbent (lb_m/lb_m)
q_1	flow rate of residue, m^3/s $(ft^3/h, cm^3/s)$
q_2	flow rate of permeate, m^3/s $(ft^3/h, cm^3/s)$
q_A	flow rate of A in permeate, cm^3 (STP)/s, m^3/s (ft^3/h)
q_c	condenser duty, kJ/h, kW (btu/h)
q_f	feed flow rate, cm^3 (STP)/s, $m^3/s/(ft^3/h)$
q_o	reject flow rate, cm^3 (STP)/s, m^3/s (ft^3/h)
q_p	permeate flow rate, cm^3 (STP)/s, m^3/s (ft^3/h)
q_R	reboiler duty, kJ/h, kW (btu/h)
Q	circulation rate, m^3/s (ft^3/h)
Q	amount absorbed, kg mol/m^2; also heat loss, W (btu/h); also heat absorbed, J/kg $(btu/lb_m, ft.lb_f/lb_m)$
r	radius, m (ft)
r_A	rate of generation, kg $A/s.m^3$ $(lb_m A/h.ft^3)$
r_H	hydraulic radius, m (ft)
$(r_2)_{cr}$	critical value of radius, m (ft)
R	rate of drying, $kg/h.m^2$ $(lb_m/h.ft^2)$
R	solute rejection, dimensionless
R	scaleup ratio, dimensionless
R	radius, m (ft); also resistance, K/W (h.°F/btu)
R_A	rate of generation, kg mol $A/s.m^3$ (lb mol $A/h.ft^3$)
R_b	radius of spindle, m (ft)
R_c	radius of outer cylinder, m (ft)
R_g	gel layer resistance, $s.m^2.atm/kg$
R_m	resistance of filter medium m^{-1} (ft^{-1})
R_i	rate of generation of i, kg/s (lb_m/h)
R_x	x component of force, N (lb_f, dyn)
s	compressibility constant, dimensionless
s	mean surface renewal factor, s^{-1}
S	conduction shape factor, m (ft)
S	solubility of a gas, m^3 solute (STP)/m^3 solid.atm [cc solute (STP)/cc solid.atm]
S	distance between centres, m (ft); also steam flow rate, kg/h, kg mol/h (lb_m/h)

S	volume of feed solution, m^3 (ft^3)
S	cross-sectional area of tower, m^2 (ft^2); also stripping factor $1/A$
S_o	specific surface area, m^2/m^3 volume (ft^2/ft^3 volume)
S_p	surface area of particle, m^2 (ft^2)
t	fin thickness m (ft)
t	time, s, min, or h
t	temperature, K, °C (°F)
t	membrane thickness, cm, m (ft)
t_b	break-point time, h
t_t	time equivalent to total capacity, h
t_T	mixing time, s
t_u	time equivalent to usable capacity up to break-point time, h
T	torque, $kg.m^2/s^2$
T	temperature, K, °C (°F, °R); also feed rate, tonne/min
T'	deviation of temperature from mean temperature $\overline{T}$, K (°F)
$\bar{u}$	average velocity, m/s (ft/s)
U	overall heat-transfer coefficient, W/m^2. K ($btu/h.ft^2.°F$)
U	internal energy, J/kg (btu/lb_m)
v	velocity, m/s (ft/s)
$\mathbf{v}$	velocity vector, m/s (ft/s)
v_A	velocity of A relative to stationary coordinates, m/s (ft/s, cm/s)
v_{Ad}	diffusion velocity of A relative to molar average velocity, m/s (ft/s, cm/s)
v_M	molar average velocity of stream relative to stationary coordinates, m/s (ft/s, cm/s)
v_H	humid volume, m^3/kg dry air (ft^3/lb_m dry air)
v_t	terminal settling velocity, m/s (ft/s)
v'_x	deviation of velocity in x direction from mean velocity $\bar{v}_x$, m/s (ft/s)
v'	superficial velocity based on cross section of empty tube, m/s (ft/s, cm/s)
v'_{mf}	velocity at minimum fluidisation, m/s (ft/s)
v'_t	terminal settling velocity, m/s (ft/s)
V_A	solute molar volume, m^3/kg mol
V	flow rate, kg/h, kg mol/h, m^3/s (lb_m/h, ft^3/s)
V	volume, m^3 (ft^3, cm^3); also specific volume, m^3/kg (ft^3/lb_m)
V	velocity, m/s (ft/s); also total amount, kg, kg mol (lb_m)
V'	inert flow rate, kg/h, kg/s, kg mol/h (lb_m/h)

w_A	mass fraction of A
$\dot{W}$	work done on surroundings, W (ft.lb$_f$/s)
W	free water, kg (lb$_m$)
$\dot{W}_s$	mechanical shaft work done on surroundings, W (ft.lb$_f$/s)
W	flow rate, kg/h, kg mol/h (lb$_m$/h)
W	work done on surroundings, J/kg (ft.lb$_f$/lb$_m$)
W	weight of wet solid, kg (lb$_m$)
W	height or width, m (ft); also power, W (hp); also mass dry cake, kg (lb$_m$)
W	width of paddle, m (ft)
W_S	mechanical shaft work done on surroundings, J/kg (ft.lb$_f$/lb$_m$)
W_p	shaft work delivered to pump, J/kg (ft.lb$_f$/lb$_m$)
x	distance in x direction, m (ft)
x	mass fraction or mole fraction; also fraction remaining of original free moisture
x_A	mole fraction of A, dimensionless
x'_B	inert mole ratio, mol B/mol inert
x_f	mole fraction of A in feed, dimensionless
x_o	mole fraction of A in reject, dimensionless
x_{oM}	mole fraction of minimum reject concentration, dimensionless
X	particle size, m (ft); also parameter $= \alpha t/x^2_1$, dimensionless
X	parameter $= Dt/x^2$, dimensionless; also free moisture, kg water/kg dry solid (lb water/lb dry solid)
y	distance in y direction, m (ft); also mole fraction
y_A	mass fraction of A or mole fraction of A; also kg A/kg solution (lb A/lb solution)
y_i	mole fraction of A in permeate at outlet of residue stream, dimensionless
y_p	mole fraction of A in permeate, dimensionless
z	distance in z direction, m (ft); also tower height, m (ft)
z	temperature range for 10:1 change in Dr, °C (°F)

Greek letters

α	correction factor $= 1.0$, turbulent flow and 1/2, laminar flow
α	absorptivity, dimensionless; also flux ratio $= 1 + N_B/N_A$
α	thermal diffusivity $= k/\rho c_p$, m^2/s (ft^2/h, cm^2/s)
α, α_{AB}	relative volatility of A with respect to B, dimensionless
α	specific cake resistance, m/kg (ft/lb$_m$)

α	angle, rad
α	kinetic-energy velocity correction factor, dimensionless
$\alpha*$	ideal separation factor $= P'_A/P'_B$, dimensionless
α_G	gas absorptivity, dimensionless
α_T	eddy thermal diffusivity, m^2/s (ft^2/h)
β	concentration polarisation, ratio of salt concentration at membrane surface to the salt concentration in bulk feed stream, dimensionless
β	momentum velocity correction factor, dimensionless
β	volumetric coefficient of expansion, $1/K$ ($1/°R$)
γ	viscosity coefficient, $N.s^{n'}/m^2$ ($lb_m/ft.s^{2-w}$)
γ	ratio of heat capacities $= c_p/c_v$, dimensionless
Γ	flow rate, $kg/s.m$ ($lb_m/h.ft$)
Γ	concentration of property, amount of property/m^3
δ	molecular diffusivity, m^2/s
δ	boundary-layer thickness, m (ft); also distance, m (ft)
Δ	difference; also difference operating-point flow rate, kg/h (lb_m/h)
ΔT_a	arithmetic temperature drop, K, °C (°F)
ΔT_{1m}	log mean temperature driving force, K, °C (°F)
ΔH	enthalpy change, J/kg, kJ/kg, kJ/kg mol (btu/lb_m, btu/lb mol)
∇	design criterion for sterilisation, dimensionless
Δp	pressure drop, N/m^2, Pa (lb_f/ft^2)
ε	roughness parameter, m (ft); or void fraction, dimensionless
ε	emissivity, dimensionless; also volume fraction, dimensionless
ε_M	mass eddy diffusivity m^2/s (ft^2/h, cm^2/s)
ε	heat-exchanger effectiveness, dimensionless
ε_G	gas emissivity, dimensionless
ε_t	momentum eddy diffusivity, m^2/s (ft^2/s)
ε_{mf}	void fraction at minimum fluidisation, dimensionless
η_f	fin efficiency, dimensionless
η	turbulent eddy viscosity, Pa.s, kg/m.s ($lb_m/ft.s$)
η	efficiency, dimensionless
θ	angle, rad
θ	cut or fraction of feed permeated, dimensionless
$\theta*$	fraction permeated up to a value of $x = 1 - q/q_f$ dimensionless
λ	wavelength, m (ft)
λ	latent heat, J/kg, kJ/kg (btu/lb_m); also mean free path, m (ft)

λ_{Ab}	latent heat of A at normal boiling point, J/kg, kJ/kg mol (btu/lb$_m$)
μ	viscosity, Pa.s, kg/m.s, N.s/m^2 (lb$_m$/ft.s, lb$_m$/ft.h, cp)
μ_a	apparent viscosity, Pa.s, kg/m.s (lb$_m$/ft.s)
v	momentum diffusivity μ/ρ, m^2/s (ft^2/s, cm^2/s)
π_1	dimensionless group, dimensionless
π	osmotic pressure, Pa, N/m^2 (lb$_f$/ft^2, atm)
ρ	density, kg/m^3 (lb$_m$/ft^3); also reflectivity, dimensionless
σ	constant, 5.676×10^{-8} W/m^2. K^4 (0.1714×10^{-8} btu/h.ft^2. °R^4); also collision diameter, Å
Σ	sigma value for centrifuge, m^2 (ft^2)
τ_{zx}	flux of x-directed momentum in z direction, kg (kg.m/s)/s.m^2 or N/m^2 (lb$_f$/ft^2, dyn/cm^2)
τ	shear stress, N/m2 (lb$_f$/ft^2, dyn/cm^2)
τ	tortuosity, dimensionless
ϕ	velocity potential, m^2/s (ft^2/h)
ϕ	angle, rad; also association parameter, dimensionless
ϕ	osmotic coefficient, dimensionless
ϕ_s	shape factor of particle, dimensionless
ψ_z	flux of property, amount of property/s.m^2
ψ	correction factor, dimensionless
ψ	stream function, m^2/s (ft^2/h)
ω	angular velocity, rad/s
ω	solid angle, sr
$\Omega_{D,\ AB}$	collision integral, dimensionless

APPENDIX II

Fundamental Constants and Conversion Factors

Gas law constant R

Numerical value	Units	Numerical value	Units
1.9872	g cal/g mol.K	8314.34	kg.m²/s².kg mol.K
1.9872	btu/lb mol.°R	10.731	ft³.lb_f/in.².lb mol.°R
82.057	cm³.atm/g mol.K	10.731	ft³.lb_f/in.².lb mol.°R
8314.34	J/kg mol.K	0.7302	ft³.atm/lb mol.°R
82.057×10^{-3}	m³.atm/kg mol.K	1545.3	ft.lb_f/lb mol.°R
8314.34	m³.Pa/kg mol.K		

Volume and density

1 g mol ideal gas at 0°C, 760 mm Hg = 22.4140 litres = 22414 cm³
1 lb mol ideal gas at 0°C, 760 mm Hg = 359.05 ft³
1 kg mol ideal gas at 0°C, 760 mm Hg = 22.414 m³
Density of dry air at 0°C, 760 mm Hg = 1.2929 g/litre
$$= 0.080711 \ lb_m/ft^3$$

Molecular weight of air = 28.97 lb_m/lb mol = 28.97 g/g mol
1 g/cm³ = 62.43 lb_m/ft³ = 1000 kg/m³
1 g/cm³ = 8.345 lb_m/U.S.gal
1 lb_m/ft³ = 16.0185 kg/m³

Length

1 in. = 2.540 cm
100 cm = 1 m (metre)
1 micron = 10^{-6} = 10^{-4} cm = 10^{-3} mm = 1 μm (micrometer)
1 Å (angstrom) = 10^{-10} m = 10^{-4} μm
1 mile = 5280 ft
1 m = 3.2808 ft = 39.37 in.

Mass

1 lb_m = 453.59 g = 0.45359 kg
1 lb_m = 16 oz = 7000 grains
1 kg = 1000 g = 2.2046 lb_m

1 ton (short) = 2000 lb_m
1 ton (long) = 2240 lb_m
1 ton (metric) = 1000 kg

Standard accleration of gravity

g = 9.80665 m/s^2
g = 980.665 cm/s^2
g = 32.174 ft/s^2
g_c (gravitational conversion factor) = 32.1740 $lb_m \cdot ft/lb_f \cdot s^2$
$$= 980.665 \ g_m \cdot cm/g_f \cdot s^2$$

Volume

1 L (litre) = 1000 cm^3 1 m^3 = 1000 L (litre)
1 in.^{3} = 16.387 cm^3 1 U.S. gal = 4 qt
1 ft^3 = 28.317 L (litre)   1 U.S. gal = 3.7854 L (litre)
1 ft^3 = 0.028317m^3 1 U.S. gal = 3785.4 cm^3
1 ft^3 = 7.481 U.S. gal   1 British gal = 1.20094 U.S. gal
1 m^3 = 264.17 U.S. gal   1 m^3 = 35.313 ft^3

Force

1 g.cm/s^2 (dyn) = 10^{-5} kg.m/s^2 = 10^{-5} N (newton)
1 g.cm/s^2 = 7.2330 × 10^{-5} lb_m.ft/s^2 (poundal)
1 kg.m/s^2 = 1 N (newton)
1 lb_f = 4.4482 N
1 g.cm/s^2 = 2.2481 × 10^{-6} lb_f

Pressure

1 bar = 1 × 10^5 Pa (pascal) = 1 × 10^5 N/m^2
1 psia = 1 lbf/in.^{2}
1 psia = 2.0360 in. Hg at 0°C
1 psia = 2.311 ft H_2O at 70°F
1 psia = 51.715 mm Hg at 0°C (ρ_{Hg} = 13.5955 g/cm^3)
1 atm = 14.696 psia = 1.01325 × 10^5 N/m^2 = 1.01325 bar
1 atm = 760 mm Hg at 0°C = 1.01325 × 10^5 Pa
1 atm = 29.921 in. Hg at 0°C

1 atm = 33.90 ft H_2O at 4°C

1 psia = 6.89476 × 10^4 g/cm.s^2

1 psia = 6.89476 × 10^4 dyn/cm^2

1 dyn/cm^2 = 2.0886 × 10^{-3} lb$_f$/ft^2

1 psia = 6.89476 × 10^3 N/m^2 = 6.89476 × 10^3 Pa

1 lb$_f$/ft^2 = 4.7880 × 10^2 dyn/cm^2 = 47.880 N/m^2

1 mm Hg (0°C) = 1.333224 × 10^2 N/m^2 = 0.1333224 kPa

Power

1 hp = 0.74570 kW   1 watt (W) = 14.340 cal/min

1 hp = 550 ft.lb$_f$/s   1 btu/h = 0.29307 W (watt)

1 hp = 0.7068 btu/s   1 J/s (joule/s) = 1 W

Heat, energy, work

1 J = 1 N.m = 1 kg.m^2/s^2

1 kg.m^2/s^2 = 1 J (joule) = 10^7 g.cm^2/s^2 (erg)

1 btu = 1055.06 J = 1.05506 kJ

1 btu = 252.16 cal (thermochemical)

1 kcal (thermochemical) = 1000 cal = 4.1840 kJ

1 cal (thermochemical) = 4.1840 J

1 cal (IT) = 4.1868 J

1 btu = 251.996 cal (IT)

1 btu = 778.17 ft.lb$_f$

1 hp.h = 0.7457 kW.h

1 hp.h = 2544.5 btu

1 ft.lb$_f$ = 1.35582 J

1 ft.lb$_f$/lb$_m$ = 2.9890 J/kg

Thermal conductivity

1 btu/h.ft.°F = 4.1365 × 10^{-3} cal/s.cm.°C

1 btu/h.ft.°F = 1.73073 W/m.K

Heat-transfer coefficient

1 btu/h.ft^2.°F = 1.3571 × 10^{-4} cal/s.cm^2.°C

1 btu/h.ft^2.°F = 5.6783 × 10^{-4} W/cm^2.°C

1 btu/h.ft^2.°F = 5.6783 W/m^2.K
1 kcal/h.m^2.°F = 0.2048 btu/h.ft^2.°F

Viscosity

1 cp = 10^{-2} g/cm.s (poise)
1 cp = 2.4191 lb$_m$/ft.h
1 cp = 6.7197 × 10^{-4} lb$_m$/ft.s
1 cp = 10^{-3} Pa.s = 10^{-3} kg/m.s = 10^{-3} N.s/m^2
1 cp = 2.0886 × 10^{-5} lb$_f$.s/ft^2
1 Pa.s = 1 N.s/m^2 = 1 kg/m.s = 1000 cp = 0.67197 lb$_m$/ft.s

Diffusivity

1 cm^2/s = 3.875 ft^2/h   1 m^2/s = 3.875 × 10^4 ft^2/h
1 cm^2/s = 10^{-4} m^2/s   1 centistoke = 10^{-2} cm^2/s
1 m^2/h = 10.764 ft^2/h

Mass flux and molar flux

1 g/s.cm^2 = 7.3734 × 10^3 lb$_m$/h.ft^2
1 gmol/s.cm^2 = 7.3734 × 10^3 lb mol/h.ft^2
1 g mol/s.cm^2 = 10 kg mol/s.m^2 = 1 × 10^4 g mol/s.m^2
1 lb mol/h.ft^2 = 1.3562 × 10^{-3} kg mol/s.m^2

Heat flux and heat flow

1 btu/h.ft^2 = 3.1546 W/m^2
1 btu/h = 0.29307 W
1 cal/h = 1.1622 × 10^{-3} W

Heat capacity and enthalpy

1 btu/lb$_m$.°F = 4.1868 kJ/kg.K
1 btu/lb$_m$.°F = 1.000 cal/g.°C
1 btu/lb$_m$ = 2326.0 J/kg
1 ft.lb$_f$/lb$_m$ = 2.9890 J/kg
1 cal (IT)/g.°C = 4.1868 kJ/kg.K
1 kcal/g mol = 4.1840 × 10^3 kJ/kg mol

Mass-transfer coefficient

$1 \ k_c$ cm/s $= 10^{-2}$ m/s

$1 \ k_c$ ft/h $= 8.4668 \times 10^{-5}$ m/s

$1 \ k_x$ g mol/s.cm^2. mol frac $= 10$ kg mol/s.m^2.mol frac

$1_, k_x$ g mol/s.cm^2. mol frac $= 1 \times 10^4$ g mol/s.m^2.mol frac

$1 \ k_x$ lb mol/h.ft^2. mol frac $= 1.3562 \times 10^{-3}$ kg mol/s.m^2.mol frac

$1 \ k_x \ a$ lb mol/h.ft^3. mol frac $= 4.449 \times 10^{-3}$ kg mol/s.m^3.mol frac

$1 \ k_G$ kg mol/s.m^2.atm $= 0.98692 \times 10^{-5}$ kg mol/s.m^2.Pa

$1 \ k_G \ a$ kg mol/s.m^3.atm $= 0.98692 \times 10^{-5}$ kg mol/s.m^3.Pa

APPENDIX III

Physical Properties of Water

Latent heat of water at 273.15 K (0°C)

Latent heat of fusion = 1436.3 cal/g mol
  = 79.724 cal/g
  = 2585.3 btu/lb mol
  = 6013.4 kJ/kg mol

Latent heat of vapourisation at 298.15 K (25°C)

Pressure (mm Hg)	Latent heat
23.75	44 020 kJ/kg mol, 10.514 kcal/g mol, 18925 btu/lb mol
760	44 045 kJ/kg mol, 10.520 kcal/g mol, 18936 btu/lb mol

Vapour pressure of water

Temperature K	°C	Vapour pressure kPa	mm Hg	Temperature K	°C	Vapour pressure kPa	mm Hg
273.15	0	0.611	4.58	323.15	50	12.333	92.51
283.15	10	1.228	9.21	333.15	60	19.92	149.4
293.15	20	2.338	17.54	343.15	70	31.16	233.7
298.15	25	3.168	23.76	353.15	80	47.34	355.1
303.15	30	4.242	31.82	363.15	90	70.10	525.8
313.15	40	7.375	55.32	373.15	100	101.325	760.0

Density of liquid water

Temperature K	°C	Density g/cm³	kg/m³	Temperature K	°C	Density g/cm³	kg/m³
273.15	0	0.99987	999.87	323.15	50	0.98807	988.07
277.15	4	1.00000	1000.00	333.15	60	0.98324	983.24
283.15	10	0.99973	999.73	343.15	70	0.97781	977.81
293.15	20	0.99823	998.23	353.15	80	0.97183	971.83
298.15	25	0.99708	997.08	363.15	90	0.96534	965.34
303.15	30	0.99568	995.68	373.15	100	0.95838	958.38
313.15	40	0.99225	992.25	NA	NA	NA	NA

Viscosity of liquid water

Temperature		Viscosity [(Pa.s) 10^3 (kg/m.s) 10^3, or cp]	Temperature		Viscosity [(Pa.s) 10^3 (kg/m.s) 10^3, or cp]
K	°C		K	°C	
273.15	0	1.7921	323.15	50	0.5494
275.15	2	1.6728	325.15	52	0.5315
277.15	4	1.5674	327.15	54	0.5146
279.15	6	1.4728	329.15	56	0.4985
281.15	8	1.3860	331.15	58	0.4832
283.15	10	1.3077	333.15	60	0.4688
285.15	12	1.2363	335.15	62	0.4550
287.15	14	1.1709	337.15	64	0.4418
289.15	16	1.1111	339.15	66	0.4293
291.15	18	1.0559	341.15	68	0.4174
293.15	20	1.0050	343.15	70	0.4061
293.35	20.2	1.0000	345.15	72	0.3952
295.15	22	0.9579	347.15	74	0.3849
297.15	24	0.9142	349.15	76	0.3750
298.15	25	0.8937	351.15	78	0.3655
299.15	26	0.8737	353.15	80	0.3565
301.15	28	0.8360	355.15	82	0.3478
303.15	30	0.8007	357.15	84	0.3395
305.15	32	0.7679	359.15	86	0.3315
307.15	34	0.7371	361.15	88	0.3239
309.15	36	0.7085	363.15	90	0.3165
311.15	38	0.6814	365.15	92	0.3095
313.15	40	0.6560	367.15	94	0.3027
315.15	42	0.6321	369.15	96	0.2962
317.15	44	0.6097	371.15	98	0.2899
319.15	46	0.5883	373.15	100	0.2838
321.15	48	0.5683			

Heat capacity of liquid water at 101.325 kPa (1 Atm)

Temperature		Heat capacity, c_p		Temperature		Heat capacity, c_p	
°C	K	cal/g.°C	kJ/kg.K	°C	K	cal/g.°C	kJ/kg.K
0	273.15	1.0080	4.220	50	323.15	0.992	4.183
10	283.15	1.0019	4.195	60	333.15	1.0001	4.187
20	293.15	0.9995	4.185	70	343.15	1.0013	4.192
25	298.15	0.9989	4.182	80	353.15	1.0029	4.199
30	303.15	0.9987	4.181	90	363.15	1.0050	4.208
40	313.15	0.9987	4.181	100	373.15	1.0076	4.219

Thermal conductivity of liquid water

Temperature			Thermal conductivity	
°C	°F	K	btu/h.ft.°F	W/m.K
0	32	273.15	0.329	0.569
37.8	100	311.0	0.363	0.628
93.3	200	366.5	0.393	0.684
148.9	300	422.1	0.395	0.684
215.6	420	588.8	0.376	0.651
326.7	620	599.9	0.275	0.476

Heat-transfer properties of liquid water, SI units

T (°C)	T (K)	ρ (kg/m^3)	c_p (kJ/kg.K)	$\mu \times 10^3$ (Pa.s, or kg/m.s)	k (W/m.K)	N_{Pr}	$\beta \times 10^4$ (1/K)	$(g\beta\rho^2/\mu^2)$ $\times 10^{-8}$ (1/K.m^3)
0	273.2	999.6	4.229	1.789	0.5694	13.3	−0.630	
15.6	288.8	998.0	4.187	1.131	0.5884	8.07	1.44	10.93
26.7	299.9	996.4	4.183	0.860	0.6109	5.89	2.34	30.70
37.8	311.0	994.7	4.183	0.682	0.6283	4.51	3.24	68.0
65.6	338.8	981.9	4.187	0.432	0.6629	2.72	5.04	256.2
93.3	366.5	962.7	4.229	0.3066	0.6802	1.91	6.66	642
121.1	394.3	943.5	4.271	0.2381	0.6836	1.49	8.46	1300
148.9	422.1	917.9	4.312	0.1935	0.6836	1.22	10.08	2231
204.4	477.6	858.6	4.522	0.1384	0.6611	0.950	14.04	5308
260.0	533.2	784.9	4.982	0.1042	0.6040	0.859	19.8	11030
315.6	588.8	679.2	6.322	0.0862	0.5071	1.07	31.5	19260

Heat-transfer properties of liquid water, english units

T(°F)	ρ $\left(\dfrac{lb_m}{ft^3}\right)$	c_p $\left(\dfrac{btu}{lb_m.°F}\right)$	$\mu \times 10^3$ $\left(\dfrac{lb_m}{ft.s}\right)$	k $\left(\dfrac{btu}{h.ft.°F}\right)$	N_{Pr}	$\beta \times 10^4$ $(1/°R)$	$(g\beta\rho^2/\mu_2) \times 10^{-6}$ $(1/°R.ft^3)$
32	62.4	1.01	1.20	0.329	13.3	−0.350	
60	62.3	1.00	0.760	0.340	8.07	0.800	17.2
80	62.2	0.999	0.570	0.353	5.89	1.30	48.3
100	62.1	0.999	0.458	0.363	4.51	1.80	107
150	61.3	1.00	0.290	0.383	2.72	2.80	403
200	60.1	1.01	0.206	0.393	1.91	3.70	1010
250	58.9	1.02	0.160	0.395	1.49	4.70	2045
300	57.3	1.03	0.130	0.395	1.22	5.60	3510
400	53.6	1.08	0.0930	0.382	0.950	7.80	8350
500	49.0	1.19	0.0700	0.349	0.859	11.0	17350
600	42.4	1.51	0.0579	0.293	1.07	17.5	30300

Heat-transfer properties of water vapour (steam) at 101.32 kPa (1 atm Abs), SI units

T (°C)	T (K)	ρ (kg/m^3)	c_p $(kJ/kg.K)$	$\mu \times 10^5$ $(Pa.s,$ or $kg/m.s)$	k $(W/m.K)$	N_{Pr}	$\beta \times 10^3$ $(1/K)$	$(g\beta\rho^2/\mu^2)$ $(1/K.m^3)$
100.0	373.2	0.596	1.888	1.295	0.02510	0.96	2.68	0.557×10^8
148.9	422.1	0.525	1.909	1.488	0.02960	0.95	2.38	0.292×10^8
204.4	477.6	0.461	1.934	1.682	0.03462	0.94	2.09	0.154×10^8
260.0	533.2	0.413	1.968	1.883	0.03946	0.94	1.87	0.0883×10^8
315.6	588.8	0.373	1.997	2.113	0.04448	0.94	1.70	52.1×10^5
371.1	644.3	0.341	2.030	2.314	0.04985	0.93	1.55	33.1×10^5
426.7	699.9	0.314	2.068	2.529	0.05556	0.92	1.43	21.6×10^5

Heat-transfer properties of water vapour (steam) at 101.32 kPa (1 Atm Abs), English units

T(°F)	ρ $\left(\dfrac{lb_m}{ft^3}\right)$	c_p $\left(\dfrac{btu}{lb_m \cdot °F}\right)$	$\mu \times 10^5$ $\left(\dfrac{lb_m}{ft.s}\right)$	k $\left(\dfrac{btu}{h.ft.°F}\right)$	N_{Pr}	$\beta \times 10^3$ (1/°R)	$(g\beta\rho^2/\mu^2)$ (1/°R.ft³)
212	0.0372	0.451	0.870	0.0145	0.96	1.49	0.877×10^6
300	0.0328	0.456	1.000	0.0171	0.95	1.32	0.459×10^6
400	0.0288	0.462	1.130	0.0200	0.94	1.16	0.243×10^6
500	0.0258	0.470	1.265	0.0228	0.94	1.04	0.139×10^6
600	0.0233	0.477	1.420	0.0257	0.94	0.943	82×10^3
700	0.0213	0.485	1.555	0.0288	0.93	0.862	52.1×10^3
800	0.0196	0.494	1.700	0.0321	0.92	0.794	34.0×10^3

References

P. G. Klemens, *Theory of the Thermal Conductivity of Solids*, Academic Press, London.

J. F. Mallory, *Thermal Insulation*, Reinhold, New York.

W. M. Rohsenow, J. P. Hartnett, and **E. N. Ganic**, *Handbook of Heat Transfer*, McGraw-Hill, New York.

H. S. Carslaw and **J. C. Jaeger**, *Conduction of Heat in Solids*, Oxford University Press, London.

L. M. K. Boelter, V. H. Chery, and **H. A. Johnson**, *Heat Transfer Notes*, University of California Press, Berkeley.

H. Gröber, S. Erk, and **U. Grigull**, *Fundamentals of Heat Transfer*, McGraw-Hill, New York.

S. V. Patankar, *Numerical Heat Transfer and Fluid Flow*, Hemisphere Publishing Corp. Washington, D.C.

F. P. Incorpera and **D. P. DeWitt**, *Fundamentals of Heat Transfer*, Wiley, New York.

A. Bejan, *Convection Heat Transfer*, Wiley, New York.

Y. Jaluria, *Natural Convection Heat and Mass Transfer*, Pergamon, New York.

W. M. Kays and **M. E. Crawford**, *Convective Heat and Mass Transfer*, McGraw-Hill, New York.

W. D. Hayes and **R. F. Probstein**, *Hypersonic Flow Theory*, Academic Press, New York.

W. M. Kays and **K. R. Perkins**, *Handbook of Heat Transfer Applications*, McGraw-Hill, New York.

W. M. Kays and **A. L. London**, *Compact Heat Exchangers*, McGraw-Hill, New York.

H. Hausen, *Heat Transfer in Counter Flow, Parallel Flow and Cross Flow*, McGraw-Hill, New York.

W. M. McAdams, *Heat Transmission*, McGraw-Hill, New York.

R. P. Stein, **J. P. Hartnett** and **T. F. Irvine**, *Advances in Heat Transfer*, Academic Press, New York.

H. B. Squire, *Modern Developments in Fluid Dynamics*, Clarendon, Oxford.

W. M. Kays and **A. L. London**, *Compact Heat Exchangers*, McGraw-Hill, New York.

W. J. Beek and **K. M. K. Muttzall**, *Transport Phenomena*, Wiley, New York.

A. H. P. Skelland, *Diffusional Mass Transfer*, Wiley, New York.

H. Hausen, *Heat Transfer in Counterflow, Parallel Flow and Cross Flow*, McGraw-Hill, New York.

E. U. Schlünder (ed.), *Heat Exchanger Design Handbook*, Hemisphere Pub., New York.

R. F. Boehm and **F. Kreith**, *Direct Contact Heat Transfer*, Hemisphere Pub. Co., New York.

M. Planck, *Theory of Heat Radiation*, Dover, New York.

F. Kreith and **W. Z. Black**, *Basic Heat Transfer*, Harper & Row, New York.

H. C. Hottel and **A. F. Sarofim**, *Radiative Heat Transfer*, McGraw-Hill, New York.

J. A. Wibelt, *Engineering Radiation Heat Transfer*, Holt, Rinehart & Winston, New York.

L. S. Tong, *Boiling Heat Transfer and Two-Phase Flow*, Wiley, New York.

A. B. Metzner, *Handbook of Fluid Mechanics*, McGraw-Hill, New York.

A. B. Metzner, *Advances in Heat Transfer*, Academic Press, New York.

W. L. Wilkinson, *Non-Newtonian Fluids*, Pergamon, New York.

A. B. Metzner, *Advances in Chemical Engineering*, Academic Press, New York.

W. M. Rohsenow, **J. P. Hartnett**, and **E. Ganic**, *Handbook of Heat Transfer Applications*, McGraw-Hill, New York.

W. H. McAdams, *Heat Transmission, 3rd edition*, McGraw-Hill Book Co.

M. Jakob, *Heat Transfer, Vol. I*, John Willey and Sons.

E. R. G. Eckert, *Introduction to the Transfer of Heat and Mass*, McGraw-Hill Book Co.

M. Fishenden and **O. A. Saunders**, *An Introduction to Heat Transfer*, Oxford at the Clarendon Press.

G. B. Wilkes, *Heat Insulation*, John Wiley and Sons

Edited by J. F. Hogerton and R. C. Grass, *Reactor Handbook: Engineering*, McGraw-Hill Book Co.

P. J. Schneider, *Conduction Heat Transfer*, Addison-Wesley Publishing Co.

F. Hildebrand, *Methods of Applied Mathematics*, Prentice Hall.

W. H. McAdams, *Heat Transmission*, McGraw-Hill Book Co.

E. R. G. Eckert, *Introduction to the Transfer of Heat and Mass*, McGraw Hill Book Co.

E. Schmidt, *The Heat Transfer by Fins*, University of Texas Publications.

M. Jakob, *Local Temperature Differences as Occurring in Evaporation, Condensation and Catalytic Reactions*, Reinhold Publishing Corporation.

W. J. Wohlenberg, *Heat Transfer by Radiation*, John Willey and Sons, New York.

Index